# Computational Group Theory

# Computational Group Theory

*Proceedings of the London Mathematical Society Symposium on Computational Group Theory*

edited by
MICHAEL D. ATKINSON
School of Computer Science,
Carleton University, Ottawa

1984

Academic Press
*(Harcourt Brace Jovanovich, Publishers)*
London Orlando San Diego San Francisco New York
Toronto Montreal Sydney Tokyo São Paulo

ACADEMIC PRESS INC. (LONDON) LTD.
24/28 Oval Road
London NW1

*United States Edition published by*
ACADEMIC PRESS INC.
(Harcourt Brace Jovanovich, Inc.)
Orlando, Florida 32887





**British Library Cataloguing in Publication Data**
Computational group theory.
1. Group, Theory of—Congresses
2. Combinatorial analysis—Congresses
I. Atkinson, M.
512′.22 QA171

ISBN 0-12-066270-1
LCCCN 83-71856

Printed in Great Britain by Thomson Litho Ltd., East Kilbride

## Symposium Participants

D. G. Arrell *Department of Mathematics and Computing, Leeds Polytechnic, Leeds LS1 3HE, U.K.*

A. O. L. Atkin *Department of Mathematics, University of Illinois at Chicago Circle, Chicago, Illinois 60680, U.S.A.*

M. D. Atkinson *School of Computer Science, Carleton University, Ottawa, Ontario, Canada K1S 5B6*

M. J. Beetham *Academic Computing Service, The Open University, Walton Hall, Milton Keynes, U.K.*

N. L. Biggs *Department of Mathematics, Royal Holloway College, Egham, Surrey TW20 0EX, U.K.*

S.C. Black *Department of Pure Mathematics, University of Birmingham, Birmingham B15 2TT, U.K.*

S. K. Burford *Department of Mathematics, Royal Holloway College, Egham, Surrey TW20 0EX, U.K.*

G. Butler *Basser Department of Computer Science, University of Sydney, Sydney 2006, Australia*

C. M. Campbell *Mathematical Institute, University of St. Andrews, St. Andrews KY16 9SS, U.K.*

J. J. Cannon *Department of Pure Mathematics, University of Sydney, Sydney 2006, Australia*

F. L. Chomse *Lehrstuhl D für Mathematik, RWTH Aachen, Templergraben 64, 5100 Aachen, West Germany*

G. Cooper *Computer Centre, Queen Mary College, Mile End Road, London E1 4NS, U.K.*

J. H. Conway *Department of Pure Mathematics and Mathematical Statistics, 16 Mill Lane, Cambridge CB2 1SB, U.K.*

A. M. Cohen *Mathematical Centre, 2e Boerhaavestraat 49, 1091 AL Amsterdam, The Netherlands*

R. T. Curtis *Department of Pure Mathematics, University of Birmingham, Birmingham B15 2TT, U.K.*

V. Felsch *Lehrstuhl D für Mathematik, RWTH Aachen, Templergraben 64, 5100 Aachen, West Germany*

A. D. Gardiner *Department of Pure Mathematics, University of Birmingham, Birmingham B15 2TT, U.K.*

R. H. Gilman *Department of Pure and Applied Mathematics, Stevens Institute of Technology, Hoboken, New Jersey 07030, U.S.A.*

R. A. Hassan *Department of Computing Mathematics, Mathematics Institute, University College, Cardiff CF1 3BB, U.K.*

G. Havas *Division of Computing Research, CSIRO, P.O. Box 1800, Canberra, ACT 2601, Australia*

M. Herzog *School of Mathematical Sciences, Tel-Aviv University, Ramat-Aviv, Tel Aviv 69978, Israel*

D. F. Holt *Mathematics Institute, University of Warwick, Coventry CV4 7AL, U.K.*

D. L. Johnson *Department of Mathematics, University of Nottingham, Nottingham, NG7 2RD, U.K.*

K. W. Johnson *Department of Mathematics, University of West Indies, Mona, Kingston 7, Jamaica*

P. E. Kenne* *Department of Mathematics, Institute of Advanced Studies, Australian National University, Canberra, ACT 2601, Australia*

A. Kerber *Lehrstuhl II für Mathematik, der Universität Bayreuth, Postfach 3008, 8580 Bayreuth, West Germany*

G. Kolesova *Department of Computer Science, Concordia University, 1455 Blvd. de Maisonneuve W., Montreal, Canada, H3G 1M8*

L. Kovacs* *Department of Mathematics, Institute of Advanced Studies, Australian National University, Canberra, ACT 2601, Australia*

R. Laue *Lehrstuhl D für Mathematik, RWTH Aachen, Templergraben 64, 5100 Aachen, West Germany*

D. W. Leavitt *Department of Computer Science, University of Nebraska-Lincoln, Lincoln, Nebraska 68588, U.S.A.*

J. Leech *Department of Computer Science, University of Stirling, Stirling FK9 4LA, U.K.*

C. R. Leedham-Green *Department of Pure Mathematics, Queen Mary College, Mile End Road, London E1 4NS, U.K.*

J. S. Leon *Department of Mathematics, University of Illinois at Chicago, Chicago, Illinois 60680, U.S.A.*

R. J. List *Department of Pure Mathematics, University of Birmingham, Birmingham B15 2TT, U.K.*

D. Livingstone *Department of Pure Mathematics, University of Birmingham, Birmingham B15 2TT, U.K.*

A. M. Macbeath *Department of Mathematics and Statistics, University of Pittsburgh, Pittsburgh, Pennsylvania 15260, U.S.A.*

J. K. S. McKay *Department of Computer Science, Concordia University, 1455 Blvd. de Maisonneuve W., Montreal, Canada H3G 1M8*

S. S. Magliveras *Department of Computer Science, University of Nebraska-Lincoln, Lincoln, Nebraska 68588, U.S.A.*

A. Mann *Department of Mathematics, Hebrew University, Jerusalem, Israel*

J. Mennicke *Fakultät für Mathematik, Universität, D-4800 Bielefeld, West Germany*

J. Neubüser *Lehrstuhl D für Mathematik, RWTH Aachen, Templergraben 64, 5100 Aachen, West Germany*

B. H. Neumann *Department of Mathematics, Institute of Advanced Studies, Australian National University, Canberra, ACT 2601, Australia*

P. M. Neumann *The Queen's College, Oxford 0X1 4AW, U.K.*

M. F. Newman *Department of Mathematics, Institute of Advanced Studies, Australian National University, Canberra, ACT 2601, Australia*

D. B. Nikolova *Bulgarian Academy of Sciences, Institute of Mathematics, 1113 Sofia, str. "Acad. G. Bonchev", block 8, Bulgaria*

S. P. Norton *Department of Pure Mathematics and Mathematical Statistics, 16 Mill Lane, Cambridge CB2 1SB, U.K.*

A. A. H. Omar *Department of Mathematics, University of El-Minya, P.O. Box 2807, Cairo, Egypt*

H. Pahlings* *Lehrstuhl D für Mathematik, RWTH Aachen, Templergraben 64, 5100 Aachen, West Germany*
R. A. Parker *Department of Pure Mathematics and Mathematical Statistics, 16 Mill Lane, Cambridge CB2 1SB, U.K.*
W. Plesken *Lehrstuhl D für Mathematik, RWTH Aachen, Templergraben 64, 5100 Aachen, West Germany*
V. Pless *Department of Mathematics, University of Illinois at Chicago, Chicago, Illinois 60680, U.S.A.*
L. Queen *Department of Pure Mathematics and Mathematical Statistics, 16 Mill Lane, Cambridge CB2 1SB, U.K.*
J. S. Richardson* *Department of Mathematics, University of Melbourne, Victoria, 3052, Australia*
E. F. Robertson *Mathematical Institute, University of St. Andrews, St. Andrews KY16 9SS, U.K.*
C. Rowley *Open University, Parsifal College, 527 Finchley Road, London NW3 7BG, U.K.*
R. Sandling *Department of Mathematics, The University, Manchester M13 9PL, U.K.*
G. Sandlöbes* *Lehrstuhl D für Mathematik, RWTH Aachen, Templergraben 64, 5100 Aachen, West Germany*
V. Schoenwaelder* *Lehrstuhl D für Mathematik, RWTH Aachen, Templergraben 64, 5100 Aachen, West Germany*
C. C. Sims *Department of Mathematics, Rutgers University, Hill Centre, New Brunswick, N.J. 08903, U.S.A.*
L. H. Soicher *Department of Pure Mathematics and Mathematical Statistics, 16 Mill Lane, Cambridge CB2 1SB, U.K.*
D. M. Solitar *Department of Mathematics, York University, 4700 Keele Street, Downsview, Ontario, Canada M3J 2R3*
M. P. Thorne* *Department of Computing Mathematics, University College, Cardiff CF1 3BB, U.K.*
M. R. Vaughan-Lee *Christ Church College, Oxford 0X1 1DP, U.K.*
A. Wagner *Department of Pure Mathematics, University of Birmingham, Birmingham B15 2TT, U.K.*

* Contributor who was not present at the symposium.

# Preface

During the last decade there has been a considerable increase in the application of computers to group theory. The contribution of machine computation to the finite simple group classification program is particularly evident. Sims' constructions of some of the sporadic groups and the Fischer–Livingstone–Thorne construction of the Monster character table are noteworthy examples. These and other successes have directed and encouraged group theorists to search for algorithms suitable for machine implementation to investigate group structure. As a result there are now several systems of group theory programs which provide a wealth of tools for studying groups. The successes of these general systems together with several notable one-off computations have led to a wide acceptance among researchers of the utility of machine computation in algebra.

Work in computational group theory has been carried out by many people, on many machines, in many parts of the world. In 1980 the time seemed ripe to begin planning an International Symposium which would bring together experts in the field and would draw together the several different strands in the subject. The London Mathematical Society readily agreed to host the symposium in Durham and Charles Leedham-Green and I took on the task of organization. In the event Charles' contribution was far greater than my own and the success of the symposium was largely due to his competence, hard work and hilarious humour.

At an early stage we decided that computing facilities should be available at the symposium so that participants could demonstrate their software. The very considerable difficulties in arranging for access to remote machines and overcoming incompatibilities between different computers were solved in a masterly way by Geoff Cooper who became the third organizer. Thanks to his expertise and the cooperation of the Durham University Computer Centre several papers in this volume describe computations done at the symposium.

Many people gave us help and advice. A complete list would include Joachim Neubüser, Chris Rowley, Mike Thorne, Tom Willmore, Steve Wilson and many others, all of whom have our hearty thanks. It is also a pleasure to acknowledge the symposium funding provided by the Science and Engineering Research Council.

The symposium took place from 30th July to 9th August 1982 and was attended by some 60 members. About half were from overseas: from Australia, Europe, the Middle East and North America. The core of the scientific programme was the contribution by the six key speakers John Cannon, John Conway, John Leech, Joachim Neubüser, Michael Newman and Charles Sims, each of whom gave several lectures. These proceedings contain accounts of most of their presentations. One omission is Charles Sims' material on algorithms related to finitely presented groups; he is planning a monograph which would include this work. Lectures were also given by many other participants, most of whom have submitted accounts for these proceedings.

Demonstrations and applications of group theoretical software were an important part of the symposium. Cannon's CAYLEY system was used by many participants. Other integrated program packages available were the CAS and SOGOS systems developed by Neubüser and his associates, and the micro-computer system of Hassan, Thorne and Atkinson. There were also many stand-alone programs. Havas and his co-workers contributed programs for coset enumeration, the nilpotent quotient algorithm, the Reidemeister–Schreier algorithm, Tietze transformations and Abelian group decomposition. Gilman, Parker, McKay, Soicher, Kolesova, Vaughan-Lee, Arrell and Robertson supplied the programs which are described in their papers. These systems and programs were available on machines at Durham, Newcastle, Cambridge, Bradford, St Andrews and Queen Mary College.

Although the primary aim of the symposium was to identify key problems and key methods in computational group theory, a secondary aim was to summarize the state of the art. Every major aspect of the subject was discussed in one form or another. These proceedings therefore comprise a fairly complete survey of computational group theory and I hope they will serve as a springboard for new workers entering the field.

In assembling the contributions for this volume I have tried to place papers on similar topics close to each other. The contents are divided into three sections: finitely presented groups, finite groups, permutation groups; most of computational group theory is within these areas. For the few papers which do not fit under just one of these headings I have made a fairly arbitrary choice in assigning it to a section.

I am most grateful to Academic Press for their advice and technical help which has made my editorial task very much easier.

MICHAEL ATKINSON
*October 1983*

# Contents

# Finitely Presented Groups

# Coset Enumeration

JOHN LEECH

## 1 Introduction

In the first two lectures at the Symposium, I gave an elementary account of coset enumeration as performed by hand and of the problems of computer implementation, together with some different interpretations of completed enumerations and some information that may be deduced from them. This was to serve as a background for subsequent speakers describing more recent work. This written account is based very freely on my lectures and includes forward references to these later lectures. I am indebted to G. Butler and V. Felsch, whose careful notes of my lectures have been most useful in preparing this account.

The original published account of coset enumeration is that of Todd and Coxeter [**13**], which of course substantially predates the advent of electronic computers. Similar accounts are more readily accessible in Coxeter and Moser [**4**]. Computer implementation was first achieved by Haselgrove in 1953 on the EDSAC 1 computer at Cambridge. (This and other early implementations are described by Leech [**7**].) This is believed to be the first application of a computer to abstract algebraic problems. There was, however, an earlier suggested application described by Newman [**12**], who proposed a calculation to enumerate the groups of order 256.

## 2 Problems

The problems with which we are concerned are the following:

(i) Given a group, to find a finite presentation for it;
(ii) Given a finitely presented group, to find its order (is it trivial, finite, infinite, otherwise known to us?).

Although (i) is of more frequent interest, our approach is via (ii) by examining generators for the group and possible sets of defining relations selected from among those satisfied by the generators.

COMPUTATIONAL GROUP THEORY
ISBN 0-12-066270-1

It is convenient to begin by describing the enumeration of *elements* of a finitely presented group rather than cosets of a non-trivial subgroup, but the term *coset* will be used for consistency with later occasions where cosets of a non-trivial subgroup are used. (Elements can be regarded as cosets of the trivial subgroup $\{I\}$.) We begin with the identity element, which we designate by the number 1. We continue by defining $2 = 1A$, $3 = 1A^{-1}$, $4 = 1B$, $5 = 1B^{-1}$, continuing through all the generators of the group and their inverses, and record this information in a *coset multiplication table* which will begin thus:

| | $A$ | $A^{-1}$ | $B$ | $B^{-1}$ | ... |
|---|---|---|---|---|---|
| 1 | 2 | 3 | 4 | 5 | |
| 2 | | 1 | | | |
| 3 | 1 | | | | |
| 4 | | | | 1 | |
| 5 | | | 1 | | |

The corresponding inverse information such as $2A^{-1} = 1$ is also recorded. We then continue with $2A, 2B, 2B^{-1}, \ldots$, omitting $2A^{-1}$ as this is known to be 1, and follow with $3A^{-1}$, $3B$, $3B^{-1}, \ldots$, omitting $3A$ similarly, etc. If our presentation had no relations, these elements would be all distinct and we would be enumerating the freely reduced elements of the free group on our generators $A, B, \ldots$ in order of increasing length and in lexicographic order for those of equal length. Of course, this enumeration will not terminate.

In cases of practical interest our presentations include *relations*, which we can take to be of the form that certain words $W_i$ in the generators and their inverses are equated to the identity; such words we call *relators*. (For relations of the form $W_i = W_j$ we write the relators as $W_i W_j^{-1}$.) In the presence of these relations, the elements or cosets enumerated as above are not all distinct. Our strategy, especially when working by hand, is to try to avoid defining apparently new cosets which the relations would show to coincide with cosets that have been defined already. Complete success, however, is not always possible.

## 3 A Worked Example

The following worked example shows how we proceed with a group with relations. We consider the presentation

$$\{A, B, C \mid AB = C, BC = A, CA = B\}, \quad (1)$$

a simple case of a *Fibonacci group* in which each generator is equated to the product of the two generators preceding it in cyclic order; we shall not assume in advance that this is a well-known presentation of the quaternion group. Our relators are thus $ABC^{-1}$, $BCA^{-1}$ and $CAB^{-1}$. In addition to the multiplication table, we form *relator tables*, one for each relator, the rows of which show the effect on each coset of multiplication, letter by letter, by the relator. (The overall effect is identity, because our relator words are equal to the identity.) We begin by defining $2 = 1A$, $3 = 1A^{-1}$ and $4 = 1B$. When inserting these into the relator tables, we see from the first relator that $3A = 1$ combines with $1B = 4$ to give $3AB = 4$, and from $AB = C$ we deduce $3C = 4$. We insert this into the multiplication table, along with the inverse entry $4C^{-1} = 3$, in the same way as an entry made by definition, but we annotate it with a subscript 1 to record that it is an entry made by deduction rather than by definition and that it is the first such entry. The corresponding place in the relator table is similarly annotated. (This annotation is not essential, but it is highly convenient should we wish to retrace the working, perhaps because of a mistake, or if we wish to use the working to prove relations as described below.) Similarly, from $BC = A$ we see that $1B = 4$ and $1A = 2$ give $4C = 2$, and we insert this into the multiplication table with the subscript 2. Our tables now appear as follows, where incomplete lines of relator tables have not been entered:

| | $A$ | $A^{-1}$ | $B$ | $B^{-1}$ | $C$ | $C^{-1}$ |
|---|---|---|---|---|---|---|
| 1 | 2 | 3 | 4 | | | |
| 2 | | 1 | | | | $4_2$ |
| 3 | 1 | | | | $4_1$ | |
| 4 | | | | 1 | $2_2$ | $3_1$ |

| | $A$ | | $B$ | | $C^{-1}$ | |
|---|---|---|---|---|---|---|
| 3 | | 1 | | $4_1$ | | 3 |

| | $B$ | | $C$ | | $A^{-1}$ | |
|---|---|---|---|---|---|---|
| 1 | | 4 | | $2_2$ | | 1 |

| | $C$ | | $A$ | | $B^{-1}$ | |
|---|---|---|---|---|---|---|

There being no further deductions to be made at this stage, we continue by defining $5 = 1B^{-1}$, from which the third relator allows us to deduce $5C = 3$, deduction number 3. Next we define $6 = 1C$, from which the first relator allows us to deduce $2B = 6$, deduction 4, the second relator allows us to deduce $5A = 6$, deduction 5, and the third relator allows us to deduce $6A = 4$, deduction 6. But now we can go further, as these deduced entries allow us to make further deductions. From $5A = 6$ and $5C = 3$ we deduce $6B = 3$ from the first relator, deduction 7, and from $5A = 6$ and $2B = 6$ we deduce $2C = 5$ from the third relator, deduction 8. Now we define $7 = 1C^{-1}$ and deduce $7A = 5$, $3B = 7$, $7B = 2$ from the respective relators and then $4A = 7$ from the first relator. Lastly we define $8 = 2A$ and deduce that the remaining blanks in lines 3–7 of the multiplication table are all to be filled with coset 8. As there are no further blank spaces for making new definitions, the enumeration is complete, and we have the following tables:

| | $A$ | $A^{-1}$ | $B$ | $B^{-1}$ | $C$ | $C^{-1}$ |
|---|---|---|---|---|---|---|
| 1 | 2 | 3 | 4 | 5 | 6 | 7 |
| 2 | 8 | 1 | $6_4$ | $7_{11}$ | $5_8$ | $4_2$ |
| 3 | 1 | $8_{16}$ | $7_{10}$ | $6_7$ | $4_1$ | $5_3$ |
| 4 | $7_{12}$ | $6_6$ | $8_{15}$ | 1 | $2_2$ | $3_1$ |
| 5 | $6_5$ | $7_9$ | 1 | $8_{13}$ | $3_3$ | $2_8$ |
| 6 | $4_6$ | $5_5$ | $3_7$ | $2_4$ | $8_{14}$ | 1 |
| 7 | $5_9$ | $4_{12}$ | $2_{11}$ | $3_{10}$ | 1 | $8_{17}$ |
| 8 | $3_{16}$ | 2 | $5_{13}$ | $4_{15}$ | $7_{17}$ | $6_{14}$ |

| | $A$ | | $B$ | | $C^{-1}$ | |
|---|---|---|---|---|---|---|
| 3 | | 1 | | 4 | 1 | 3 |
| 1 | | 2 | 4 | 6 | | 1 |
| 5 | | 6 | 7 | 3 | | 5 |
| 7 | 9 | 5 | | 1 | | 7 |
| 4 | 12 | 7 | | 2 | | 4 |
| 2 | | 8 | 13 | 5 | | 2 |
| 6 | | 4 | | 8 | | 6 |
| 8 | | 3 | | 7 | | 8 |

| | $B$ | | $C$ | | $A^{-1}$ | |
|---|---|---|---|---|---|---|
| 1 | | 4 | 2 | 2 | | 1 |
| 5 | | 1 | | 6 | 5 | 5 |
| 6 | | 3 | | 4 | | 6 |
| 3 | 10 | 7 | | 1 | | 3 |
| 7 | | 2 | | 5 | | 7 |
| 2 | | 6 | 14 | 8 | | 2 |
| 8 | | 5 | | 3 | 16 | 8 |
| 4 | | 8 | | 7 | | 4 |

| | $C$ | | $A$ | | $B^{-1}$ | |
|---|---|---|---|---|---|---|
| 5 | 3 | 3 | | 1 | | 5 |
| 1 | | 6 | 6 | 4 | | 1 |
| 2 | 8 | 5 | | 6 | | 2 |
| 7 | | 1 | | 2 | 11 | 7 |
| 3 | | 4 | | 7 | | 3 |
| 4 | | 2 | | 8 | 15 | 4 |
| 8 | 17 | 7 | | 5 | | 8 |
| 6 | | 8 | | 3 | | 6 |

## 4 Some Comments and Definitions

It will be seen that several lines of the relator tables above do not show deduction numbers. These are *consistent* lines whose filling shows that the eight elements are truly distinct and that the whole working is free from contradiction. Each relator table must finally contain the equivalent of a line beginning (and therefore ending) with each of the cosets. However, where a relator is a power of a generator or of a word in the generators (as does not happen in this case), a line of the table may contain the equivalent of several beginnings. In particular the relator table for a power of a generator needs only to contain every coset number somewhere in some line.

In this enumeration, the deduction $2C = 5$ (deduction 8) was not made until after coset 6 had been defined. It is an example of a *deep deduction*—one which relates cosets with smaller numbers than that most recently defined—but which involves this latter in the working leading to the deduction.

An *inconsistent* line of a relator table is of one of two forms. First, we may begin a line with a certain coset number and work straight through, without encountering an undefined coset, but find that the line, though complete, has different coset numbers at its beginning and end. Second, we may fill a line, with the same coset number at its beginning and end, and find an intermediate place which should be filled with two different coset numbers when it is approached from either side. In either case we discover a *coincidence*, establishing that two differently numbered cosets are identical. Discovery of a coincidence may lead to the exhibition of other pairs of coincident cosets, either because the products by a generator of the coincident cosets are both defined and bear different numbers, or because there are other lines of relator tables which can now be completed, but inconsistently. The aggregate of such coincidences is termed a *collapse*; it is *total* if all cosets are found to coincide, otherwise *partial*. The absence of inconsistent lines in the foregoing example shows the freedom from coincidences and the distinctness of the eight elements.

The occurrence of coincidences is closely related to that of deep deductions. A typical coincidence occurs when we define a new coset to fill a space in the multiplication table which later working would have shown by a deep deduction to have been due for filling with a smaller coset number. This is sometimes the consequence of defining cosets in an inefficient sequence, but there are occasions when the working unavoidably uses both versions of a coset before exhibiting the deduction that they coincide. This is usual when a collapse is total, several cosets having to be defined before these can be shown to be coincident. Details of an enumeration involving coincidences and a partial collapse are given by Coxeter [**3**].

## 5 Enumeration of Cosets

It is tedious and laborious to have to enumerate all the separate elements of a group, and the size of group that can be thus handled is seriously limited. To proceed more expeditiously, we enumerate *cosets* of a subgroup, availing ourselves of Lagrange's theorem that these cosets comprise equal numbers of elements, and that any two are either disjoint or coincident. Thus the order of the group is just the product of the order of the subgroup by its *index*, the number of distinct cosets. To do this, we precede the foregoing enumeration by the insertion of initial entries in the multiplication table which specify that coset 1 is fixed by the generators of the subgroup; this may involve the definition of further cosets in addition to coset 1.

An example is the presentation (1) above. To enumerate cosets of $\{A\}$ we insert the entries $1A = 1A^{-1} = 1$ in the multiplication table; this specifies that 1 now designates this subgroup. After this insertion we define $2 = 1B$, and find that the respective relators allow us to deduce $1C = 2$, $1C^{-1} = 2$ and $1B^{-1} = 2$, followed by $2A = 2$ to complete the working and two consistent rows to exhibit overall consistency. We now have the following tables:

| | $A$ | $A^{-1}$ | $B$ | $B^{-1}$ | $C$ | $C^{-1}$ |
|---|---|---|---|---|---|---|
| 1 | 1 | 1 | 2 | $2_3$ | $2_1$ | $2_2$ |
| 2 | $2_4$ | $2_4$ | $1_3$ | 1 | $1_2$ | $1_1$ |

| | $A$ | | $B$ | | $C^{-1}$ | |
|---|---|---|---|---|---|---|
| 1 | | 1 | | 2 | $_1$ | 1 |
| 2 | $_4$ | 2 | | 1 | | 2 |

| | $B$ | | $C$ | | $A^{-1}$ | |
|---|---|---|---|---|---|---|
| 1 | | 2 | $_2$ | 1 | | 1 |
| 2 | | 1 | | 2 | | 2 |

| | $C$ | | $A$ | | $B^{-1}$ | |
|---|---|---|---|---|---|---|
| 2 | | 1 | | 1 | $_3$ | 2 |
| 1 | | 2 | | 2 | | 1 |

The subgroup $\{A\}$ is thus found to be of index 2 in the group. Unfortunately it tells us little about the subgroup, and is thus somewhat uninformative about the group. In particular we cannot instantly deduce that the subgroup is of order 4, as we could have got exactly the same tables if it had been of order 2 and the group the Klein 4-group which satisfies the same relations. We shall see later how to enlarge on this limited information.

## 6 Analysis of Operation

We now examine the operations performed in coset enumeration, both by hand and as implemented on computers. After setting up the presentation of the group and specifying the chosen subgroup, we have to perform the following operations:

(i) defining new cosets, by making appropriate entries in the multiplication table;
(ii) inserting new entries from the multiplication table (whether made by definition or deduction) in the relator tables;
(iii) detecting closure of rows of relator tables and inserting corresponding deduced entries in the multiplication table;
(iv) completing consistent and inconsistent rows of relator tables, detecting coincidences and eliminating them by replacing all references to the larger

numbered coset by references to the smaller and dealing with any consequential coincidences.

The last three of these can be dealt with fairly briefly. Concerning (ii) we remark that it is the normal practice for computer implementation to store the relators but not the corresponding relator tables. (Thus the only extensive table retained is the coset multiplication table, and this is sometimes called *the coset table.*) Instead, each line of each relator table is constructed anew whenever a new multiplication table entry is made in a column corresponding to a letter in that relator. There are three possibilities. The line may fail to close, and no action is taken. It may just close, leading to a new deduction for insertion into the multiplication table. Or the line may be filled, either consistently or inconsistently; in the former event we take no action, but in the latter event a place in the line can be filled with two different coset numbers and we have a coincidence. Dealing with coincidences is straightforward in principle, but computer implementation is complicated in detail and will not be described here (see [**7**, **9**, **10**]).

The sequence of definition of cosets in (i) is at our disposal and we have considerable scope for choice. This has been the subject of much investigation (see especially [**2**]).

## 7 Termination of Process

We remark first that in principle termination will always be reached for any presentation of a finite group, provided that cosets are defined in a sequence which ensures that the multiplication table is "filled", in the sense that no place in the table can escape receiving an entry, by definition or deduction, not exceeding some function of its row number such as a simple multiple. For then every relation used in some proof of finiteness will eventually be spelled out, and the proof of finiteness of the number of elements (and *a fortiori* of cosets) will exhibit the closure of the working. But there is no algorithm for bounding the length of the working for a given presentation in terms of the final order, even for presentations of the trivial group. So our recipes for sequences of definitions, though usually termed *algorithms*, are not strictly algorithms in the sense of mathematical logic.

## 8 Hand Working

When working by hand, we usually define cosets in a sequence that "gives most value for money" at each stage, by causing or bringing nearer the closure of lines of relator tables and deduction of further multiplication table entries, while keeping in mind the need for "filling" the table as mentioned above. For example, when enumerating elements of the quaternion group in presentation (1) above, having defined $2 = 1A$, we would not (as above) go straight on to define $1A^{-1}$, but would notice that definition of $1B$ or $1C$ or $1C^{-1}$ leads to a deduction, while definition of $1A^{-1}$ does not. Other things being equal, we choose the earliest of these and define $3 = 1B$, deducing $3C = 2$. There are now several choices for a next definition giving a further deduction, including some which allow two further deductions; of these latter, the earliest is $1C$, so we define $4 = 1C$ and deduce $2B = 4$ and $4A = 3$. Now whatever definition we make next will lead to two further deductions, and continuing in this way we easily complete the working without encountering any coincidences or even deep deductions.

The following presentation will be used to illustrate three points. It has seven involutory generators $A, B, C, D, E, F, G$, and the first remark is that with involutory generators we do not need separate columns for their inverses in the multiplication table and we do not need relator tables for $A^2$ etc. as the corresponding deductions are made immediately. Thus as soon as we define $2 = 1A$ we deduce and insert the inverse entry $2A = 1$ at once. These relators are not presented to the machine. The full presentation is:

$$\{A, B, C, D, E, F, G \mid A^2 = B^2 = C^2 = D^2 = E^2 = F^2 = G^2 = I,$$
$$(AC)^2 = (BD)^2 = (CE)^2 = (DF)^2 = (EG)^2 = (FA)^2 = (GB)^2 = I,$$
$$(ABC)^2 = (BCD)^2 = (CDE)^2 = (DEF)^2 = (EFG)^2 = (FGA)^2 = (GAB)^2 = I,$$
$$ABCDEFG = AFDBGEC = I\}. \qquad (2)$$

Our next observation is a warning on the need for filling the multiplication table. Suppose we begin to enumerate cosets of the subgroup $\{C, E\}$ of order 4, and naively make definitions, each of which will fill a line of a relator table and allow a deduction to be made. We have $1C = 1E = 1$, defining the subgroup. Then defining $2 = 1A$ we deduce $2C = 2$ from $(AC)^2 = I$, defining $3 = 1G$ we deduce $3E = 3$ from $(EG)^2 = I$, defining $4 = 2E$ we deduce $4C = 4$ from $(CE)^2 = I$, defining $5 = 3C$ we deduce $5E = 5$ from $(CE)^2 = I$, and so on, filling only columns $A, C, E, G$ of the multiplication table and only rows of the relator tables for $(AC)^2 = (CE)^2 = (EG)^2 = I$. It is easily seen that it is never at any stage possible to make any other one definition that will itself close a line of a relator table, and that the foregoing sequence will continue indefinitely and will never lead to the correct conclusion that there are only 16 distinct cosets (the group is in fact the elementary abelian group of order 64). So we must not confine our definitions of cosets to those which close lines of relator tables without regard to filling the multiplication table, or we could get into an endless cycle.

The third remark is that in hand work we often use information from lines of relator tables which are not closed. Suppose for example that we had defined cosets in this presentation (2), including $2 = 1A$, $3 = 2B$, $4 = 3C$ and $5 = 4D$. Then from the relator $ABCDEFG$ we can deduce $5EFG = 1$. We can insert this in the relator table for $(EFG)^2$ and deduce $1EFG = 5$, which in turn we can insert in the relator table for $ABCDEFG$, deducing $5ABCD = 1$. In hand work our definitions are often directed towards making deductions in this way. This is difficult to emulate on a computer, though work described by Sims at the Symposium, and promised in a forthcoming monograph, indicates progress in this direction. One could introduce redundant generators, equal to products such as $EFG$ above which appear in more than one relator, but this is likely to produce a greater expansion of the multiplication table in one dimension than it would save in the other.

## 9 Computer Implementation

Computer implementations are most readily characterized by the choice of sequence of definition of new cosets, and fall broadly into two classes. That most closely resembling the hand work, as described above, was first used by Bandler [**1**], but is usually associated with the name of Felsch [**5**] who gave the first published account. Bandler did not deal with coincidences by program, but sought to avoid them by suitable sequences of definitions of cosets, such as always filling the earliest gap in the multiplication table or the earliest gap in the next column after that in which the previous definition was made (or, if this column was full, the next column which was not full). Although successful at completing many enumerations without coincidences, Bandler's work was seriously restricted by the absence of the facility for handling coincidences and the limited computer power available. Felsch, with a more powerful computer, used a similar sequence based on filling the earliest gaps in the multiplication table. His programs were more powerful, as they handled coincidences automatically. His, and similar later, programs have done much valuable work in subsequent years.

A disadvantage of this sequence of definition of cosets is that, in many enumerations involving coincidences, a large number of cosets are defined which play no part in the final collapse. In an example quoted by Leech [**10**], the collapse, in this case total, was precipitated by the definition of coset number 1544, but the next highest numbered coset to play a part in the working was numbered 1142. Some four hundred intermediate cosets were unnecessary to the working. One way of attempting to alleviate this inefficiency is to keep a record of "small gaps" in relator tables, places where a line is not yet closed but could be closed by the definition of one further coset (or a few more for very long relators), and giving these priority over the consecutive filling of the multiplication table. Havas reported to the

Symposium on some experiments with this modified sequence of definitions. But, as noted above, this priority must not be absolute, as the multiplication table may not be "filled". One possibility is to give priority to filling spaces in the multiplication table within some specified initial fraction of its current length, if there are any, otherwise to closing gaps in relator tables. Some "initial gap filling" can be achieved by including some or all of the group relators, perhaps in cyclically permuted forms, as (redundant) subgroup generators. This will ensure the filling of lines of relator tables with coset 1 at the beginning and at positions to which it is cyclically permuted.

The implementations of Haselgrove (in 1953 but unpublished) and his successors, use a different strategy. Recognizing that ultimately each relator table must contain rows beginning with each coset, they examine the rows of the relator tables beginning with each coset in turn. Whenever one is found that is not complete, they always define sufficient new cosets to complete it. But these new multiplication table entries are not then tried in other relators. While this results in many more redundant cosets to be eliminated by coincidence, it presents a much simpler algorithm for the programmer. It is often faster running than the Felsch algorithm in spite of the additional redundancy, and much useful work has been done with such programs.

The inefficiency of this algorithm is conspicuous with group presentations including long relators. Suppose, for example, we wish to enumerate elements (cosets of $\{I\}$) of the group:

$$\{A, B \mid A^2 = B^3 = (AB)^7 = (AB^{-1}AB)^8 = I\}. \tag{3}$$

The algorithm requires us to fill a row, beginning and ending with 1, for each relator, before considering rows beginning with 2 etc. This results in the definition of elements such as $(AB^{-1}AB)^4$ before elements such as $BA$, and elements beginning with $BA$ will not be defined until a very late stage.

This inefficiency can be partially alleviated by means of "look-ahead", [**2**]. Instead of insisting on completing every line of each relator table when we come to it, we can on suitable occasions examine lines of relator tables, ignoring any that are incomplete, to see if any of them produce deductions or coincidences without definition of new cosets. We can adopt this search mode when the number of cosets reaches specified limits (perhaps corresponding to blocks of store), reverting to ordinary working when any coincidences have been eliminated, or we might use it all the time on certain long relators, making definitions only when working on shorter relators. Each alternative has advantages over the unmodified algorithm.

## 10 Choice of Presentation

We now consider the effects of choice of presentation of the group, first as it affects the performance of the programs. With algorithms of the Haselgrove type, it is clear that the sequence in which the relators are presented affects the sequence of definition of cosets. Cannon *et al.* [2] have examined this in some detail and conclude that it is advantageous (i) to present the relators in order of increasing length, with the shortest first, and (ii) where possible to permute relators cyclically so that some of the relators (or their inverses) have common initial segments.

With algorithms of the Felsch type, the ordering of the relators is of no consequence but the sequence of generators and inverses heading the columns of the multiplication table can have a substantial effect. It is usually advantageous to place involutory generators first, followed by others, with their inverses, in increasing order of period (where these periods are known from the relators or otherwise). In an example where the generators are of the same period, Havas [6] describes enumerations of one-generator subgroups of the Fibonacci group,

$$\{A, B, C, D, E, F, G \mid AB = C, BC = D, CD = E, DE = F, EF = G, FG = A, GA = B\}$$

(cf. (1) above), in which choice of subgroup is equivalent to cyclic permutation of the columns of the multiplication table. He found that the different enumerations needed from 327 to 742 cosets to be defined before the total collapse set in. In this example a hand enumeration, making definitions so as to maximize the deductions made from each definition, collapsed after the definition of 129 cosets. Cutting out the dead wood (parts of the working playing no part in the final collapse) reduced this number to 55. Of course, such a minimal number (if indeed it is minimal) will usually be found only retrospectively, but the initial hand number of 129 will serve as a target for "intelligent" computer algorithms. (Use of this enumeration halves the length of Havas's formal proof of the order of the group, but the proof is still so opaque that I am content to state the existence theorem: "There exists a formal proof half as long as Havas's proof".) But one cannot expect perfection in a general purpose computer algorithm. It seems unlikely that such an algorithm would enumerate the five cosets of $\{A\}$ in $\{A, B \mid A^4 = A^2BAB^{-2} = I\}$ without coincidences. This would require that, after the insertion of $1A = 1$ to specify the subgroup, the coset $1B$ must not be defined, but other working must continue until $1B$ is deduced.

We now look at group theoretical considerations affecting our choices of subgroup and of group presentation. Usually we choose the subgroup to be as large as possible, subject to our ability to ascertain its order. Sometimes the order of the subgroup is deducible retrospectively. In the Fibonacci group last mentioned, the order of the subgroup is most readily deduced after we have found that the enumeration collapses. This collapse implies that the subgroup is the whole group, which is therefore cyclic and Abelian; it is then readily seen to be of order 29. In some cases we are not interested in the order of the group or subgroup, provided that the enumeration terminates and exhibits the finite index.

The ease or difficulty of enumeration can vary considerably with the choice of subgroup, even from among those of equal order. Thus in the group

$$\{A, B \mid A^8 = B^7 = (AB)^2 = (A^{-1}B)^3 = I\} \tag{4}$$

enumeration of the 448 cosets of the octahedral subgroup $\{A^2, A^{-1}B\}$ is difficult and needs the definition of many redundant cosets. But enumeration of the 448 cosets of $\{BA, AB^3\}$ is straightforward. The problem now is that of showing that this latter subgroup is of order 24, which depends on showing that the product of the subgroup generators is of period 6, i.e. $(A^2B^4)^6 = I$, and this is a surprisingly difficult problem [**11**].

Another influence on the difficulty of an enumeration is the choice of relations for the presentation. Often addition of redundant relations effects substantial simplification. Thus, if to the presentation (4) above we adjoin the redundant relation $(A^2B^4)^6 = I$, enumeration of the 448 cosets of the octahedral subgroup ceases to present any difficulty. Change of generators may also be helpful. For example the presentations (3) and (4) above are isomorphic, but the latter is easier to work with as it has no long relators comparable with $(AB^{-1}AB)^8$ in the former. The following example illustrates an experience which might not be foreseen. The relations $B^7 = (AB)^2 = (A^{-1}B)^3 = I$, together with any one of the eight possible relations $A^{\pm 4}BA^{\pm 4}B^2A^{\pm 4}B^4 = I$ obtained by choice of signs, define a group of order 1344. With any choice of signs the relation $A^8 = I$ is implied, so these are all equivalent. But enumeration of the 192 cosets of $\{B\}$ is of widely varying difficulty, depending on the choice of signs, the version $A^{-4}BA^4B^2A^4B^4 = I$ being especially troublesome. Cannon *et al.* [**2**] have discussed several examples of the effect on enumeration of the choice of presentation.

## 11 Interpretations

A completed coset enumeration may be interpreted in several ways in addition to the obvious one. We can remove a relation from the presentation of the group, and interpret its use in lines of relator tables as adjoining conjugates of the relator to the set of elements generating the subgroup. This was done by Leech [**8**] when finding generators for normal subgroups of $(2, 3, 7)$ with specified quotient groups. If we do this with all the relations, we obtain a set of generators for a subgroup of finite index in the free group on the original generators. For example, our second enumeration for the presentation (1) above shows that the elements $A$, $BC^{-1}$, $BC$, $B^{-1}C$, $C^{-1}AB$ generate a subgroup of index 2 in the free group $\{A, B, C\}$.

## 12 Permutation Representations

In the first instance, the information obtained from the completion of a coset enumeration is no more than the bare minimum, namely the index of the subgroup in the group. We can, however, glean further information. The action of the generators on the cosets, as displayed in the multiplication table, gives a permutation representation of the group. This, though, may not be faithful, as is evident from the differing interpretations mentioned above. What it gives is a faithful representation of the quotient group obtained by factoring out the intersection of all conjugates of the subgroup by elements of the group. In many cases, such as simple groups, this intersection is trivial and the group must be faithfully represented, but this may need checking in other cases. The two cosets of $\{A\}$ in our second enumeration of (1) above supply an easy example where the representation is not faithful.

We can work this process backwards to find whether a group has a permutation representation of fairly small degree by seeking a subgroup of the specified index, which we do by attempting to construct a multiplication table of that number of cosets which is consistent with the relators for the group. Often many small numbers can be immediately excluded, and it may not take long to find the smallest permutation representation for the group (while bearing in mind that this may not be faithful).

## 13 Proof of Relations

Another use to which coset enumerations can be put is the proof of relations in the group, and in particular the derivation of sets of defining relations for the subgroup. A method for this latter problem was given by Moser in a lecture at the 1967 Oxford Conference but was not then published; an account is given in the fourth edition, 1980, of Coxeter and Moser [**4**].

The method of Leech [**10**] is different in organization but substantially equivalent in effect. It is illustrated in the following example of the group presented by $\{A, B \mid A^2 = B^2 = (AB)^{-2}\}$. Direct enumeration to obtain the order can be done only by the laborious enumeration of elements (cosets of $\{I\}$), as the obvious non-trivial subgroups contain the central element $Z = A^2 = B^2 = (AB)^{-2}$ and the enumerations do not directly exhibit the order of $Z$. So we proceed as follows, with annotations as before, to enumerate cosets of the subgroup $\{A\}$. After inserting $1A = 1$ to specify the subgroup, we define

$2 = 1B$. From $A^2B^{-2} = I$, in the form $B^{-1} = A^{-2}B$, we deduce $1B^{-1} = 2$, and we write this line of the relator table in the form

$$1 \ B^{-1} \ 2 = 1 \ A^{-1} \ 1 \ A^{-1} \ 1 \ B \ 2.$$

(For typographical convenience we write the generators and coset numbers on the same line.) Then from $A^3BAB = I$, in the form $A = B^{-1}A^{-3}B^{-1}$, we deduce $2A = 2$ and write this line of the relator table as

$$2 \ A \ 2 = 2 \ B^{-1} \ 1 \ A^{-1} \ 1 \ A^{-1} \ 1 \ A^{-1} \ 1 \ B^{-1} \ 2 = 2 \ B^{-1}A^{-5}B \ 2,$$

where we have substituted for $1 \ B^{-1} \ 2$ using the first result. At each stage we construct an expression not involving deduced entries by substituting for these where they occur. (Lines of working such as this should be read as correct statements in the group with the coset numbers added as punctuation and indicating the substitutions being made.) This completes the enumeration, and we have the following tables, including the consistent lines:

| | $A$ | $A^{-1}$ | $B$ | $B^{-1}$ |
|---|---|---|---|---|
| 1 | 1 | 1 | 2 | $2_1$ |
| 2 | $2_2$ | $2_2$ | $1_1$ | 1 |

| | $A$ | $A$ | $B^{-1}$ | $B^{-1}$ |
|---|---|---|---|---|
| 1 | 1 | 1 | $_1$ 2 | 1 |
| 2 | 2 | 2 | 1 | 2 |

| | $A$ | $A$ | $A$ | $B$ | $A$ | $B$ |
|---|---|---|---|---|---|---|
| 1 | 1 | 1 | 1 | 2 | $_2$ 2 | 1 |
| 2 | 2 | 2 | 2 | 1 | 1 | 2 |

We now operate similarly on the consistent lines. From $A^2B^{-2} = I$ we get

$$2 \ I \ 2 = 2 \ A \ 2 \ A \ 2 \ B^{-1} \ 1 \ B^{-1} \ 2 =$$

$$2 \ B^{-1}A^{-5}B \ 2 \ B^{-1}A^{-5}B \ 2 \ B^{-1} \ 1 \ A^{-2}B \ 2 = 2 \ B^{-1}A^{-12}B \ 2,$$

so we deduce the relation $A^{12} = I$. From $A^3BAB = I$ we get

$$2 \ I \ 2 = 2 \ A \ 2 \ A \ 2 \ A \ 2 \ B \ 1 \ A \ 1 \ B \ 2 =$$

$$2 \ B^{-1}A^{-5}B \ 2 \ B^{-1}A^{-5}B \ 2 \ B^{-1}A^{-5}B \ 2 \ B^{-1}A^2 \ 1 \ A \ 1 \ B \ 2 = 2 \ B^{-1}A^{-12}B \ 2,$$

giving redundantly the same relation $A^{12} = I$. As this completes all the lines of the relator tables, we infer that this is the complete set of defining relations for the subgroup, so this is of order 12 and the whole group is of order 24.

Similar working for the other cases shows that the order of the group $\{A, B \mid A^l = B^m = (AB)^{-n}\}$ (when finite, i.e. $l^{-1} + m^{-1} + n^{-1} > 1$) is equal to the order of the factor group $\{A, B \mid A^l = B^m = (AB)^n = I\}$, dihedral or polyhedral, multiplied by the order of the central element $Z = A^l = B^m = (AB)^{-n}$, and this latter exceeds the order of the factor group by 2. Thus from the enumeration of the 12 cosets of $\{AB\}$ in $\{A, B \mid A^2 = B^3 = (AB)^{-5}\}$ we can calculate the order of the group to be $60.62 = 3720$, a considerable saving of effort over the enumeration of this number of elements.

The procedure is more complicated if there are coincidences. Some examples of how to proceed in such cases are given in [**10**]. One of these, the non-Hopfian group $\{A, B \mid A^{-1}B^2A = B^3\}$ with subgroup $\{A, B^2\}$, is also given in [**4**] (1980 edition), affording a comparison of the two methods.

A more elaborate example is the group presentation (2) above. Here it is difficult to enumerate the 16 cosets of subgroups such as $\{C, E\}$, of evident order 4, coincidences being unavoidable. But if we enumerate the 16 cosets of $\{F, G\}$, we can achieve this without coincidences. With the Felsch sequence of definitions we can "only just" achieve this: after defining $1A, 1B, \ldots$ to fill the gaps in the multiplication table in sequence, we reach the definition $13 = 3E$ before making the deep deduction $3F = 3$, just in time to avoid defining $14 = 3F$. (Had we chosen $\{A, B\}$ as our subgroup, this equivalent problem would have led to coincidences with the Felsch sequence; this is another example of the effect of permuting the columns of the multiplication table. Defining cosets to close lines of relator tables also leads to coincidences.) In this enumeration, the order of the subgroup is not known at the outset, and we have to deduce it. It turns out that no deduction was made from the relation $AFDBGEC = I$ (and if this relation is omitted, both group and subgroup are doubled in size—another interpretation of the same enumeration). The consistent line $1 \;\; I \;\; 1 = 1 \;\; FDBGECA \;\; 1$ is found, on substituting for deduced entries as above, to lead to $(GF)^2 = I$, showing that the subgroup $\{F, G\}$ is of order 4 and the group of order 64. (This group is the normal subgroup of the group $(8, 7 \mid 2, 3)$ (presentation (4) above) generated by conjugates of $A^4$ [**11**], and the problem of proving $(FG)^2 = I$ in (2) is equivalent to that of proving $(A^2B^4)^6 = I$ in (4).)

## 14 Schreier Coset Diagrams

The information contained in a coset multiplication table may be portrayed graphically in the form of a Schreier coset diagram. The nodes of this graph are the cosets, and directed joins, of types corresponding to the generators of the group, join each coset to its image under the action of the generator. (For involutory generators it is convenient to use a single undirected join in place of each pair of directed joins which would join a pair of cosets in the two directions.) Some examples are given in [**4**]. Where feasible, it is attractive to draw the graph so as to portray its automorphisms as geometrical symmetries.

Some drawings by the author were shown at the Symposium, the most elaborate being of 240 cosets and 756 cosets in the automorphism groups of the quadratic forms $E_8$ and $K_{12}$ respectively. Unfortunately these are not suitable for reproduction in these Proceedings. In cases of multi-generator groups such as these, it is desirable to preserve as far as possible the symmetries of the subgraphs formed by subsets of the generators, and indeed these large drawings were built up from copies of drawings of the subgraphs. They are essentially isometric, with all joins of equal length (except for those few which cross the centre of symmetry of the diagram for $K_{12}$ which are twice the length of others), but some distortion from this is needed to avoid exact superposition of nodes for distinct cosets. It appears to be possible to preserve this isometry if expansion into three dimensions is allowed, but the author has not yet faced the task of planning and constructing such three-dimensional models.

## References

1. Bandler, P. A. (1956). M.A. Thesis, Manchester University.
2. Cannon, J. J., Dimino, L. A., Havas, G. and Watson, J. M. (1973). Implementation and analysis of the Todd–Coxeter algorithm, *Mathematics of Computation* **27**, 463–490.
3. Coxeter, H. S. M. (1970). *In* "Twisted Honeycombs", 33–34. American Mathematical Society.
4. Coxeter, H. S. M. and Moser, W. O. J. (1957, 1965, 1972, 1980). "Generators and Relations for Discrete Groups". Springer-Verlag, Berlin.
5. Felsch, H. (1961). Programmierung der Restklassenabzählung einer Gruppe nach Untergruppen, *Numerische Mathematik* **3**, 250–256.
6. Havas, G. (1976). Computer aided determination of a Fibonacci group, *Bull. Austral. Math. Soc.* **15**, 297–305.
7. Leech, J. (1963). Coset enumeration on digital computers, *Proc. Camb. Phil. Soc.* **59**, 257–267.
8. Leech, J. (1965). Generators for certain normal subgroups of (2, 3, 7), *Proc. Camb. Phil. Soc.* **61**, 321–332.
9. Leech, J. (1970). Coset Enumeration. *In* "Computational Problems in Abstract Algebra" (Ed. J. Leech), 21–35. Pergamon Press, Oxford.
10. Leech, J. (1977). Computer proof of relations in groups. *In* "Topics in Group Theory and Computation" (Ed. M. P. J. Curran), 38–61. Academic Press, London and New York.
11. Leech, J. and Mennicke, J. (1961). Note on a conjecture of Coxeter, *Proc. Glasgow Math. Assoc.* **5**, 25–29.
12. Newman, M. H. A. (dated 1951, but privately issued 1953). The influence of automatic computers on mathematical methods. Manchester University Computer Inaugural Conference 1951, p. 13.
13. Todd, J. A. and Coxeter, H. S. M. (1936). A practical method for enumerating cosets of a finite abstract group, *Proc. Edinburgh Math. Soc.* (*2*) **5**, 26–34.

# Space Saving in Coset Enumeration

M. J. BEETHAM

## 1 Introduction

Many programs for enumerating the cosets of a finitely generated subgroup $H$ of a finitely presented group $G$ have been described. [**1**, **2**, **3**, **4**, **5**]. The algorithm we are about to describe differs from these in the method in which it handles collapses or coincidences. That is the action taken when it is discovered that two labels given in the execution of the algorithm turn out to describe the same coset.

The method of using the relations and subgroup generators to define new cosets and discover collapses used in the implementation described below is standard. It is of the type described in [**1**]. We will refer to this as the SEARCH routine. The procedure of interest in the space saving connection is the collapse handling routine which we refer to as COLLAPSE. Section 2 of the paper describes the theoretical storage requirements of the program. Section 3 describes the COLLAPSE routine. The program was originally implemented on an IBM 360/44 at St. Andrews and now runs on the IBM 3081 at Cambridge University Computing Laboratory.

The idea for the present algorithm arose following a conversation with I. D. Macdonald who suggested that it might be possible to use the space created by collapses to store the collapses remaining to be processed. This is not possible using conventional methods but can be done by modifying the collapse processing procedure.

## 2 The Multiplication Table

In a hand enumeration, one uses tables corresponding to the subgroup generators and group relations having one row for each coset label defined. In a machine implementation, one dispenses with these tables and uses instead a multiplication table with a column corresponding to each group generator and a row corresponding to each coset label.

To avoid having to search the table it is also convenient to keep in the same

COMPUTATIONAL GROUP THEORY
ISBN 0-12-066270-1

table, by adding further columns, a record of the products by the inverses of the generators. Unknown products are represented by zero entries. In the following, we will assume that the first two columns of the multiplication table correspond to a group generator and its inverse.

The size of the multiplication table is the limiting factor in the implementation of the algorithm. The algorithm can terminate incompletely as it may require to define a new label with no space available in the machine to do this. Various techniques are used to delay a final failure of the algorithm, such as packing the table to remove redundant entries. Such methods are discussed in more detail in section 4.

## 3 The Algorithm for Handling Collapses

As we remarked above, the essential difference between the present algorithm and previous enumeration algorithms is in the handling of collapses or coincidences. The usual procedure when discovering a coset with two different labels is to compare the corresponding rows of the multiplication table and note in a store any further collapses discovered in the comparison. All references to the larger of the two labels can then be removed from the table, either by replacing by the lower label and entering the inverse information or by deleting if the corresponding product of the lower label is also defined. The remaining collapses, if any are left in the store, are then processed in the same way before returning to the search routine.

The aim of the present algorithm is to save the additional storage required in the above process. The idea behind the method is that a collapse results in space being created in the multiplication table. In the above method, the whole row of the multiplication table corresponding to the larger label becomes available. However, it is easy to see that this does not provide enough space for storing consequential collapses if the multiplication table itself is efficient. To overcome this difficulty we process collapses in the following way.

On discovering the first of a series of collapses, the corresponding entries of the first two columns of the multiplication table are compared. Any collapses resulting from this comparison are then entered in a four word storage area. The first column of the multiplication table is then used to record the collapses being processed. To enable this to be done, we have to remove some of the information at present held in this position to a vacant space in the table. Suppose the lower of the two labels being processed is LOW and the higher one is HIGH. Then suppose the parts of the table we are considering are as shown in Table 1.

There is, of course, no intention to imply that the labels occur in the order shown in the table and, if some of the entries are zero, then the corresponding rows will not occur. We now enter the information in the first two columns of the

*Table 1*

| | Column 1 | Column 2 |
|---|---|---|
| | ... | ... |
| HIGH | HIGH1 | HIGH2 |
| | ... | ... |
| LOW | LOW1 | LOW2 |
| | ... | ... |
| HIGH2 | HIGH | *** |
| | ... | ... |
| HIGH1 | *** | HIGH |
| | ... | ... |
| LOW2 | LOW | *** |
| | ... | ... |
| LOW1 | *** | LOW |

rows labelled HIGH or LOW into general registers so that we can re-allocate the usage of the corresponding table locations. We convert the positions of Table 1 to the form shown in Table 2. In this table the alternative entries in the HIGH1 row arise in the following way. If LOW1 is zero then, provided HIGH1 is non zero, LOW1 is replaced by HIGH1 and now appears in column 2 of row HIGH. The column 2 entry for HIGH1 then becomes LOW. Otherwise this particular entry is zero. In the same way the alternative entries occur in row HIGH2 with consequent changes in column 2 of rows LOW and HIGH.

In this way we now have a multiplication table still containing the same information as before but which has space, in column 1 of the rows corresponding to cosets involved in collapses, for containing information about the relationship between such labels.

Column 1 of these rows is now used to store pointers producing a linked list

*Table 2*

| | Column 1 | Column 2 |
|---|---|---|
| | ... | ... |
| HIGH | LOW | LOW1 or HIGH1 |
| | ... | ... |
| LOW | HIGH$ | LOW2 or HIGH2 |
| | ... | ... |
| HIGH2 | 0 or LOW | *** |
| | ... | ... |
| HIGH1 | *** | 0 or LOW |
| | ... | ... |
| LOW2 | LOW | *** |
| | ... | ... |
| LOW1 | *** | LOW |

of labels all corresponding to the same coset. The flag, denoted by $, on the column 1 entry of row LOW shows that this label, LOW, is the lowest of such a linked list. The multiplication table must now be read in the following way.

We will assume, as happens in practice, that we will only need to look up the products, by the first generator and its inverse, of labels which either do not belong to a collapse or, alternatively, are the first label in such a linked list. Which of the two alternatives is determined by examining the flag corresponding to the label. If no flag is present then the table is read in the normal way. Otherwise the column 1 entry is used to find the required entry in column 2.

As Havas has pointed out, since seeing an earlier version of this paper, it is possible to use this technique with the type of collapse routine described in [**1**]. Indeed this technique is now used in his implementation of the algorithm.

The method we use processes collapses in a different way, making use of a more complex data structure. This ensures that the multiplication table remains consistent, in the sense that the occurrence of a non zero label $x$ in a row $y$ means that the entry in the $x$ row corresponding to the inverse generator will be $y$.

Subsequent collapses are processed in substantially the same way as the first pair of labels. The next collapse to be considered is obtained from the four word storage area mentioned above. This frees two words of that area and in processing this collapse only one, not two, collapses can occur from the first two columns. For example, in the case shown in the above tables, we would next process the collapse LOW1 = HIGH1, assuming this is a collapse, when it is clear that no collapse results from an examination of column 2.

In practice, the linked lists of collapses are themselves joined together to save having to search the table for the positions of collapsed labels later in the process. For this reason, we will call the sequence of labels all corresponding to the same coset a link. We combine the links to form a chain, the top entry of one link pointing to the bottom of the next link which, if the chain is one link long, will be the same link again. This is illustrated in the example of Table 2.

We actually keep two such chains, one involving links of length two and the other links of length at least three. The reason for this will become clear from the following discussion of the method of adding new links to a chain.

For each chain, we store in another four word storage area the top and bottom of each chain. If a new link to be added has length two, we simply insert it between the top and bottom and change the top entry accordingly. (Notice that each chain is in fact a loop. The entry in column 1 of the top row points to the bottom entry, as in Table 2.) If the collapse being processed involves labels which have been involved in collapses before then, we have to carry out a more complicated adjustment of the chains.

Essentially, what we have to do is to remove any links involving the new labels from the existing chains, connect two links together, or add a new label to one link if there is only one, and then insert this new link into the chain whose links have length greater than two.

To remove a link from the chain with links of length two, we have no alternative but to search all the way round the chain, recall that this is a loop, to find the previous link in the chain. Having done this we can remove the link and rejoin the chain.

To facilitate the removal of links of length greater than two from their chain, we make a further modification. It is clear that in such a link we are carrying no information in column 2 of the third and subsequent rows. Accordingly we make use of the column 2 position of the third row in a link to store a pointer to the top of the previous link in the chain. (This is the position in the link which we must alter when rejoining the chain after removal of a link.) This avoids us having to search all round the chain when processing links from the second chain.

The collapses we have considered so far have resulted firstly from a collapse found in the SEARCH routine and then subsequently from the consequential collapses arising from consideration of the first two columns of the multiplication table. When all such collapses have been processed, we then use the chains to enable us to compare the other columns of rows corresponding to the same coset.

If a further collapse results from the comparison of two rows, then we simply re-enter the COLLAPSE routine at its start. If no further collapse is discovered in the comparison, then we delete the higher of the two rows by adding a flag to its column 1 entry and adjusting the link in the chain to remove references to the deleted row.

It will be clear that this can result in the length of a link of length three reducing to length two, or a link of length two reducing to length one. In the first case it is necessary to move the resultant link to the other chain. In the second case we remove the link from the chains altogether, remove the flag from the column 1 entry and restore the information in columns 1 and 2 to its normal form.

Processing of collapses and then search for new collapses proceeds in the above manner until both chains have disappeared. The program then returns to the search routine.

## 4 Other Features of the Algorithm

The difficulty of coset enumeration lies in the fact that there is no bound in terms of the final result of the amount of space required to carry out the algorithm. Thus, even when working with presentations of finite groups, even small finite groups such as the trivial group, the algorithm can run out of the space available. When this happens, as mentioned in section 2, we first pack the table to make available the space released by collapsed labels. If no space is available in this way, then we continue the search without defining new labels in the hope of finding further collapses which will either terminate the algorithm successfully or allow extra space, and so extra cosets, to be created.

Eventually, however, the process can terminate incompletely with no space to define new coset labels and with no collapses resulting from the information present in the multiplication table.

One method of coping with this is to use secondary storage, [**1**]. Another is to search for representations on larger subgroups.

One way, which has been incorporated into St. Andrews algorithms since the early days of coset enumeration in the University, is to replace the subgroup $H$ by a larger subgroup in which $H$ is normal. In fact it is easy to see that the normalizer of $H$ is the union of those cosets which are fixed by the generators of $H$.

Thus, to carry out a further search from the point at which space runs out, we only need to find those labels which are fixed by the generators of $H$ and process the collapses which make them equal to 1, the label of $H$. In practice, so that we can repeat this operation several times, we also find coset representatives of these cosets so that we can add them to the generators of $H$.

It will be clear that this can be useful even when the process does terminate successfully. For example it can be used to find normalizers of subgroups, subgroups in which a given group is subnormal and representations of smaller degree of given groups.

Another method which was developed in connection with a problem of I. D. Macdonald, is to modify the algorithm in the following way.

Each time that a collapse is discovered in the SEARCH routine, we calculate coset representatives, $x$ and $y$ say, corresponding to the two labels and note on secondary storage that $xy^{-1}$ is an element of $H$.

When the algorithm runs out of space we simply run it again using the extra, redundant, generators of $H$ in addition to the original ones. This results in the collapses discovered in the first run being processed without so many cosets being defined and so leaves room to discover more information about $G$ and $H$. In practice, it has been found that this method can be repeated several times before the algorithm finds the required representation or, alternatively, discovers no essentially new information.

Another technique, which has proved surprisingly powerful, is to run the algorithm with a subset of the relations until it terminates incompletely. At this stage, further relations are searched, producing collapses which allow the algorithm to terminate successfully.

A final method which has been used with success to find a presentation of a simple group, is to use an incomplete representation obtained from an incomplete multiplication table to calculate the least common multiple of the lengths of the cycles representing a given word, $w$ say, in the group. This number, $n$ say, is then used to suggest the additional relation $w^n = 1$. This either is true or collapses the group to the trivial group and so it is easy to check that the method is successful.

One final point which is of interest in the search algorithm is the fact that although this is a normal HLT search algorithm as used in other implementations, the relation formed by taking the product $aa^{-2}a$ over all group generators $a$, is always added to the relations of the presentation. This is done automatically by the program and has the effect of ensuring that the product of $n$ by any generator or inverse is always defined when the $n$th row of the multiplication table is being checked.

## 5 Acknowledgements

I am grateful to many former colleagues in the University of St. Andrews and in particular to the staff of both the Computing Laboratory and the Department of Computational Science, for their help in transforming the algorithm into a working assembly language program. I should also like to thank Dr. I. D. Macdonald for his interest and encouragement in the early stages of the work. Finally I must thank Dr. G. Havas for encouraging me to publish an account of this work.

## References

1. Cannon, J. J., Dimino, L. A., Havas, G. and Watson, J. M. (1973). Implementations and analysis of the Todd–Coxeter algorithm, *Mathematics of Computation* **27**, 463–490.
2. Felsch, H. (1961). Programmierung der Restklassenabzählung einer Gruppe nach Untergruppen, *Numerische Mathematik* **3**, 250–256.
3. Guy, M. J. T. (1967). Coset enumeration. Lecture delivered at the conference on computational problems in abstract algebra, Oxford.
4. Leech, J. (1970). Coset Enumeration. *In* "Computational Problems in Abstract Algebra" (Ed. J. Leech), 21–35. Pergamon Press, Oxford.
5. Trotter, H. T. (1964). A machine program for coset enumeration, *Canad. Math. Bull.* **7**, 357–368.

# A Modified Todd-Coxeter Algorithm

D. G. ARRELL AND E. F. ROBERTSON

## 1 Introduction

Several different modifications of the Todd-Coxeter algorithm have been described which enable a presentation of a subgroup of finite index to be determined on a specified set of subgroup generators; see for example [**1**], [**2**], [**4**], [**6**] and [**7**]. All these methods tend to produce numerous and lengthy relations for the given subgroup and the authors of [**1**] give some thought to the way in which tree structures can be used to store the words used to produce the relations. They comment that their method of implementation leads more readily to the effective use of Tietze transformations than merely using a Tietze transformation program on the final subgroup presentation. Computer implementations to find presentations of finite index are most commonly based on the Reidemeister-Schreier method which relies heavily on Tietze transformations to eliminate redundant generators and shorten the longer relations.

The algorithm and its implementation, which we describe below, use Tietze transformations to eliminate generators and shorten relations in a similar way to the Reidemeister-Schreier method but, in common with the modified version of Todd-Coxeter referred to above, guarantees a presentation on the specified set of subgroup generators. The idea for the algorithm we describe originates in the work of [**1**] and indeed the first half of our implementation uses the program described in [**1**].

## 2 Description of the Algorithm

Suppose $G$ is a finitely presented group $\langle X \mid R \rangle$ and $H = \langle Y \rangle$ is a subgroup of finite index in $G$. We use essentially the Todd-Coxeter method to enumerate the cosets of $H$ in $G$ but instead of $1, 2, 3, \ldots$ representing cosets, we shall use these numbers to represent coset representatives. So, for $x \in X$ and coset representatives $i, j$ we obtain information of the type

$$i \,.\, x = h \,.\, j \tag{1}$$

COMPUTATIONAL GROUP THEORY
ISBN 0-12-066270-1

where $h$ is a word in the subgroup generators $Y$. This method is nicely described in §13 of [**5**]. However we make one change which at first sight appears to have only technical significance. We do not allow products of length greater than one to be built up.

When, in the course of the algorithm, a product

$$y_{i_1}^{\varepsilon_1} y_{i_2}^{\varepsilon_2} (y_{i_1}, y_{i_2} \in Y, \ \ \varepsilon_i = \pm 1)$$

would normally occur as we attempt to calculate the value of $h$ in (1) (by the method described in [**5**]) we introduce a new subgroup generator

$$\bar{y} = y_{i_1}^{\varepsilon_1} y_{i_2}^{\varepsilon_2} \text{ and put the relator } \bar{y}^{-1} y_{i_1}^{\varepsilon_1} y_{i_2}^{\varepsilon_2}$$

into a set $T$ and form an augmented set of subgroup generators $\bar{Y} = Y \cup \{\bar{y}\}$. In the same way, whenever a product

$$\bar{y}_{i_1}^{\varepsilon_1} \bar{y}_{i_2}^{\varepsilon_2} \text{ with } \bar{y}_{i_1}, \bar{y}_{i_2} \in \bar{Y}$$

occurs, we proceed as before to define this to be a new subgroup generator which is added to $\bar{Y}$ and the corresponding relator is added to $T$.

We define an order on $\bar{Y}$ as follows. Order the elements of $Y$ arbitrarily and then, at each stage, when we add a generator $\bar{y}$ to $\bar{Y}$ we define $\bar{y}$ to be the greatest element of the new set $\bar{Y}$. Now the relators in $T$ express each generator as a product of two smaller generators or their inverses and hence the relators in $T$ allow us to write any element of $\bar{Y}$ as a word in the subgroup generators $Y$.

When the coset enumeration is complete and $H$ has been found to have index $n$ in $G$, then our monitor table, which records $i.x = h.j$ for each $x \in X$ and each coset representative $i$ with $1 \leqslant i \leqslant n$, will have $h \in \bar{Y}$, $h^{-1} \in \bar{Y}$ or $h = 1$. Construct the subset $\bar{B}$ of $\bar{Y}$ consisting of those elements $h$ of $\bar{Y}$ such that $h$ or $h^{-1}$ appears in this monitor table. Put $B = \bar{B} \cup Y$. Relators $R_1(B)$ for $H$ in terms of the generators $B$ of $H$ are of two types. There are those obtained by applying each relator $r \in R$ to each coset representative $i$, $1 \leqslant i \leqslant n$, and also those obtained when each $y \in Y$ is equated to the word resulting from applying $y$ to coset 1.

If at this stage we use the relators in $T$ to write each element of $B$ as a word in $Y$ and substitute these into the relations $R_1(B)$, we have a presentation for $H$ identical to that described in [**5**] and, were we to do this, our introduction of $\bar{Y}$ and $T$ would only be for the technical purpose of computational efficiency. However our algorithm proceeds as follows.

If there is a relator in $R_1(B)$ which allows the largest generator in $B$, say $g$, to be expressed in terms of smaller generators, eliminate $g$ using a Tietze transformation. If $g$ cannot be eliminated in this way, add to $R_1(B)$ the relator in $T$ which expresses $g$ in terms of smaller generators and then eliminate $g$ using the Tietze transformation. Our set of relators becomes $R_1(A)$, where $A$ is a subset of $\bar{Y}$ involving generators smaller than $g$. Continue in this way to remove the largest generator occurring in $A$. Because each generator is being replaced by words involving smaller generators, the ordering on $\bar{Y}$ requires that after a finite number of steps we reach $R_1(Y)$, i.e. the relators involve only the subgroup generators $Y$.

At any stage we may attempt to simplify the relators $R_1$ by searching for relators of the form $w_1 w_2^{-1}$ with $w_1$ shorter than $w_2$ and $w_2$ occurring in another relator. Then replace $w_2$ in this second relator by $w_1$; we call this process *substring searching*. Duplicate relators may occur in $R_1$ and we then discard one.

The final relations $R_1(Y)$ give a presentation for $H$ on the given generators $Y$,

$$H = \langle Y \mid R_1(Y) \rangle$$

and we give a proof of this in the next section. We remark at this point that many of the relators in $T$ may never be used; in fact it is possible that no relators in $T$ are used in the above method. This is in contrast with the usual modified algorithm which would, in effect, mean rewriting $R_1(B)$ in terms of the subgroup generators $Y$ and this involves the use of every relator of $T$.

## 3 Proof of the Algorithm

Let $U$ be a transversal of $H$ in $G$ and denote by $\bar{f}$ the element of $U$ representing coset $Hf$. Let $s_{u,x} = ux\overline{ux}^{-1}$ be a Schreier generator. Then the Reidemeister-Schreier theorem (see §12 of [**5**]) gives the presentation for $H$

$$H = \langle s_{u,x} \mid t(uru^{-1}), \ s_{u,x}^{-1} t(ux\overline{ux}^{-1}) \rangle$$

where $t$ is the rewriting function of Lemma 2.3 of [**5**]. Now this relates to the sets defined in §2 in that the transversal $U$ has been built up in the usual way by the monitor table of the modified algorithm, the generators $\bar{B}$ are Schreier generators and the relators $\{t(uru^{-1})\}$ are the first type of relators in $R_1(B)$. Let us denote by $R_2(B)$ the relators $\{s_{u,x}^{-1} t(ux\overline{ux}^{-1}\}$ and put $R_3(B) = \varnothing$, the empty set. Then,

$$H = \langle B \mid R_1(B), R_2(B), R_3(B) \rangle .$$

Consider what happens to $H$ when we eliminate the largest generator $g \in B$ by the method described in §2. If $g$ is eliminated using a relator in $R_1(B)$ then clearly

$$H = \langle A \mid R_1(A), R_2(A), R_3(A) \rangle$$

where $A = B \backslash \{g\}$, $R_2(A)$ and $R_3(A)$ are $R_2(B)$ and $R_3(B)$ with $g$ eliminated by the same Tietz transformation that eliminated it from $R_1(B)$. Suppose now that $g$ has to be eliminated by adding to $R_1(B)$ the necessary relator from $T$. Three cases must be considered.

(i) $g = g_1 g_2$ *where* $g_1, g_2 \in B$.

In this case $g = g_1 g_2$ is theoretically deducible from the relators $R_1(B)$, $R_2(B)$, $R_3(B)$, so adding it to $R_1(B)$, then eliminating $g$ from $R_1(B)$, $R_2(B)$, $R_3(B)$ as before, gives again

$$H = \langle A \mid R_1(A), R_2(A), R_3(A) \rangle \text{ where } A = B \backslash \{g\}.$$

(ii) $g = g_1 g_2$ *where* $g_1 \in B$ *and* $g_2 \notin B$.

Now $g_2 = g_1{}^{-1} g$ and adding $g_2$ to $B$ and $g_2 = g_1{}^{-1} g$ to $R_1(B)$ is a Tietze transformation. Now eliminate $g$ as in case (i) to get

$$H = \langle A \mid R_1(A), R_2(A), R_3(A) \rangle \text{ where } A = (B \backslash \{g\}) \cup \{g_2\}.$$

(The case $g_1 \notin B$ and $g_2 \in B$ is similar.)

(iii) $g = g_1 g_2$ *where* $g_1, g_2 \notin B$.

Put into $R_3(B)$ two relators $g_i{}^{-1} w_i(Y)$, $i = 1, 2$, which express $g_1$ and $g_2$ as words in $Y$. We do not specify at this stage what these words are but merely note that sufficient information to do this is contained in $T$. Now proceed as in (i) and eliminate $g$ from

$$\langle B \cup \{g_1, g_2\} \mid R_1(B \cup \{g_1, g_2\}), R_2(B \cup \{g_1, g_2\}), R_3(B \cup \{g_1, g_2\}) \rangle$$

to get a presentation

$$H = \langle A \mid R_1(A), R_2(A), R_3(A) \rangle \text{ where } A = (B \backslash \{g\}) \cup \{g_1, g_2\}.$$

Continue to follow the algorithm as described in §2 until the presentation is in terms of $Y$, i.e. $H = \langle Y \mid R_1(Y), R_2(Y), R_3(Y) \rangle$. Now note that each relator of $R_3$ started life as $g_i{}^{-1} w_i(Y)$ and must now have become $\bar{w}_i(Y)^{-1} w_i(Y)$ where $\bar{w}_i(Y)$ is the result of applying to $g_i$ the successive Tietze eliminations. We can now assume that $\bar{w}_i(Y)$ was our original choice of $w_i(Y)$ so $R_3(Y)$ is empty. In [**8**], Neubüser compares the Reidemeister-Schreier and modified algorithms and deduces from Theorem 3 and Theorem 3′ in the paper that, when produced by the modified algorithm, the relators $\{s_{u,x}^{-1} t(ux\overline{ux}^{-1})\}$ become consequences of the others when written in terms of the original subgroup generators. Hence the relators $R_2(Y)$ are consequences of $R_1(Y)$ and we have proved that $H = \langle Y \mid R_1(Y) \rangle$ as required.

## 4 An Example using a Computer Implementation

We consider an example of the type of result which our computer implementation of the algorithm will produce. First we must say a little about this implementation. The first point to note is that we introduce a new group generator equal to each given subgroup generator and add the corresponding relators to the given group relators. This results in more relators in $R_1(B)$ than one would have otherwise expected.

The subgroup generators $Y$ are denoted by positive even integers, being ordered by the natural order on the integers. Additional generators in $\bar{Y}$ are denoted by the next available even integer, say $n$, and if the new generator $n$ is defined by

$$n = n_1^{\varepsilon_1} . n_2^{\varepsilon_2},$$

we add the relator $n^{-1}n_1^{\varepsilon_1}n_2^{\varepsilon_2}$ to $T$ by entering $\pm n_1$ and $\pm n_2$ (+ if $\varepsilon = 1$, − if $\varepsilon = -1$) into records $n$ and $n+1$ respectively, in a file which we shall also call $T$. The monitor tables are constructed in the usual way and the relators $R_1(B)$ are written to a file.

The second phase of the program begins by eliminating trivial and duplicate relators in $R_1(B)$. Then the elimination of the highest numbered generators starts using the relators $R_1$ if possible. If at some stage a generator $m$ cannot be eliminated using $R_1$, its expression as a product of two lower numbered generators is found by looking up records $m$ and $m+1$ in the file $T$. After adding this relator to $R_1$, then $m$ is eliminated. The frequency of substring searching is set at the beginning of this second phase. Several techniques are available to simplify the final presentation $\langle Y | R_1(Y) \rangle$ and are best used interactively.

As an example consider

$$G = \langle a, b \,|\, ab^2a^{-1}b^{-1}a^3b^{-1} = ba^2b^{-1}a^{-1}b^3a^{-1} = 1 \rangle$$

and put $x = [a^{-1}, b^{-1}]$, $y = [a^{-1}, b]$ and $z = [a, b]$. Our notation here is such that $[a, b] = a^{-1}b^{-1}ab$. Consider first the subgroup $H = \langle x, y, z \rangle$ which has index 18 in $G$. In [**3**] this enumeration was performed using a machine implementation of the usual version of the modified Todd-Coxeter algorithm. A presentation for $H$ consisting of 20 relators of total length about 1200 was found. After much labour with hand calculations this was reduced to 14 relators of total length 246, this final presentation being given in [**3**].

Our computer implementation of the algorithm described in this paper produces 89 relators in $R_1(B)$ and 147 relators in $T$. After eliminating duplicate and trivial relators, 73 relators remain in $R_1(B)$. If we use substring searching after every 6 eliminations, we obtain a presentation for $H$ on the given generators $x$, $y$, $z$ with 11 relators of total length 72. Relators are as follows:

$$\begin{aligned}[x^2, y] &= [x^2, z] = [y^2, x] = [z^2, x] = [z^2, y] = y^{-1}(z^{-1}x)y(z^{-1}x) \\ &= y^{-1}(xz^{-1})y(xz^{-1}) = y^{-2}(zx)^2 = x^4z^{-4} = y^4z^4 \\ &= y^{-1}(zy^{-1}x)y(zy^{-1}x) = 1.\end{aligned}$$

During the course of the eliminations 14 of the 147 relators in file $T$ were used.

As a second example, consider the subgroup $K = \langle x, y \rangle$ of index 72 in the group $G$. The algorithm produces 275 relators in $R_1(B)$ which reduces to 234 after duplicates and trivial relators have been eliminated. There are 405 relators in $T$. Substring searching after every 20 eliminations gives a presentation for $K$ on the given generators with the following 7 relations:

$$[x^2, y] = [x, y^2] = x^4y^{-4} = (xy)^2(yx)^{-2} = (xy^{-1})^4 = x^8y^4 = (x^2y)^4 = 1.$$

Only 4 of the 405 relators in $T$ are used.

Of course, small changes can have a considerable effect on the data we have quoted for these examples. If we change the order of the group relations when finding a presentation for $K$ and do substring searching after every 12 eliminations, 41 of the relators in $T$ are used although the final presentation obtained for $K$ does not differ greatly from the one given, above.

## References

1. Arrell, D. G., Manrai, S. and Worboys, M. F. (1982). A procedure for obtaining simplified defining relations for a subgroup. *In* "Groups St. Andrews 1981" (Eds. C. M. Campbell and E. F. Robertson), L.M.S. Lecture Notes **71**, 155–159. Cambridge University Press.
2. Beetham, M. J. and Campbell, C. M. (1976). A note on the Todd-Coxeter coset enumeration algorithm, *Proc. Edinburgh Math. Soc.* **20**, 73–79.
3. Campbell, C. M. and Robertson, E. F. (1975). Remarks on a class of 2-generator groups of deficiency zero, *J. Austral. Math. Soc.* **19**, 297–305.
4. Coxeter, H. S. M. and Moser, W. O. J. (1980). "Generators and Relations for Discrete Groups", 4th ed. Springer-Verlag, Berlin.
5. Johnson, D. L. (1980). Topics in the theory of group presentations. Cambridge University Press.
6. McLain, D. H. (1977). An algorithm for determining defining relations of a subgroup, *Glasgow Math. J.* **18**, 51–56.
7. Mendelsohn, N. S. (1964). An algorithmic solution for a word problem in group theory, *Canad. J. Math.* **16**, 509–516.
8. Neubüser, J. (1982). An elementary introduction to coset table methods in computational group theory. *In* "Groups St. Andrews 1981" (Eds. C. M. Campbell and E. F. Robertson), L.M.S. Lecture Notes **71**, 1–45. Cambridge University Press.

# An Algorithm for Double Coset Enumeration?

J. H. CONWAY

## 1 Introduction

In many theoretical investigations, large groups $G$ have been constructed or described in terms of their double coset decompositions with respect to a pair of smaller groups $H$ and $K$. (Usually $H = K$, but in the present paper we found no need to lose this generality.) Sims has pioneered the use of computers in such work, and has used the double coset technique successfully in the construction of several large sporadic groups, and in particular in that of the "Baby Monster" group $F_{2+}$, jointly with Leon. So the main message of this paper is not the underlying method, but the attempt to formulate it as a direct generalization of ordinary single coset enumeration that might be programmed for use in a routine way.

I have described the ideas below to several interested groups of people over the last 15 years (notably to J. McKay and J. Neubüser) but, so far as I know, the method has never been programmed for a computer. The first implementation will certainly not be without problems! On the other hand, I have used it manually to provide working tables for calculation with several groups, and in particular with the Janko groups $J_1$ (which I later learnt was originally constructed in almost exactly the same way), and $J_3$ (where I suspect that my calculations may have repeated some of G. Higman's).

The algorithm is best regarded as a way of performing the calculations of an ordinary coset enumeration (of the cosets of $H$ in $G$), but condensing them to some extent by using known properties of $K$. We shall in fact suppose that $K$ is very well known indeed (mnemonic: $K$ = $K$nown), as will appear from the required subroutines listed below. Also, in the present description, we consider the only saving to be one of *space*—the *time* taken for the double coset enumeration will theoretically be essentially the same as that for the corresponding single coset enumeration. But, as we shall see, the algorithm will often have obtained the correct coset table long before it runs to completion, and in a number of applications it can then be stopped without loss.

COMPUTATIONAL GROUP THEORY
ISBN 0-12-066270-1

The known subgroup $K$ is regarded in much the same way that one regards a ground field elsewhere in mathematics, and we think of its elements as constants. Thus, for instance, if $G = \langle K, x \rangle$ we call $G$ a *1-generator group over* $K$. For simplicity we shall first discuss only this case, and we shall also suppose that $x^2 = 1$, so that each relation of $G$ can be written in the form

$$k_1 x k_2 x \ldots k_n x = 1 \quad (k_1, k_2, \ldots, k_n \in K)$$

Then $x$ will normalize the subgroup $L = K \cap x^{-1}Kx$ and the *gain factor* obtained by performing the double coset enumeration rather than the corresponding single coset enumeration will be $|L|$. Unless $L$ is large, this gain is likely to be swamped by the extra administrative costs, but there are plenty of cases where $|L|$ is very large indeed, rendering an impossibly large single coset enumeration quite practicable as a double coset one. We suppose that a particular transversal $T$ is chosen for the cosets $aL$ of $L$ in $K$.

It is handy at this point to make a note about single coset enumeration. Normally this is viewed as a process in which there are two basic operations—*defining* new cosets and *identifying* old ones. Mathematically it is simpler to suppose that all the defining operations have already been done (producing an infinite table which initially describes the regular representation of a free group) and regard the entire process as one of identification. This is obviously computationally possible using a "servant" subroutine whose function is to arrange that, whenever a coset is called for, it is hurriedly defined, if necessary. It is said that Queen Victoria never needed to look behind her when she sat down—there was always a chair there! We shall emulate this practice by supposing that whenever we examine an entry in one of our double coset tables, the corresponding information is there, and has been "consistently" filled in—in other words, whenever we assert that coset $Ha$ times some generator is coset $Hb$, then $Hb$ times the inverse generator is $Ha$.

Each element of $K$ is required to have a *name*. The algorithm does not concern itself with the form of these names, but we imagine that they will often be explicit permutations or matrices and may themselves have been produced by some mechanical means, perhaps using CAYLEY. Here are the required subroutines:

MULT—given the names of $k_1, k_2 \in K$, finds that of $k_1 k_2$.
SPLIT —given the name of $k \in K$, expresses it as $tl$ $(t \in T, l \in L)$.
ACT —given the name of $l \in L$, finds that of $l^x$.
LIST —produces the successive elements of $K$ in order.

## 2 A Double Coset Table

Let us suppose that the double cosets of $H$ and $K$ in $G$ are

$Ha_1K, Ha_2K, \ldots, Ha_NK$, and that $T = \{t_1, \ldots, t_M\}$. Then when the enumeration is complete there will be, for the generator $x$, a table whose rows correspond to $a_1, \ldots, a_N$ (we shall actually refer to "the row $Ha_i$") and whose columns correspond to $t_1, \ldots, t_M$ ("the column $t_j$"). The entry in the $Ha_i, t_j$ space will be of one of two forms, as follows.

*$\alpha$-entry.* The name of another row, say the row $Ha_I$, and the name of an element of $K$, say $k$. This will be described as the $\alpha$-entry $Ha_I, k$, and indicates the relation

$$Ha_it_jx = Ha_Ik.$$

In some sense it is the normal case.

*$\beta$-entry.* In this case the entry will consist of the name of an *earlier* column $t_J$ ($J < j$) and an element $l$ of $L$, and indicates that there is a coincidence between the two $(H, L)$ double cosets $Ha_it_jL$ and $Ha_it_JL$, and in fact

$$Ha_it_j = Ha_it_Jl.$$

When the algorithm has run to completion, the $\alpha$-entries will correspond to a complete system of representatives for the double cosets of $H$ and $L$ in $G$ and the rows will be in 1–1 correspondence with the double cosets of $H$ and $K$ in $G$. Before then, however, there will often be many names for the same double coset and it is the job of the algorithm to find these coincidences by using the relations of $G$, and then to take appropriate action.

Here is how this is done. For each $a_i$ and each $k \in K$, we must arrange to identify the cosets

$$Ha_ik \quad \text{and} \quad Ha_ikk_1xk_2x \ldots k_nx$$

(for each relation $k_1x \ldots k_nx = 1$ of $G$). What we do is actually to find the current name of $Ha_ikk_1x \ldots k_nx$ in the form $Ha_{i'} . k'$ for some $k' \in K$), and then move into the IDENTIFY phase.

We appeal in order to

LIST, to give us the next element $k$ of $K$;
MULT, to form the element $kk_1$;
SPLIT, to express this as $t_jl_1$ for some $t_j \in T$, $l_1 \in L$;
ACT, to find the $l_2$ for which $l_1x = xl_2$.

We now know that

$$Ha_ikk_1x = Ha_it_jxl_2$$

and we LOOK in the table to see that $Ha_it_jx = Ha_Ik^*$, say. (For the moment we ignore the possibility that we find a $\beta$-entry.) We are now faced with the simpler problem of finding

$$Ha_I k^* l_2 k_2 x k_3 x \ldots k_n x,$$

and we continue inductively.

Had we found a $\beta$-entry indicating that $Ha_i t_j = Ha_i t_J l_3$, then we should instead ACT on $l_3$ to produce $l_4$, say, and would be faced with the problem

$$Ha_i t_J x l_4 l_2 k_2 x k_3 x \ldots k_n x$$

which this time would correspond to an $\alpha$-entry.

So we must now face the problem of what to do in the IDENTIFY phase. The information we will be given has the form

$$Ha_{i_1} k_{i_2} = Ha_{j_1} k_{j_2},$$

say, where $k_{i_2}, k_{j_2}$ are elements of $K$ which may be SPLIT as

$$t_{i_3} l_{i_4} \quad \text{and} \quad t_{j_3} l_{j_4}.$$

If $i_1 \neq j_1$, we can suppose $i_1 < j_1$, and we are faced with the problem of identifying two previously distinct rows of the table. We delete the row $Ha_{j_1}$, after translating all the information it contains into information about $Ha_{i_1}$ (this will probably generate new coincidences). This process is directly analogous to the identification procedure in ordinary coset enumeration, although now, of course, there will be new headaches.

If instead $i_1 = j_1$, we have an identification inside the $Ha_{i_1}$ row which probably indicates that the double coset $Ha_{i_1} K$ does not have the full size. We proceed as follows. If $i_3 \neq j_3$, say $i_3 < j_3$, the fact that

$$Ha_{i_1} t_{i_3} l_{i_4} = Ha_{i_1} t_{j_3} t_{j_4}$$

allows us to place a $\beta$-entry in the place $Ha_{i_1}, t_{j_3}$. (Of course if this place is non-empty yet further identifications may be produced.) If, finally, $i_3 = j_3$, then what we have shown (if $i_4 \neq j_4$) is that the coset $Ha_{i_1} t_{i_3}$ is fixed under multiplication by the element $l_{i_4} l_{j_4}^{-1}$.

This interesting information should plainly be part of the output of the algorithm, but I am completely unsure what to do about it. This may not matter, since the algorithm itself makes no use of such identities! However, it is probably best to consider keeping this information in the form of a list of elements of $L$ that fix a given coset. Since such elements actually form a subgroup of $L$, there may be more efficient ways to hold it—for instance if $L$ is a cyclic group, any subgroup of it is conveniently specified by its order. It is easy to see in any case that the information is only needed for the *first* element $Ha_i t_1$ in any given row, and so we shall imagine a list $M(a_i)$, which for sentimental reasons we shall call the *muddle list* for $a_i$.

We briefly survey the entire process. Identifications are found, in the first instance, by *applying* relations $k_1 x, \ldots k_n x$ to each coset $Hak$. Since the table only holds information about the particular cosets $Hat$ for which $t \in T$, the action of $x$ on a more general coset $Hak$ is found using SPLIT, and MULT is also required, since we must be able to multiply $Hak$ by $k' \in K$. When a relation has been applied to a coset, we find an identification which can be put into a canonical form $Hatl = Ha't'l'$, say. If the two $a$ factors are distinct, we can delete one row in favour of the other, and record a number of consequent identifications. If the $a$ factors are the same, but the $t$ factors differ, we can insert a $\beta$-entry (and again may have more identifications to do). Finally, if the $a$ and $t$ factors are the same, but the $l$ factors differ, we can produce an element of $L$ to throw into the corresponding muddle list.

The final number of cosets of $H$ in $G$ is obtained by summing $|K|/|M|$ for each $\alpha$-entry, where $|M|$ is the size of the corresponding muddle list.

As the reader can see, the algorithm can hardly be said to be in final form! But I have handled a number of cases, often involving quite large groups, entirely by hand, and have seen no problems which appear to be insuperable. Since the gain over single coset enumeration can be tremendous, it is surely worthwhile attempting a machine version, whatever the problems.

I add some comments about the expected behaviour of the algorithm in "early-closing" situations. We shall take as our example an enumeration for the group $J_1$, with $H = K = L_2(11)$, of order 60, and $x$ an involution which normalizes (in fact centralizes) a group $L = A_5$ in $K$. In this case there are just 5 double cosets and $|T| = 11$, so the final output will be a $5 \times 11$ table (and some muddle lists). Discounting the relations in $K$, and the relation $x^2 = 1$, $G$ can be defined by just two relations $k_1xk_2xk_3xk_4xk_5x = 1$ (I believe), so that officially our procedure will perform $5 \times 660 \times 2$ tests, involving $5 \times 660 \times 10$ table lookups, to find 55 entries. In fact, the table will have been cut down to size and all 55 entries completed long before all these 33 000 lookups have been performed. If, therefore, we already know that $G$ exists, and has 5 double cosets of the appropriate sizes, we can expect to be able to stop the algorithm long before its official completion date. In larger cases one might well have an official procedure involving *billions* of operations to construct and check a table with only a few *thousand* entries! For these reasons, if one intends to stop the algorithm when the table is complete, one should replace LIST by a subroutine that generates *random* elements of $K$, since a systematic listing procedure will probably cause the algorithm to traverse the same segment of table too many times.

We now comment, as promised, on the more general situation in which $G$ may need more than one generator over $K$ and in which the generators may be non-involutory. In this case, it is best to consider a finite set of generators $x_1, \ldots, x_n$, *closed under inversion*. There will be a double coset table, as described, for each of these generators and also a group $L_i = x_iKx_i^{-1} \cap K$ which again depends on the generator $x_i$. Consequently, the action of subroutines SPLIT and ACT will depend on the particular generator. Otherwise the generalization is obvious, although we suspect the extra complexity will mean that initial implementations stick to our special case.

# Groups of Exponent Six

M. F. NEWMAN

## 1 Introduction

In 1958 Marshall Hall [**1**] gave an affirmative answer to a case of the Burnside question by proving that *every finitely generated group of exponent 6 is finite*. His proof "involves long calculations" (§18.4 of [**2**]). Because of developments since then it is possible to give a comparatively simple proof.

## 2 Proof

The crux of the proof is the following lemma. Its proof is given later.

*Lemma.* Let $H$ be a group of exponent 6 which has a generating set $\{x\} \cup A$ such that $x^2 = \phi$ (the identity) and $a^3 = (xa)^2 = \phi$ for all $a$ in $A$. The subgroup generated by $A$ has exponent 3.

Let $G$ be a finitely generated group of exponent 6. Let $M$ be the subgroup generated by the cubes of all elements of $G$. Then $M$ is normal in $G$ and is generated by elements of order 2. The quotient group $G/M$ is a finitely generated group of exponent 3 and so is finite (see §18.2 of [**2**] or §5.12 of [**3**]). Hence, by Schreier's Theorem (Theorem 2.9 in [**3**]), $M$ is finitely generated. It follows that there is a finite set $X$ of elements of order 2 which generates $M$. For $x$ in $X$ let $x^M$ be the normal closure of $x$ in $M$. As subgroup $x^M$ is generated by $\{x\} \cup A(x)$ where $A(x) = \{x^{-1}x^w : w \in M\}$. Observe that $x^2 = \phi$ and $(x[x, w])^2 = \phi$ where, as usual, $[x, w] = x^{-1}x^w$. So to apply the lemma, it suffices to prove $[x, w]^3 = \phi$ for all $w$ in $M$. This is done by induction on the length of $w$ as a product of elements of $X$. If $w = \phi$, it is obvious. Otherwise $w = vy$ with $y$ in $X$ and, by inductive assumption, $[x, v]^3 = \phi$. Moreover, since $x, y$ have order 2,

$$[x, y] = x^{-1}y^{-1}xy = (xy)^2,$$

COMPUTATIONAL GROUP THEORY
ISBN 0-12-066270-1

so $[x,y]^3 = \phi$. Hence, $([x,y]^{-1}[x,v])^3 = \phi$ by applying the lemma to $\langle x, [x,y], [x,v]\rangle$ where $\langle\ldots\rangle$ denotes the subgroup generated by $\{\ldots\}$. It follows that $[x,w]^3 = \phi$ because

$$[x,w] = xyv^{-1}xvy = ([x,y]^{-1}[x,v])^y.$$

Applying the lemma to $\{x\} \cup A(x)$ in $x^M$ gives that $\langle A(x)\rangle$ has exponent 3. Clearly each $\langle A(x)\rangle$ is normal in $M$. So their product $P$ is normal in $M$, and has 3-power exponent. Since $P$ also has exponent 6, it has exponent 3. In $M/P$ each generator $xP$ is central and of order 2, so $M/P$ is finite. Hence $P$ is finitely generated, and therefore finite. The proof is complete.

*Proof of Lemma.* The core of the proof is to show

(1) $[a,b,c,d] = \phi$ for all $a,b,c,d$ in $A$.

Given (1) it follows that $\langle A\rangle$ is nilpotent (of class 3). Since every element of $A$ has order 3, every finite subset of $A$ generates a group of 3-power order. Therefore it has exponent 3 because it also has exponent 6. Hence $\langle A\rangle$ has exponent 3 as required.

A fairly straightforward argument (see the proof of Lemma 5.11 in [**3**] or the relevant proof in §18.2 of [**2**]) shows that (1) follows from

(2) $\langle a,b,c\rangle$ and $\langle a,b,[c,d]\rangle$ have exponent 3 for all $a,b,c,d$ in $A$.

This can be proved by coset enumeration as follows (there is a remark on this in the next section). Consider the group $U$ generated by $\{y,u,v\}$ with defining set of relations

$$y^2 = \phi, \quad (yu)^2 = u^3 = (yv)^2 = v^3 = \phi,$$
$$(uv)^6 = (uv^{-1})^6 = (yuv)^6 = (y(uv)^3)^6 = \phi.$$

Coset enumeration shows that $\langle y,u\rangle$ has index 9 in $U$. Moreover, $\langle y,u\rangle$ is a quotient group of the group defined by $y^2 = (yu)^2 = u^3 = \phi$ which has order 6. Hence $\langle u,v\rangle$ has order dividing 27. For $a,b$ in $A$ the subgroup $\langle a,b\rangle$ is a quotient of $\langle u,v\rangle$ and has exponent 6, hence

(3) $\langle a,b\rangle$ has exponent 3 for all $a,b$ in $A$.

Since $(ab^{\pm}a)^3 = (xab^{\pm}a)^2 = \phi$, the same argument gives

(4) $\langle ab^{\pm}a, c\rangle$ has exponent 3 for all $a,b,c$ in $A$.

Similarly,

(5) $\langle ab^{\pm}a, cd^{\pm}c\rangle$ has exponent 3 for all $a,b,c,d$ in $A$.

Now consider the group $V$ generated by $\{u,v,w\}$ subject to the sixteen relations

$$u^3 = v^3 = w^3 = \phi, \quad (uv^{\pm})^3 = (vw^{\pm})^3 = (wu^{\pm})^3 = \phi,$$
$$(uv^{\pm}uw^{\pm})^3 = \phi, \quad (u^{-1}vw)^6 = (uv^{-1}w)^6 = (uvw^{-1})^6 = \phi.$$

Coset enumeration shows that $\langle u, v\rangle$ has index 81 in $V$. Since $\langle u, v\rangle$ has order dividing 27, it follows that $V$ has 3-power order. For $a, b, c$ in $A$, using (3) and (4) gives that $\langle a, b, c\rangle$ is a quotient of $V$; since it also has exponent 6,

(6) $\langle a, b, c\rangle$ has exponent 3 for all $a, b, c$ in $A$.

Moreover, using (5) and (6), this argument gives

$$\langle c, d, ab^{\pm}a\rangle \text{ has exponent 3 for all } a, b, c, d \text{ in } A.$$

In particular, $(ab^{\pm}a[c, d]^{\pm})^3 = \phi$, so, using (6) again,

$$\langle a, b, [c, d]\rangle \text{ has exponent 3 for all } a, b, c, d \text{ in } A.$$

This completes the proof.

## 3 Remark

Since the term "coset enumeration" covers a number of (related) algorithms which may behave quite differently on a given input, a proof quoting coset enumeration should be accompanied by information about the algorithm actually used. In this case, I used the computer program Todd-Coxeter version 2.2A (written by Alford and Havas) defining new cosets with its default Felsch method. Both enumerations were easy to run. They required little space—room for 300 cosets was adequate, and little time—about a second on a VAX 11/780.

## References

1. Hall, Jr., M. (1958). Solution of the Burnside problem for exponent six, *Illinois J. Math.* **2**, 764–786.
2. Hall, Jr., M. (1959). "The theory of groups". Macmillan, New York.
3. Magnus, W., Karrass, A. and Solitar, D. (1966). "Combinatorial group theory". Interscience, New York, London, Sydney.

# On a Class of Groups Related to SL(2, $2^n$)

C. M. CAMPBELL AND E. F. ROBERTSON

## 1 Introduction

In [2] we gave a presentation for SL(2,8) with 2 generators and 2 relations. It became natural to investigate whether SL($2,2^n$), $n \geqslant 4$, could be presented with a small number of relations, in particular does SL($2,2^n$) have a presentation with an equal number of generators and relations for all $n \geqslant 3$?

SL($2,2^n$), n $\geqslant$ 4, might have such a presentation since it has trivial Schur multiplier. In general, for a group $G$ with Schur multiplier $M(G)$ having rank $m$, the best one could hope for, see [6], would be a presentation with $n$ generators and $m+n$ relations. $G$ is called *efficient* if it has such a presentation.

Higman and McKay [5] give a nice presentation for SL(2,16), namely,

$$\langle a, b, c \mid a^2 = b^{15} = c^2 = (bc)^2 = (ac)^3 = aba^{-4}ab^3 = 1 \rangle.$$

This presentation motivated our definition of the class of groups $\theta(n,k)$ defined by

$$\theta(n,k) = \langle a,b,c \mid a^2 = b^{2^n-1} = c^2 = (bc)^2 = (ac)^3 = abab^{-k}ab^{k-1} = 1 \rangle.$$

The class of groups $\theta(n,k)$ is a reasonable one to study when searching for nice presentations of SL($2,2^n$) in view of the following lemma.

*Lemma 1.* There is an epimorphism $\phi:\theta(n,k) \rightarrow \mathrm{SL}(2,2^m)$ when there is an $\alpha \in \mathrm{GF}(2^m)$ with $\alpha^k+\alpha+1 = 0$ and the order of $\alpha$ divides $2^n-1$.

*Proof.* Define

$$\phi(a) = \begin{bmatrix} 1 & 1 \\ 0 & 1 \end{bmatrix}, \quad \phi(b) = \begin{bmatrix} \alpha^{t-1} & 0 \\ 0 & \alpha^t \end{bmatrix}, \quad \phi(c) = \begin{bmatrix} 0 & 1 \\ 1 & 0 \end{bmatrix},$$

where $t = 2^{n-1}$ and $\alpha^k+\alpha+1 = 0$. It is a straightforward task to check that these matrices satisfy the relations of $\theta(n,k)$ and the lemma is proved.

In this paper we use both algebraic techniques and computer implementations of group theory algorithms to investigate the class $\theta(n,k)$. We give a

COMPUTATIONAL GROUP THEORY
ISBN 0-12-066270-1

2-generator, 2-relation efficient presentation for SL(2,16) again using both algebraic and computing results. Finally we give some results on PSL$(2, p^n)$, $p$ an odd prime.

We would like to thank J. M. Cohen and P. D. Williams for many helpful discussions. In particular, P. D. Williams is largely responsible for the proof of Theorem 6 and we thank him for allowing us to quote the result here.

## 2 The Groups $\theta(n, k)$

It is not hard to extend the result of Lemma 1 to show that, under certain conditions, $\theta(n, k)$ has a direct product of groups SL$(2, 2^{n_i})$ as a homomorphic image.

*Theorem 2.* Suppose $x^k + x + 1 = p_1(x)p_2(x)\dots p_m(x)$ is the decomposition as a product of distinct irreducible factors over GF(2). Let $p_i(x)$ have degree $n_i$ and suppose

$$p_{i_1}(x), p_{i_2}(x), \dots, p_{i_j}(x)$$

are those polynomials having a root of order dividing $2^n - 1$. Then there is an epimorphism

$$\phi: \theta(n, k) \rightarrow \mathrm{SL}(2, 2^{n_{i_1}}) \times \mathrm{SL}(2, 2^{n_{i_2}}) \times \dots \times \mathrm{SL}(2, 2^{n_{i_j}}).$$

It is interesting to know under what conditions $\phi$ is an isomorphism or indeed whether $\phi$ is always an isomorphism. We can show that in many cases $\phi$ is indeed an isomorphism and as a first step we investigate the subgroup $\langle a, b \rangle$ of $\theta(n, k)$ which we can show always has an order which is consistent with $\phi$ being an isomorphism.

Let us define a group $H$ by

$$H = \langle a, b \mid a^2 = b^l = abab^{-k}ab^{k-1} = 1 \rangle$$

where $l = 2^n - 1$. The subgroup $\langle a, b \rangle$ of $\theta(n, k)$ is a homomorphic image $\bar{H}(n, k)$. With the aim of calculating the order of $\bar{H}(n, k)$, we first examine the structure of $H$. Note that $a \in H'$, the derived group of $H$, so if we put $x_i = b^{-i+1}ab^{i-1}$ then $K = \langle x_i | 1 \leqslant i \leqslant l \rangle \subseteq H'$. However, enumerating the cosets of $K$ in $H$ shows that $|H:K| = l$ and since $H/H' \cong C_l$, the cyclic group of order $l$, we see $K = H'$.

Since $b^l = 1$ we have $x_i = x_{i+\lambda l}$ for any integer $\lambda$, so we can assume that the suffices on the $x_i$ are reduced modulo $l$. The Reidemeister-Schreier algorithm may now be used to find a presentation for $H'$ on the generators $\{x_i : 1 \leqslant i \leqslant l\}$ using a similar argument to that given in §3 of [**1**]. The following presentation is obtained:

$$\langle x_1, x_2, \dots, x_l \mid x_{k+i} = x_{k+i-1}x_i, x_i^{\ 2} = 1, 1 \leqslant i \leqslant l \rangle.$$

Now Theorem 3.3 of [**1**] shows that the relations $[x_i, x_j] = 1, 1 \leqslant i \leqslant j \leqslant l$, can be deduced from the relations $x_{k+i} = x_{k+i-1}x_i$ so $H'$ is abelian and thus an elementary abelian 2-group. Since $x_{k+1} = x_k x_1, x_{k+2} = x_{k+1}x_2, \ldots$ we can write $x_i$ for $i \geqslant k+1$ in terms of $x_1, x_2, \ldots, x_k$ so $H'$ is generated by $x_1, x_2, \ldots, x_k$. Thus $H'$ has order dividing $2^k$ showing that $H$ is a metabelian group of order dividing $l \,.\, 2^k$.

We can give the order of $H$ precisely. If we let $M$ be the circulant $l \times l$ matrix with first row

$$(1 \quad \underbrace{1 \quad 0 \quad \ldots \quad 0 \quad 1}_{k+1} \quad 0 \quad \ldots \quad 0),$$

then $H'$ has order $2^{l-r}$, where $r$ is the rank of $M$. For, the relation matrix of the abelian group $H'$ is the $2l \times l$ matrix

$$\begin{pmatrix} M \\ 2I \end{pmatrix}, \text{ where } I \text{ is the } l \times l \text{ identity matrix.}$$

But the number of roots of $x^k + x + 1$ in GF($2^n$) is $l - r$ where $r$ is the rank of $M$. This can be proved by forming the product of the Vandermonde matrix with rows $1, \beta, \ldots, \beta^{l-2}$ where $\beta \in \text{GF}(2^n) \backslash \{0\}$ with the matrix $M$ (see p. 211 of [**3**]). We have therefore proved the following result.

*Theorem 3.* $|H| = (2^n - 1) \,.\, 2^v$ where $v$ is the number of roots of $x^k + x + 1$ in GF($2^n$).

It is now easy to calculate the order of $\bar{H}(n, k)$. Let $f(k)$ be the period of the polynomial $x^k + x + 1$, i.e. the least integer $j$ with $x^k + x + 1$ dividing $x^j - 1$. Now $x_{f(k)+i} = x_i$. But $x_{l+i} = x_i$ so, if $t = $ h.c.f. $(f(k), l)$, then $x_{t+i} = x_i$, giving

$$[b^t, a] = 1. \tag{1}$$

Since $c^2 = 1$, $cbc = b^{-1}$ gives

$$cb^t = b^{-t}c. \tag{2}$$

Using $(ac)^3 = 1$ we have $b^t acacac = b^t$, which, after using (1) and (2), becomes $acacacb^{-t} = b^t$. Hence $b^{2t} = 1$. But $l = 2^n - 1$ is odd, so $b^l = 1$ and $b^{2t} = 1$ give $b^t = 1$.

We have therefore shown that the order of $\langle a, b \rangle$ in $\theta(n, k)$ divides $t \,.\, 2^v$, but examining Theorem 2 shows that $|\langle a, b \rangle|$ is at least this value. We have proved:

*Theorem 4.* $|\langle a,b \rangle|$ in $\theta(n,k)$ is $t \,.\, 2^v$ where $t$ is the h.c.f. of $l$ and the period of $x^k + x + 1$ while $v$ is the number of roots of this polynomial in GF($2^n$). Moreover, the epimorphism $\phi$ of Theorem 2 is an isomorphism when restricted to $\langle a, b \rangle$.

From Theorem 4, we see that to check whether the epimorphism $\phi$ in Theorem 2 is an isomorphism for particular values of $n$ and $k$, it is sufficient to determine the index of $\langle a, b\rangle$ in $\theta(n, k)$. One would expect to be able to gain a considerable insight into whether $\phi$ might always be an isomorphism by using a machine implementation of the Todd-Coxeter algorithm. Although to some extent this is the case, it is unfortunate that most of the significant examples are outside our computing range. Since $b^l = 1$ in $\theta(n, k)$ (where $l = 2^n - 1$), we can clearly consider $k$ reduced modulo $l$. However we have a further isomorphism.

*Lemma 5.* $\theta(n, k) \cong \theta(n, 2^n - k)$.

*Proof.* Replace each of the generators $a, b, c$ by their inverses to get $\theta(n, k) \cong \theta(n, -k+1)$. Since $-k+1 \equiv 2^n - k \pmod{l}$, we have the required result.

Clearly, $\theta(n, k)$ is trivial if $n = 1$ or if $k = 0$ or if $k = 1$. A computer implementation of the Todd-Coxeter algorithm verifies that the index of $\langle a, b\rangle$ in $\theta(n,k)$ has the correct value to prove $\phi$ of Theorem 2 an isomorphism for $n = 2, 3, \ldots, 7$ and any $k$. This is not too surprising in view of the following theorem, a proof of which will appear elsewhere.

*Theorem 6.* The epimorphism of Theorem 2 is an isomorphism if each of the distinct irreducible factors $p_i(x)$ of $x^k + x + 1$ with a root of order dividing $2^n - 1$ is primitive and their degrees $n_i$ are coprime.

The Todd-Coxeter checking of $\phi$ being an isomorphism up to $n = 7$ has already shown that Theorem 6 does not give necessary conditions since it has verified one case which is not covered by Theorem 6, namely $\theta(6,8)$, where the degrees of the irreducible factors of $x^8 + x + 1$ are 2 and 6, so not coprime. We can also use Todd-Coxeter to prove that the primitive condition of Theorem 6 is not necessary since $\theta(9, 9) \cong SL(2,2^9)$. In this case, $x^9 + x + 1$ is irreducible but not primitive, so not covered by Theorem 6.

To use Theorem 6, one needs to know properties of the polynomial $x^k + x + 1$ over GF(2) and several authors have indeed studied this polynomial, see, for example, [**6**].

Perhaps the most interesting case not covered by Theorem 6 and which is beyond our available computing range, is $\theta(8,16)$. In order to check that $\theta(8,16) \cong SL(2,2^8) \times SL(2,2^8)$, it would be necessary to show that $|\theta(8,16):\langle a, b\rangle| = (2^8 - 1)(2^8 + 1)^2$. This example is at the beginning of the series

$$\theta(2^n, 2^{2^{n-1}})$$

which, since $x^{2^{2^{n-1}}}+x+1$ is a product of $2^{2^{n-1}-n}$ irreducible factors of degree $2^n$, would be, if $\phi$ were an isomorphism, isomorphic to the direct product of

$$2^{2^{n-1}-n} \text{ copies of } \mathrm{SL}(2,2^{2^n}).$$

This result would be so surprising that further investigation would appear very profitable. Another series of the same type is

$$\theta(3^n, 2^{3^{n-1}}+1)$$

which, if $\phi$ were an isomorphism, would be isomorphic to

$$(2^{3^{n-1}}+1)/3^n \text{ copies of } \mathrm{SL}(2, 2^{3^n}).$$

The smallest case where both non-primitive factors and non-coprime degrees occurs is $\theta(12,17)$ which, if $\phi$ were an isomorphism, would be $\mathrm{SL}(2,2^2)\times\mathrm{SL}(2,2^3)\times\mathrm{SL}(2,2^{12})$.

In view of the above results we state the following problems.

*Problem 1.* Given $n$, does there exist a $k$ such that $\theta(n,k)\cong\mathrm{SL}(2,2^n)$?

*Problem 2.* Is $\phi$ in Theorem 2 always an isomorphism?

Irrespective of the answers to these problems the question of the efficiency of $\mathrm{SL}(2,2^n)$ is still not completely resolved. It is not hard to reduce by one, the number of relations of $\theta(n,k)$ and in the case where $\theta(n,k)$ has trivial Schur multiplier, we can reduce the number by two as the following lemma shows.

*Lemma 7.* Let $l=2^n-1$. Then:

(i) $\theta(n,k)\cong\langle a,b,c\,|\,a^2=1, b^l=1, (ac)^3=c^2, (bc)^2=1, abab^{-k}ab^{k-1}=1\rangle$;

(ii) if $\theta(n,k)$ has trivial Schur multiplier,

$$\theta(n,k)\cong\langle a,b,c\,|\,a^2=b^l, a^2=(bc)^2, (ac)^3=c^2, abab^{-k}ab^{k-1}=1\rangle.$$

*Proof.* (i) $(ac)^3=c^2$ gives $cac^{-1}=(aca)^{-1}$, so $(aca)^2=(cac^{-1})^{-2}=1$, since $a^2=1$. But $(aca)^2=1$ gives $c^2=1$, since $a^2=1$, and the result follows. (ii) If we denote by $G$ the group with presentation given in the statement of (ii), then it is not hard to check that $G$ is perfect. It remains to show that $a^2$ is central in $G$ since then $a^2\in M(G)$ which is the trivial group. But $[a^2,b]=1$, since $a^2=b^l$ and $[a^2,c]=1$, since $a^2=(bc)^2$ gives $[a^2,bc]=1$.

$\theta(4,4)$ therefore gives a 3-generator 4-relation presentation for SL(2,16). In the final section we improve this to give an efficient presentation for SL(2,16).

## 3 Efficient Presentations

In this section we give efficient presentations for the simple groups SL(2,16), PSL(2,25), PSL(2,27) and PSL(2,49). From a presentation for SL(2,16) given by Coxeter and Moser in [4], Sandlöbes [7] gives the following presentation for SL(2,16):

$$\langle a,b \mid a^{15} = b^3 = (a^7b)^2 = (a^7(ab)^9)^2 = (a^{11}(ab)^2)^2 = 1\rangle.$$

A machine implementation of Todd-Coxeter shows that the fourth relation may be omitted. Now put $x = a^7b$, $y = b$ to obtain the following presentation:

$$\langle x,y \mid x^2 = 1, y^3 = 1, (xy)^{15} = 1, ((xy^{-1})^5xyx^{-1}yx^{-1}y)^2 = 1\rangle.$$

Again use Todd-Coxeter to show that the third and fourth relations may be replaced by a single relation to obtain the presentation for SL(2,16):

$$\langle x,y \mid x^2 = 1, y^3 = 1, (yx^{-1})^9y(yx^{-1})^2y(xy^{-1})^6y(yx^{-1})^2y = 1\rangle.$$

Finally an argument similar to that in [2] for SL(2,8) shows that an efficient presentation for SL(2,16) is

$$\langle x,y \mid x^2y^{-3} = 1, (yx^{-1})^9y(yx^{-1})^2y(xy^{-1})^6y(yx^{-1})^2y = 1\rangle.$$

Williams [8] has generalized the results of §2 on the groups SL$(2,2^n)$ to the groups PSL$(2,p^n)$, $p$ an odd prime. In general, his presentations contain a fixed but greater number of relations. However, using a machine implementation of Todd-Coxeter in a similar way to that described above, he has obtained the following efficient presentations for PSL(2,25), PSL(2,27) and PSL(2,49):

$$\mathrm{PSL}(2,25) = \langle a,b \mid a^3 = (aba^{-1}b)^2, b^5 = (ab^{-1}ab^2)^4, (ab^2a^{-1}b^{-2})^2 = a^3b^5\rangle;$$

$$\mathrm{PSL}(2,27) = \langle a,b \mid a^2 = b^{13} = (ab^4)^3, (ab^{-1}ab^5)^3 = (ab^{-1}ab^2ab^5)^2a^2\rangle;$$

$$\mathrm{PSL}(2,49) = \langle a,b \mid a^3(ab^{-1})^7 = (ab^2ab)^3b^{-5} = ab(ab^2)^2a^2b^{-1}(b^2a)^{-2} = 1\rangle.$$

Finally we note that we have also obtained efficient presentations for PSL(2,25) and PSL(2,27), again using computational techniques. Starting from the presentations given by Sandlöbes [7], namely:

$$\mathrm{PSL}(2,25) = \langle a,b \mid a^2 = 1, b^3 = 1, [a,b]^{13} = 1, (ab)^4[a,b]^4(ab)^3(ab^{-1})^2 = 1\rangle$$

and

$$\mathrm{PSL}(2,27) = \langle a,b \mid a^2 = 1, b^3 = 1, (ab)^7 = 1, [a,b]^{13} = 1,$$
$$(ab)^2[a,b]^2(ab(ab^{-1})^2)^2[a,b^{-1}]^4[a,b](ab^{-1})^2 = 1\rangle$$

where $[a,b]$ denotes $a^{-1}b^{-1}ab$, we obtain the efficient presentations for PSL(2,25) and PSL(2,27), respectively,

$$\langle a,b \mid a^2b^3 = 1, [a,b]^{13} = 1, a^{-6}(ab)^4[a,b]^4(ab)^3(ab^{-1})^2 = 1\rangle$$

and

$$\langle a,b \mid a^2b^3 = 1, (ab)^7[a,b]^{13} = 1,$$
$$(ab)^2[a,b]^2(ab(ab^{-1})^2)^2[a,b^{-1}]^4[a,b](ab^{-1})^2 = 1\rangle.$$

## References


1. Campbell, C. M. and Robertson, E. F. (1978). Classes of groups related to $F^{a,b,c}$, *Proc. Roy. Soc. Edinburgh* **78A**, 209–218.
2. Campbell, C. M. and Robertson, E. F. (1980). A deficiency zero presentation for SL(2,$p$), *Bull. London Math. Soc.* **12**, 17–20.
3. Cohn, P. M. (1977). "Algebra", Vol. 2. Wiley, London.
4. Coxeter, H. S. M. and Moser, W. O. J. (1972). "Generators and Relations for Discrete Groups", 3rd edn. Springer-Verlag, Berlin.
5. Higman, G. and McKay, J. (1969). On Janko's simple group of order 50, 232, 960, *Bull. London Math. Soc.* **1**, 89–94.
6. Huppert, B. (1967). "Endliche Gruppen I", Springer-Verlag, Berlin.
7. Sandlöbes, G. (1977). Perfekte Gruppen bis zur Ordnung $10^4$, *Schriften zur Informatik und angewandten Mathematik* **36**, RWTH Aachen.
8. Williams, P. D. (1982). Ph.D. Thesis, University of St. Andrews.
9. Zierler, N. (1970). On $x^n+x+1$ over $GF(2)$, *Information and Control* **16**, 502–505.

# Enumerating Infinitely Many Cosets

ROBERT H. GILMAN

## 1 Introduction

The Todd-Coxeter method for determining the cardinality of a group $G$ given by a finite presentation works when $G$ is finite. There is another procedure, $P$, which succeeds whenever $G$ is finite and in some additional cases as well. In this article we give some applications of $P$, and we discuss a possible improvement. A description of $P$ appears in [**2**], and [**3**] contains an extension applicable to enumeration of double cosets.

We give a brief description of $P$. Let $F$ be a finitely generated free monoid. Well-order $F$ first by length and then lexicographically among words of the same length. Suppose $R = \{w_i = v_i\}$ is a finite presentation of a group $G$ as a quotient of $F$. We assume $w_i > v_i$ for all $i$ where $>$ is the well-ordering defined above. $G$ is isomorphic to the quotient of $F$ by the congruence generated by $R$. $P$ tests $R$ for a condition called confluence (completeness in [**2**]). If $R$ is confluent, then $L$, the set of words in $F$ not divisible by any $w_i$, is a cross-section for the congruence. $L$ is a regular subset of $F$ (see chapter 2 of [**5**] for an introduction to regular sets), and it is straightforward to determine the cardinality of $L$ and hence of $G$. If $R$ is not complete, $P$ generates new relations which are added to $R$. $P$ continues alternately testing $R$ and adding new relations to it. Either $P$ never terminates, or $P$ terminates with a confluent set of relations which give a presentation for $G$. If $P$ could run forever, it would always generate a confluent set of relations for $G$.

A confluent set of relations, $R$, for $G$ embodies a solution to the word problem for $G$. Namely, if $w$ is any word in $F$, reduce $w$ by replacing any subword $w_i$ by $v_i$ and continue this process as long as possible. The process must terminate with a word $w^*$ in $L$. Thus $w$ and $v$ represent the same element of $G$ if, and only if, $w^* = v^*$. The use of procedures like $P$ to generate confluent presentations of algebras and thereby solve the word problem, is well known in mathematics and computer science. For two articles with extensive and almost disjoint lists of references, see [**1**] and [**6**].

COMPUTATIONAL GROUP THEORY
ISBN 0-12-066270-1

## 2 Experimental Evidence

In this section we give a few applications of $P$ worked out using a computer program prepared for demonstration at the Durham Symposium.

*Example 1.* $G = \langle a, b; a^4 = b^2 = [a, b]^2 = 1 \rangle$.

The actual input to the program was a presentation of $G$ as a quotient of the free monoid on generators $a, b, c, d$ by relations

$$R = \{ab = 1, ba = 1, cd = 1, dc = 1, (acbd)^2 = 1, a^4 = 1, c^2 = 1\}.$$

The program determined that $G$ is infinite with growth of degree 2. That is, the number of elements of $G$ represented by words of length at most $n$ in the generators is of order a constant times $n^2$. The program gave the following confluent set of relations for $G$:

$$\begin{array}{ccc} cbca = bcac & cacaa = acbcb & cacb = acbc \\ d = c & c^2 = 1 & b^2 = a^2 \\ ba = 1 & a^3 = b & ab = 1 \end{array}$$

By [4] we know that $G$ has a nilpotent subgroup of finite index. We will find such a subgroup. Let $N$ be the normal closure of $(\overline{ac})^2$ in $G$ (where $\bar{w}$ is the element of $G$ represented by $w$). $G/N$ is dihedral of order 8. Since we can solve the word problem in $G$, it is a straightforward calculation to check that $(\overline{ac})^2$ has 4 conjugates in $G$, namely $(\overline{ac})^{-2}$, $(\overline{bc})^2$, $(\overline{bc})^{-2}$ in addition to $(\overline{ac})^2$, and that $N$ is abelian of rank at most 2. Because $G$ has growth of degree 2, $N$ must be free abelian of rank 2.

*Example 2.* $G(i, j) = \langle a, b; a^j = b^i = [a, b]^2 = 1 \rangle$.

Since $G(i, j) \simeq G(j, i)$, we assume $i \leqslant j$. The program gave the information shown in Table 1.

The program could not determine the cardinality of $G(4,5)$ or $G(4,6)$ directly, although since these groups have $G(2,5)$ and $G(2,6)$ respectively as

*Table* 1 The Order of $G(i,j)$

| | j | 2 | 3 | 4 | 5 | 6 |
|---|---|---|---|---|---|---|
| i | 2 | 8 | 24 | ∞ | ∞ | ∞ |
| | 3 | | 288 | ∞ | ∞ | ∞ |
| | 4 | | | ∞ | ? | ? |

quotients, we know that they are infinite. Also $G(2,4)$ is the only group in Table 1 with polynomial growth. The other infinite groups have exponential growth.

The behaviour of $P$ can vary with small changes in the input. Consider the following presentations of the $(2,3,7)$ group as quotients of free monoids with generators $a, b$ and $a, b, c, d$ respectively.

*Example 3.* $G = \langle a, b; a^7 = 1, b^3 = 1, (ab)^2 = 1 \rangle.$

*Example 4.* $G = \langle a, b, c, d; a^7 = 1, c^3 = 1, d^2 = 1, acd = 1, ab = 1, ba = 1 \rangle.$

The program mentioned above fails for Example 3 but gives the right answer for Example 4. Moreover, it seems that the failure occurs not because of peculiarities of the program but because Procedure $P$ does not terminate with the input of Example 3. The reason is that, after a while, the relations produced by the program fall into a pattern which appears to go on indefinitely. The same phenomenon occurs for the $(2,3,8)$ group with the presentation below.

*Example 5.* $G = \langle a, b; a^8 = 1, b^3 = 1, (ab)^2 = 1 \rangle.$

We have not been able to get our program to work on the $(2,3,8)$ group by making small changes in the presentation of Example 5 as we did in going from Example 3 to Example 4. The method of small changes works for $(2,3,n)$, $n = 7,9,11,13$, but not yet for $n = 6,8,10,12$. What happens with Example 5 is that after a while the program generates only the relations

$$bab^2a^2(b^2ab^2a^2b^2a^2)^m b^2ab^2a^2b^2ab = a^5(b^2a^3b^2a^4)^{m+1}a \tag{1}$$

for successive values of $m$, starting at $m = 2$. It seems likely that the infinite family of relations (1), $m \geqslant 2$ together with the finite set of other relations generated by the program, forms a confluent set of relations, $R$, for $G$. If this is the case, the set $L$ of all words not divisible by $w_i$ for any relation $w_i = v_i$ in $R$ is still a cross-section for the congruence defining $G$ and also still a regular set. Thus, if $P$ could be extended to deal with families like (1) the way it deals with single relations, its performance would be improved.

## 3 An Extension of Procedure $P$

We indicate briefly how $P$ might be modified as discussed in the last section. A formal description would be tedious, so we give an illustration.

*Example 6.* $G = \langle a, b, c, d; ba = ab, ac = 1, ca = 1, bd = 1, db = 1 \rangle.$

Clearly $G$ is a free abelian group on two generators. However, as is shown in Example 3 of [**2**], $P$ does not terminate. In fact it is not hard to see that $P$ produces the following relations,

$$\begin{aligned} ab^nc = b^n \quad n \geqslant 1 \qquad\qquad & bc^nd = c^n \quad n \geqslant 1 \\ dc = cd \qquad\qquad da = ad \qquad & cb = bc \end{aligned} \tag{2}$$

and that the addition of (2) to the input yields a confluent set of relations for $G$.

Let $P^*$ be a modification of $P$ which works in the following way, (here we assume the reader is familiar with the description of $P$ in [**2**]). Whenever $P$ compares two single relations by computing two different reductions of a word $w$, $P^*$ computes reductions of all powers $w^n$, $n \geqslant 1$, simultaneously. For example, when comparing relations $ba = ab$ and $ac = 1$, $P$ computes two reductions of $w = bac$. Namely (using arrows to indicate the effect of substituting the right hand side of a relation for the left hand side),

$$bac \longrightarrow abc \qquad\qquad bac \rightarrow b.$$

Thus, $P$ adds the relation $abc = b$ to the input. $P^*$ computes

$$(bac)^n \rightarrow (abc)^n = a(bca)^{n-1}bc \rightarrow ab^{n-1}bc = ab^nc$$

$$(bac)^n \longrightarrow b^n$$

and adds the infinite family $ab^nc = b^n$, $n \geqslant 1$. When comparing an infinite family of relations to a single relation or to another infinite family, $P^*$ proceeds in analogy with $P$.

The preceding vague description can be extended to a definition of a procedure $P^*$ which will cope with Example 6. $P^*$ has not yet been programmed on a computer or tried on more complicated problems like Examples 4 and 5. Several issues arise in specifying $P^*$ precisely; we mention one. In checking for confluence, $P$ reduces a word $w$ in two ways to get two irreducible words. $P^*$ cannot always reach irreducible words and so may not be able to tell that it has reached a confluent set of relations, when in fact it has. In this connection it is interesting that Theorem 4.1.4 of [7] shows that, for certain benign infinite sets of relations, confluence is recursively undecidable. We do not know if confluence is decidable for infinite sets of relations involving finite numbers of families like (1).

## References

1. Bergman, G. M. (1978). The diamond lemma for ring theory, *Adv. in Math.* **29**, 178–218.

2. Gilman, R. H. (1979). Presentations of groups and monoids, *J. Alg.* **57**, 544–554.
3. Gilman, R. H. (1982). Enumeration of double cosets, *J. Pure and Appl. Alg.* **26**, 183–188.
4. Gromov, M. (1981). Groups of polynomial growth and expanding maps, *Publ. Maths. I.H.E.S.* **53**, 53–73.
5. Harrison, M. A. (1979). "Introduction to Formal Language Theory". Addison-Wesley, Reading, Mass.
6. Huet, G. and Oppen, D. C. (1980). Equations and rewrite rules: A survey. *In* "Formal Language Theory: Perspectives and Open Problems" (Ed. R. V. Book), 349–405. Academic Press, London, Orlando and New York.
7. O'Dúnlaing, C. (1981). Infinite regular Thue systems. *In* "Topics in Theoretical Computer Science: Five Papers on String Replacement". Dept. of Math., University of California at Santa Barbara.

*Note*

The dissertation of Susanne Kemmerich (Unendliche Reduktionssysteme, Aachen 1983) contains another treatment of infinite families of relations.

# Presentations for Cubic Graphs

N. L. BIGGS

## 1 Introduction

The process of coset enumeration is formally equivalent to the construction of a graph, often called the Schreier coset diagram. In this paper I shall exploit this relationship in two ways. First, I shall explain how a problem in graph theory can be reduced to coset enumerations, and then I shall show how graphical constructions can throw some light on the results of the computations.

Many people have given me invaluable help: John Conway, who developed the elegant (and practical) theory which underlies the whole paper; John Cannon, who constructed the Cayley language; Kaye Burford and Miles Hoare, who did many computations; and George Havas and Edmund Robertson, who showed me how to do many more (see [**8**]).

## 2 Cubic Graphs

A *cubic graph* is an undirected graph which is regular, with valency 3. If $\Gamma$ is a cubic graph, an *r-arc* in $\Gamma$ is a sequence of vertices $(v_0, v_1, \ldots, v_r)$ such that $\{v_i, v_{i+1}\}$ is an edge $(0 \leqslant i \leqslant r-1)$ and $v_i \neq v_{i+2}$ $(0 \leqslant i \leqslant r-2)$.

Suppose that $G$ is a group of automorphisms of $\Gamma$ which acts transitively on the 1-arcs (ordered pairs of adjacent vertices). Then we can define

$$s = \sup\{r \mid G \text{ acts transitively on the } r\text{-arcs of } \Gamma\}.$$

The following basic theorems were discovered by Tutte [**11**].

*Theorem 1.* $G$ acts regularly on the $s$-arcs.

*Theorem 2.* If $\Gamma$ is finite, then $s \leqslant 5$.

Tutte's original paper was published in 1947, and since that time many improvements have been made in the proof of Theorem 2. In particular, the

COMPUTATIONAL GROUP THEORY
ISBN 0-12-066270-1

work of Sims [**10**], Djoković [**6**], and Weiss [**12**] should be mentioned. Also there is a paper by Miller [**9**], which deals with a simple case of the problem outlined below. However, a much more profound influence on the present work comes from unpublished investigations started by Conway in the 1950s and carried on by Conway and Guy in the 1960s. Their work was generously made available to the author in March 1981. It provides a set of canonical presentations for groups acting $s$-regularly on cubic graphs, and, as will appear, it reduces the problem of determining such graphs to a series of coset enumerations. A more theoretical treatment of the subject has recently been published by Djoković and Miller [**7**].

## 3 Presentations

Suppose that $G$ acts $s$-regularly on $\Gamma$. Choose an $s$-arc $(v_0, v_1, \ldots, v_s)$ in $\Gamma$, and let $v_{s-1}, w', w''$, be the three vertices of $\Gamma$ which are adjacent to $v_s$. Since $G$ acts regularly on the $s$-arcs, there are unique automorphisms $a, b, \sigma$ in $G$, such that

$$a(v_0, v_1, \ldots, v_{s-1}, v_s) = (v_1, v_2, \ldots, v_s, w'),$$

$$b(v_0, v_1, \ldots, v_{s-1}, v_s) = (v_1, v_2, \ldots, v_s, w''),$$

$$\sigma(v_0, v_1, \ldots, v_{s-1}, v_s) = (v_s, v_{s-1}, \ldots, v_1, v_0).$$

If $\Gamma$ is connected, then $a$ and $b$ generate $G$; $\sigma$ is introduced in order to describe the underlying geometry more easily.

Since $\sigma^2$ fixes the basic $s$-arc, we must have $\sigma^2 = 1$. Similarly, examining the action of $\sigma a \sigma$ we see that it must be either $a^{-1}$ or $b^{-1}$. For the purposes of this paper we shall confine our attention to the case (called $s^+$) when first possibility holds, so that we have the relations

$$\sigma a \sigma = a^{-1}, \quad \sigma b \sigma = b^{-1}.$$

It can be shown that the $s^+$ case can occur only for $s = 2,3,4,5$. In each of these cases, an analysis of the action of suitable combinations of $a, b$, and $\sigma$ on the basic $s$-arc provides a set of $s+3$ relations which must hold in $G$. For the $4^+$ and $5^+$ cases there is no loss of generality in choosing these relations as follows:

$$4^+ : \sigma^2 = 1, (\sigma a)^2 = 1, (\sigma b)^2 = 1, (a^{-1}b)^2 = 1,$$
$$(a^{-2}b^2)^2 = 1, a^3 b^{-3} a^3 = bab, a^3 b \sigma a^4 = ba^2 b.$$

$$5^+ : \sigma^2 = 1, (\sigma a)^2 = 1, (\sigma b^2 = 1, (a^{-1}b)^2 = 1,$$
$$(a^{-2}b^2)^2 = 1, (a^{-3}b^3)^2 = 1, a^4 b^{-4} a^4 = ba^2 b, a^4 b \sigma a^5 = ba^3 b.$$

The relations will be satisfied in any group of type $4^+$ or $5^+$ acting on a cubic graph. In particular, they are satisfied in suitable groups acting on the infinite cubic tree, and this shows that the groups defined by these relations alone are infinite. However, when the graph has cycles (and this must happen when it is finite) there are further relations to be satisfied. It can be shown that each cycle of length $l$ corresponds to a word of length $l$ in non-negative powers of $a$ and $b$ which represents the identity. So, if we adjoin such a relation to the relations for $4^+$ or $5^+$, then we shall obtain a quotient of $4^+$ or $5^+$, which may be finite. Indeed, if our extra relation is ill-chosen, then the quotient group may collapse.

## 4 Coset Enumerations

For the purposes of this paper we shall confine our attention to the situation where some $l$-cycle in $\Gamma$ is represented by the word $a^l$. We shall write $4^+$ $(a^l)$ and $5^+(a^l)$ for the groups defined by the relations for $4^+$ and $5^+$, respectively, with $a^l = 1$ adjoined.

If $G$ acts $s$-regularly, then the elements $a^{-i}b^i$ $(1 \leqslant i \leqslant s)$ fix the vertex $v_0$, and in fact they generate the stabilizer of $v_0$, which has order $3.2^{s-1}$. Thus, if we define

$$H = \langle a^{-i}b^i \,|\, i = 1, 2, \ldots, s\rangle,$$

then the index $|G:H|$ is the cardinality of a cubic graph on which $G$ acts. This graph is the Schreier coset diagram of $G$ over $H$, with edges defined by the elements $a, a^{-1}$ and $b^{-1}$.

The determination of the index by coset enumeration techniques was begun by Conway and Guy. Recently, the availability of the Cayley language and programs has enabled their calculations to be checked and extended, and this has been done by Burford using the implementation of Cayley on an ICL 2980 machine at Queen Mary College, London. Some results for $n = |G:H|$ when $G = s^+(a^l)$ and $s = 4,5$ are given in Table 1. A cross indicates that the index is 1 or 2, and a question mark indicates that the order of the graph was unknown when this paper was delivered at the Symposium.

*Table 1*

| $l$ | $\leqslant 5$ | 6 | 7 | 8 | 9 | 10 | 11 | 12 | 13 | 14 | 15 | 16 |
|---|---|---|---|---|---|---|---|---|---|---|---|---|
| $s = 4$ | × | 14 | × | 14 | 102 | 90 | × | ? | 10270 | ? | ? | ? |
| $s = 5$ | × | × | × | 30 | × | × | × | 650 | 234 | 2352 | × | ? |

## 5 Graphical Constructions

In the rest of this paper we shall use graphical methods to throw some light on the results of the coset enumerations. For example, we can explain the collapses for small values of $l$, since it is well-known that the length of cycle in an $s$-regular graph must satisfy $l \geqslant 2s-2$ [**11**].

A set of basic building blocks for cubic graphs with $s = 4$ or 5 is provided by the sextet construction of [**5**]. For each odd prime $p$ there is a *sextet graph* $S(p)$: it admits a $4^+$ group for all $p$, and also a $5^+$ group when $p \equiv 3$ or 5 (mod 8). The order $n$ of $S(p)$ depends on the congruence class of $p$ modulo 16, as follows:

$$n = \frac{1}{48}p(p^2-1) \quad \text{when} \quad p \equiv 1, 15 \pmod{16},$$

$$n = \frac{1}{24}p(p^2-1) \quad \text{when} \quad p \equiv 7, 9 \pmod{16},$$

$$n = \frac{1}{24}p^2(p^4-1) \quad \text{when} \quad p \equiv 3, 5, 11, 13 \pmod{16}.$$

The full automorphism group of $S(p)$ is $PSL(2,p)$, $PGL(2,p)$, $P\Gamma L(2,p^2)$ in the respective cases. The generators $a$ and $b$ can be represented in a concrete way, and their orders calculated for each $p$. (Unfortunately, we have no formulae for these orders as functions of $p$.)

Using this information we can identify some of the groups and graphs arising from the coset enumerations.

(i) $S(7)$ is Heawood's graph on 14 vertices, occurring as the graph of $4^+(a^6)$. The group $4^+(a^6)$, with order $14 \cdot 24 = 336$, is therefore $PGL(2,7)$. In $S(7)$ we also have $b^8 = 1$, and since the interchange of $a$ and $b$ induces an automorphism of $4^+$, this accounts for the occurrence of $S(7)$ as the graph of $4^+(a^8)$.

(ii) $S(17)$ has 102 vertices and it is one of the "three remarkable graphs" discussed in [**1**]. Since $a^9 = 1$ in $S(17)$, we see that this is the graph of $4^+(a^9)$, and the group is $PSL(2,17)$.

(iii) In a similar way we find that $S(79)$ is the graph of $4^+(a^{13})$, and the group is $PSL(2,79)$.

(iv) The graph $S(3)$ is Tutte's 8-cage on 30 vertices; it is the graph of $5^+(a^8)$ and the group is $P\Gamma L(2,9)$, or Aut $S_6$.

(v) The graph $S(5)$ on 650 vertices is the graph of $5^+(a^{12})$ and the group is $P\Gamma L(2,25)$.

We can derive more graphs with $4^+$ and $5^+$ groups from the sextet graphs, by using the covering construction given in Chapter 19 of [**2**].

Let $K$ be any group, and suppose that $G$ is a group of automorphisms of a connected graph $\Gamma$ which also acts as a group of automorphisms of $K$. In other words, for each $v$ in $V\Gamma$, $\kappa$ in $K$, and $g$ in $G$ we have $g(v)$ in $V\Gamma$ and $\hat{g}(\kappa)$ in $K$, and these actions preserve the respective structures. The split extension $K \rtimes G$ is the group whose elements are the ordered pairs $(\kappa, g)$ in $K \times G$, with composition defined by

$$(\kappa_1, g_1)(\kappa_2, g_2) = (\kappa_1 \cdot \hat{g}_1(\kappa_2), g_1 g_2).$$

The group $K \rtimes G$ acts on the set $K \times V$ in the obvious way:

$$(\kappa_1, g_1)(\kappa, v) = (\kappa_1 \hat{g}_1(\kappa), g_1(v)).$$

Now we shall take $\tilde{V} = K \times V$ to be the set of vertices of a graph $\tilde{\Gamma}$. Suppose that each 1-arc $(v, v')$ in $\Gamma$ is assigned an element $\lambda(v, v')$ of $K$, in such a way that the following compatibility condition (*) is satisfied:

$$\lambda(g(v), g(v')) = \hat{g}(\lambda(v, v')). \qquad (*)$$

Let us say that two vertices $(\kappa, v)$, $(\kappa', v')$ are adjacent in $\Gamma$ if $\kappa' = \kappa\lambda$, where $\lambda = \lambda(v, v')$. It is easy to check that the action of $K \rtimes G$ on $\tilde{V} = K \times V$ defines automorphisms of $\tilde{\Gamma}$.

The proof of the following theorem may be found on p. 129 of [**2**].

*Theorem 3.* If $G$ acts transitively on the $s$-arcs of $\Gamma$, then $K \rtimes G$ acts transitively on the $s$-arcs of $\tilde{\Gamma}$.

In fact, if $a, b$ are the basic "shunt" automorphisms associated with the $s$-arc $(v_0, v_1, \ldots, v_s)$ and $\lambda(v_0, v_1) = \lambda$, then the basic shunts for the action of $K \rtimes G$ are given by

$$\tilde{a} = (\lambda, a), \tilde{b} = (\lambda, b).$$

Of course, it must be remembered that $\tilde{\Gamma}$ need not be connected. In the worst cases, $\tilde{\Gamma}$ will be just the union of $|K|$ copies of $\Gamma$, and nothing will be gained from the construction. However, there are at least three useful methods of choosing $K$ and the labels $\lambda$ so that the components of $\tilde{\Gamma}$ are strictly larger than $\Gamma$. We shall examine these methods and their implications for our particular problem.

*Method A.* Let $K$ be the cyclic group of order 2 with generator $t$, and define $\lambda(v, v') = t$ for all 1-arcs in $\Gamma$. Any automorphism group of $\Gamma$ may act trivially on $K$ so that (*) is satisfied.

The graph $\tilde{\Gamma}$ is a double covering of $\Gamma$, and it is trivial (two components, both $\Gamma$) if and only if $\Gamma$ is bipartite. If the shunt $a$ has odd order, then $\Gamma$ has an odd cycle and $\tilde{\Gamma}$ is non-trivial; the order of $\tilde{a}$ is twice that of $a$. Applying this construction to S(17) and its group $4^+(a^9)$, we deduce that the group $4^+(a^{18})$ has a quotient which is the group of a graph with 204 vertices. Consequently, the order of $4^+(a^{18})$ is at least 204.24. (Probably the group is much larger.)

*Method B.* (Biggs, [**4**]). Suppose that $G$ is a group of $4^+$ type acting on $\Gamma$. Associated with each edge $e$ in $\Gamma$ there is a unique involution $j_e$ which fixes the vertices of $e$ and their immediate neighbours in $\Gamma$. Let $K$ be the subgroup of $G$ generated by the involutions $j_e$, for all $e$ in the edge-set $E$ of $\Gamma$. Then $K$ is normal in $G$, and in fact:

$$|G:K| = \begin{cases} 2, \text{ if } \Gamma \text{ is bipartite;} \\ 1, \text{ otherwise.} \end{cases}$$

Consequently, $G$ acts by conjugation as a group of automorphisms of $K$. If we label the 1-arcs of $\Gamma$ so that

$$\lambda(v, v) = j_e, \text{ when } e = \{v, v'\},$$

then the condition $(*)$ is satisfied.

In this case there is a special bonus. It is shown in [**4**] that $\tilde{\Gamma}$ admits, not only the $4^+$ group $K \rtimes G$, but also a 2-fold extension of it acting in the $5^+$ manner. If $(ab)^r = 1$ in $\Gamma$, then the shunt $\tilde{a}_5$ (with respect to the $5^+$ group) has order $2r$.

When $\Gamma$ is S(7) and $G$ is its group $4^+$ $(a^6)$, we find that $(ab)^7 = 1$. In this case it can be shown directly, [**3**], that $\tilde{\Gamma}$ is connected, and so it admits a $5^+$ group with $\tilde{a}_5{}^{14} = 1$. Since $\Gamma$ is bipartite, we have $|K| = 168$ and $|\tilde{V}| = 168 \cdot 14$. Thus $\tilde{\Gamma}$ is the graph of $5^+$ $(a^{14})$.

*Method C.* (Conway). Let $K$ be the free $\mathbb{Z}_2$-module on the set $E$ of edges of $\Gamma$. For each edge $e = \{v, v'\}$ we may define $\lambda(v, v')$ to be the generator $e$ of $K$, so that $G$ acts naturally on $K$ and condition $(*)$ is trivially satisfied. The covering graph $\tilde{\Gamma}$ has $2^{|V|-1}$ components, each with $2^{|E|-|V|+1}|V|$ vertices. Also, the order of $\tilde{a}$ is twice the order of $a$.

Applying this construction to S(7) with $G = 4^+(a^6)$ we conclude that a quotient of $4^+(a^{12})$ is the group of a $2^8$-fold covering of S(7). Hence the order of $4^+(a^{12})$ is at least $2^8 . 14 . 24 = 86\,016$.

Applying the construction to S(3) with $G = 5^+(a^8)$ we conclude that a quotient of $5^+(a^{16})$ is the group of a $2^{16}$-fold covering of S(3). Hence the order of $5^+(a^{16})$ is at least $2^{16} . 30 . 48 = 94\,371\,840$.

The only graphs in our short table not yet accounted for, are those with 90 and 234 vertices. The 90-vertex graph is a triple covering of S(3), not of the type obtained by Methods *A*, *B*, *C*. The 234-vertex graph is the unique primitive 5-regular graph described by Wong [**13**]. At the time of writing we know of no 4-regular or 5-regular graphs which are not coverings (possibly trivial) of sextet graphs or Wong's graph.

## 6 Postscript

After this paper was read at the Symposium, Havas and Robertson [**8**] showed, using clever computational techniques, that both $4^{+}(a^{12})$ and $5^{+}(a^{16})$ are "practically" infinite.

## References


1. Biggs, N. L. (1973). Three remarkable graphs, *Canad. J. Math.* **25**, 397–411.
2. Biggs, N. L. (1974). "Algebraic Graph Theory". Cambridge University Press.
3. Biggs, N. L. (1982a). A new 5-arc-transitive cubic graph, *J. Graph Theory* **6**, 447–451.
4. Biggs, N. L. (1982b). Constructing 5-arc-transitive cubic graphs, *J. London Math. Soc.* **26**, 193–200.
5. Biggs, N. L. and Hoare, M. J. (1982c). The sextet construction for cubic graphs, *Combinatoria* (to appear).
6. Djoković, D. Z. (1972). On regular graphs II, *J. Combinatorial Theory* (*B*) **12**, 252–259.
7. Djoković, D. Z. and Miller, G. L. (1980). Regular groups of automorphisms of cubic graphs, *J. Combinatorial Theory* (*B*) **29**, 195–230.
8. Havas, G. and Robertson, E. F. (1983). Two groups which act on cubic graphs, (these Proceedings).
9. Miller, R. C. (1971). The trivalent symmetric graphs of girth at most six, *J. Combinatorial Theory* (*B*) **10**, 163–182.
10. Sims, C. C. (1967). Graphs and finite permutation groups, *Math. Zeitschr.* **95**, 76–86.
11. Tutte, W. T. (1947). A family of cubical graphs, *Proc. Camb. Phil. Soc.* **43**, 459–474.
12. Weiss, R. (1978). s-transitive graphs. *In* "Algebraic Methods in Graph Theory" (Eds. L. Lovasz and V. T. Sos), 827–847. North–Holland, Amsterdam.
13. Wong, W. J. (1967). Determination of a class of primitive permutation groups, *Math. Zeitschr.* **99**, 235–246.

# Two Groups which Act on Cubic Graphs

GEORGE HAVAS AND EDMUND F. ROBERTSON

## 1 Introduction

Biggs [1] studies automorphism groups which act on cubic graphs. In particular he describes two families of groups:

$$4^+(a^l) = \langle a, b, \sigma \mid \sigma^2 = (\sigma a)^2 = (\sigma b)^2 = (a^{-1}b)^2 = (a^{-2}b^2)^2 = 1,\\ a^3b^{-3}a^3 = bab, a^3b\sigma a^4 = ba^2b, a^l = 1\rangle;$$

$$5^+(a^l) = \langle a, b, \sigma \mid \sigma^2 = (\sigma a)^2 = (\sigma b)^2 = (a^{-1}b)^2 = (a^{-2}b^2)^2 = (a^{-3}b^3)^2 = 1,\\ a^4b^{-4}a^4 = ba^2b, a^4b\sigma a^5 = ba^3b, a^l = 1\rangle.$$

In the tables presented in his talk at the Symposium, Biggs indicated that the orders of $5^+(a^{16})$ and $4^+(a^{12})$ were unknown but that the groups have homomorphic images with orders $2^{16}.30.48$ and $2^8.14.24$. He asked for more information about these groups. We show that both these groups have considerably larger quotients than those quoted above and present here details of computations done at the Symposium which prove this. We expect that these computations may form the basis of a proof that both these groups are infinite.

## 2 The Group $5^+(a^{16})$

In Biggs [1], it is shown that $G = 5^+(a^{16})$ has a quotient of order $2^{16}.30.48$, and the question of the finiteness of $G$ is raised. Initial attempts to prove finiteness involved coset enumerations in $G$ over the subgroups

$$K_1 = \langle a^{-1}b, a^{-2}b^2, a^{-3}b^3, a^{-4}b^4, a^{-5}b^5\rangle$$

of order 48 and

$$K_2 = \langle a^8, b^{10}, (ab)^6, (a^2b)^8, (ab^2)^4, (a^2b^2)^5, (a^3b)^6, (ab^3)^4, (a^4b)^8,\\ (a^3b^2)^4, (a^2b^3)^{10}, (ab^4)^8, (a^2b^4)^5, (a^3b^3)^6, (a^4b^2)^5, (a^5b)^6\rangle$$

COMPUTATIONAL GROUP THEORY
ISBN 0-12-066270-1

of index expected to be a multiple of 1440. Both of these enumerations were unsuccessful, for reasons which will become obvious. These coset enumerations were performed with computer programs designed and implemented by Havas and Alford which are descendants of that described in [**4**].

Because of the difficulty of these coset enumerations, we decided to try to show that $G$ was infinite, or at least very large. The approach taken was to find a subgroup which can be shown to have either an infinite or a very large quotient. There was no obvious subgroup to try, so we started by looking for possible candidates. There are two approaches which we used here. On the one hand Cayley (see [**2**]) was used to find low index subgroups directly from the presentation for $G$. On the other hand we looked for subgroups by using the random coincidence technique combined with coset enumeration [**3**].

Cayley (used by Cannon himself) provided a subgroup of index 90 with a concise generating set, namely

$$K_3 = \langle a, \sigma, b^2abab^2, ba^7b, baba^5bab \rangle.$$

The random coincidence technique found subgroups with index 144, 180, 240, 288, 360, 480 and 720 in the quotient of $G$ obtained by adding the relation $b^{20} = 1$. These subgroups are all extensions of the image of $K_2$ in this quotient of $G$.

Since $K_3$ was derived with by far the neatest generating set of all these subgroups, we elected to study it. A presentation for $K_3$ was found using the Reidemeister-Schreier program described in [**5**]. This presentation on 181 generators with 504 relators, was used as input to the Tietze transformation program described in [7]. The program reduced the presentation for $K_3$ to the following 4-generator 22-relator presentation

$$\begin{aligned} K_3 = \langle a, b, c, d \mid\, & a^2, (ab)^2, (ac)^2, (ad^{-1})^2, (adc^{-1}d)^2, (bc^{-1})^4, \\ & bc^{-1}bd^2c^{-1}d^2, (cd^3)^2, (bc^{-1}d^{-1}bdc^{-1})^2, (bdb^{-1}d^3)^2, (bd^5)^2, \\ & (cd^2c^{-1}d^{-2})^2, (acb^{-1}d^{-5})^2, (ad^2b^{-1})^4, (bd^{-2}c^{-1}d^{-1}c^{-1}d^{-2})^2, \\ & d^{16}, (bc^{-1}d^{-1}bc^{-1}d^{-1}cdc^{-1}d)^2, (ab^{-1}cb^{-1}cd^2c^{-1}d^{-1}c^{-1}d)^2, \\ & (ad^3bc^{-1}d^{-1}bc^{-1}d^{-2})^2, (cdcdcd^{-1}c^{-1}d^{-1}cd^{-1}c^{-1}d^{-1})^2, \\ & (ad^{-1}cd^2cd^{-2}c^{-1}b)^4, (acdc^{-1}dbc^{-1}d^{-2}cd^{-1}adcb^{-1}d^{-1}cd^{-1})^4 \rangle. \end{aligned}$$

The abelian quotient of $K_3$ is elementary abelian of order 16. The maximal 2-class 13 quotient of $K_3$, obtained by the nilpotent quotient algorithm program (see [**6**]), has order $2^{130}$. This shows that $G$ has order which is a multiple of $90 . 2^{130}$.

One plausible approach, within this framework, to showing $G$ infinite is to find a subgroup which has an infinite cyclic quotient. In an effort to do this, which was to prove unsuccessful, we obtained a presentation for $K_4 = K_3'$. The derived quotient of $K_4$ is elementary abelian of order $2^6$ and $K_4$ has a maximal 2-class 6 quotient of order $2^{138}$, showing that $G$ has order at least $90 . 16 . 2^{138}$.

## 3 The Group $4^+(a^{12})$

Biggs showed that the group $H = 4^+(a^{12})$ has a quotient of order $2^8 . 14 . 24$ and asked whether $H$ is finite. We studied $H$ with the same overall techniques as applied to $G$. However we found the initial subgroup in a different way.

Consider the quotient $L$ of $H$ obtained by adding the relation $b^{16} = 1$. Then, by coset enumeration, the subgroup

$$\langle a^{-1}b, a^{-2}b^2, a^{-3}b^3, a^{-4}b^4 \rangle$$

has index 3584. Random coincidences in this coset table yield that

$$\langle a^{-1}b, a^{-2}b^2, a^{-3}b^3, a^{-4}b^4, bab^2ababa^2 \rangle$$

has index 14 in $L$. By coset enumeration we see also that

$$H_1 = \langle a^{-1}b, a^{-2}b^2, a^{-3}b^3, a^{-4}b^4, bab^2ababa^2 \rangle$$

has index 14 in $H$.

Using the Reidemeister-Schreier and Tietze programs, we obtain the following presentation for $H_1$:

$$\langle a, b, c, d \mid a^2 = b^2 = c^2 = d^2 = (ab)^2 = (cd)^2 = (ac)^3 = (ad)^3 = (bc)^4 = abcbcacbc = (bdc)^4 = (adb)^6 = 1 \rangle.$$

$H_1/H_1'$ is cyclic of order 2 and, using the Reidemeister-Schreier and Tietze programs again, we obtain a presentation for $H_2 = H_1'$:

$$\langle a, b, c \mid a^2 = b^3 = c^3 = (bc^{-1})^2 = (ab)^3 = (ab^{-1}c)^4 = (acb^{-1})^4 = (ac^{-1})^6 = 1 \rangle.$$

Now $H_3 = H_2'$ has index 3 in $H_2$ and the same method as above yields

$$H_3 = \langle a, b, c, d \mid a^2 = b^2 = c^2 = d^2 = (ac)^2 = (bd)^2 = (ab)^4 = (bc)^4 = (cd)^4 = (acd)^4 = (adb)^4 = (abcdacbd)^2 = (acdb)^4 = 1 \rangle.$$

$H_3/H_3'$ is elementary abelian of order 16 and the nilpotent quotient algorithm program shows that $H_3$ has a maximal 2-class 19 quotient of order $2^{76}$ (and in fact for $n < 20$ the maximal 2-class $n$ quotient has order $2^{4n}$). Thus $H$ has order which is a multiple of $14 . 2 . 3 . 2^{76}$.

Continuing down the derived series of $H_1$, we obtained a presentation for $H_4 = H_3'$. Now $H_4/H_4'$ is elementary abelian order $2^9$ and $H_4$ has a maximal 2-class 9 quotient of order $2^{73}$ (for $n < 10$ the maximal 2-class $n$ quotient has order $2^{8n+1}$). This shows that $H$ has order at least $14 . 2 . 3 . 16 . 2^{73}$.

### References


1. Biggs, N. L. (1983). Presentations for cubic graphs, (these Proceedings).
2. Cannon, J. J. (1982). (Preprint). A Language for Group Theory. University of Sydney.

3. Cannon, J. J. and Havas, G. Implementation and application of the Todd-Coxeter algorithm, (to appear).
4. Cannon, J. J., Dimino, L. A., Havas, G. and Watson, J. M. (1973). Implementation and analysis of the Todd-Coxeter algorithm, *Mathematics of Computation* **27**, 463–490.
5. Havas, G. (1974). "A Reidemeister-Schreier program", Proc. Second Internat. Conf. Theory of Groups (Canberra 1973), Lecture Notes in Mathematics, Vol. **372**, 347–356. Springer-Verlag, Berlin.
6. Havas, G. and Newman, M. F. (1980). Application of computers to questions like those of Burnside. *In* "Burnside Groups" (Ed. J. L. Mennicke). Lecture Notes in Mathematics, Vol. **806**, 211–230. Springer-Verlag, Berlin.
7. Havas, G., Kenne, P. E., Richardson, J. S. and Robertson, E. F. (1983). A Tietze Transformation Program, (these Proceedings).

# A Tietze Transformation Program

GEORGE HAVAS, P. E. KENNE, J. S. RICHARDSON
AND E. F. ROBERTSON

## 1 Introduction

A Reidemeister-Schreier program which yields a presentation of a subgroup $H$ of finite index in a finitely presented group $G$ is described in [**3**]. The program has two stages; first, Schreier generators and Reidemeister relators for $H$ are computed, then the resulting presentation is simplified by eliminating redundant generators and by using a substring searching technique. The Tietze transformation program which we describe in this paper was originally designed to improve the simplification stage of that Reidemeister-Schreier program and now also forms part of the implementation of the modified Todd-Coxeter method [**1**]. The program described here is written in a reasonably portable superset of FORTRAN 66, and was available at the Symposium.

## 2 Program Description

The three main principles used by the Tietze program in simplifying the presentation are as follows.

(i) All relators of length 1 and non-involutory relators of length 2 are used to eliminate generators.

(ii) Long substrings of relators are replaced by shorter equivalent strings. First substring searching is performed. A relator $r_1$ is chosen (either by the program or by the user), and other relators are searched for a matching substring $v$ in a cyclic permutation $uv$ of $r_1$ and in a cyclic permutation $wv$ of $r_2$ or its inverse, with the length of $v$ greater than the length of $u$. Then, when a suitable match is found, the relator $r_2$ is replaced by a canonical representative of the shorter relator $wu^{-1}$. One substring replacement pass involves the application of this process with $r_1$ running through all relators.

COMPUTATIONAL GROUP THEORY
ISBN 0-12-066270-1

(iii) Redundant generators (i.e. generators which occur only once in some relator) are eliminated using relators with length greater than 2.

Notice that (i) and (ii) cannot increase the total length of the presentation but (iii) can, and probably does, increase the length.

The relators are stored as circular doubly-linked lists. This enables rapid access to all cyclic permutations of a relator and its inverse. A list of relators is maintained, sorted into ascending order (the canonical order defined in [**3**]). This assists the elimination of duplicate relators and also helps to reduce the work in substring searching. Information about a relator, such as its length and exponent, are maintained in a separate array.

A list of generators is also kept, containing for each generator such information as the total number of occurrences of the generator. This information is useful since it is often a good strategy to eliminate a generator which occurs infrequently.

## 3 Substring Searching

A substring search with two relators takes time dependent on the length of the longer relator but relatively independent of the length of the shorter relator, by virtue of the following observation. It suffices to check initially for occurrences of at most two letters of the shorter relator in the longer relator. If the short relator has unit exponent, any "long" common substring must contain the first letter of the short relator or a letter half way round that relator. If the short relator has non-trivial exponent then a long common substring must contain the first letter of the short relator. This makes the substring search process in the Tietze program take time dependent on $N^2 l$, where $N$ is the number of relators and $l$ is the total relator length.

Substring replacement passes can be applied with potential beneficial effect after any significant alteration to a presentation. In particular, they are often effective after generator eliminations and after substring replacement passes which have led to changes. In calculations with lengthy presentations containing many hundreds of relators, the amount of substring searching may have to be curtailed to save time.

Two ways of reducing substring searching are to do several eliminations between substring replacement passes and to reduce the number of consecutive passes. The largest saving in length is usually made on the first pass and interrupting the search after one or two passes may save much time. Rather surprisingly, better final presentations are often obtained by making several eliminations between substring replacement passes, particularly when the redundant generators are being replaced by reasonably short words.

Sims pointed out, during his lectures at the Symposium, that substring searching can be done in time proportional to $Nl$. Some interesting practical comparisons of string matching algorithms are made in [**6**].

## 4 Other Simplification Techniques

The latest version of the Tietze program contains several features to help simplify presentations in addition to (i), (ii) and (iii) described in §2. The first of these is similar to the substring replacement technique except that substitutions are only made for matching substrings of equal length. After equal length substitutions have been made, a substring replacement pass may again shorten the presentation. A separate technique deals with involutory generators.

Another feature allows the introduction of a new generator equal to the substring of length 2 (not a square) which occurs most often in the presentation. The generator from the relator of length 3 introduced, which occurs least often in the presentation, is then eliminated. This technique, like the others described in this section, seems to be most effective if used only in the final stages of reducing a presentation.

## 5 Applications

The Tietze transformation program is run with the user selecting an appropriate level of interaction. On the one hand the user may control the program closely (selecting which simplification techniques to apply in which order, monitoring the presentation as the program runs) or, alternatively, the user may leave it all to automatic running.

The program is mainly used to simplify Reidemeister-Schreier presentations, so that a resulting presentation for the subgroup may be input to a Todd-Coxeter coset enumeration program or a Reidemeister-Schreier program. Output may also be input to a nilpotent quotient algorithm or an abelian decomposition program. Typical applications of the version of the program discussed in this paper are described in [**5**] and [**2**].

Applications of an earlier version of the Tietze program are described in several papers, for example [**4**].

## 6 Test Examples

In order to give an indication of the improvements that the Tietze program gives, we examine the same 6 test examples discussed in [**3**].

(a) $G_1 = \langle a, b \mid a^3 = b^6 = (ab)^4 = (ab^2)^4 = (ab^3)^3 = 1, ab^2a^2b^2ab^2 = b^2ab^2a^2b^2a \rangle,$

$H_1 = \langle a, b^2 \rangle, |G_1 : H_1| = 26.$

(b) $G_2 = \langle a,b \mid a^4 = b^4 = (ab)^4 = (a^{-1}b)^4 = (a^2b)^4 = (ab^2)^4 = (a^2b^2)^4 = [a,b]^4 = (a^{-1}bab)^4 = 1 \rangle$,
$H_2 = \langle a,b^2 \rangle, |G_2 : H_2| = 64$.

(c) $G_3 = \langle a,b,c \mid a^{11} = b^5 = c^4 = (bc^2)^2 = (abc)^3 = (a^4c^2)^3 = b^2c^{-1}b^{-1}c = a^4b^{-1}a^{-1}b = 1 \rangle$,
$H_3 = \langle a,b,c^2 \rangle, |G_3 : H_3| = 12$.

(d) $G_4 = \langle a,b,c \mid a^{11} = b^5 = c^4 = (ac)^3 = b^2c^{-1}b^{-1}c = a^4b^{-1}a^{-1}b = 1 \rangle$,
$H_4 = \langle a,b,c^2 \rangle, |G_4 : H_4| = 12$.

(e) $G_5 = \langle a,b,c \mid a^3 = b^7 = c^{13} = (ab)^2 = (bc)^2 = (ca)^2 = (abc)^2 = 1 \rangle$,
$H_5 = \langle ab,c \rangle, |G_5 : H_5| = 42$.

(f) $G_6 = \langle a,b,c \mid a^3 = b^7 = c^{14} = (ab)^2 = (bc)^2 = (ca)^2 = (abc)^2 = 1 \rangle$,
$H_6 = \langle ab,c \rangle, |G_6 : H_6| = 78$.

The notation in Tables 1 and 2 is:

$a$ = number of generators, $b$ = number of relators
$c$ = length of longest relator, $d$ = total length of relators
(where length ignores overall exponent on relators);
I = Reidemeister-Schreier presentation (for which we omit column $d$),
II = final output presentation from Havas's program, [**3**].
III = output from Tietze using alternate eliminations and long substring replacement automatically,
IV = output from Tietze using all simplification techniques interactively.

*Table 1*

| | I | | | II | | | | III | | | | IV | | | |
|---|---|---|---|---|---|---|---|---|---|---|---|---|---|---|---|
| | $a$ | $b$ | $c$ | $a$ | $b$ | $c$ | $d$ | $a$ | $b$ | $c$ | $d$ | $a$ | $b$ | $c$ | $d$ |
| $H_1$ | 27 | 156 | 14 | 2 | 29 | 168 | 1664 | 2 | 9 | 12 | 44 | 2 | 7 | 8 | 26 |
| $H_2$ | 65 | 576 | 16 | 2 | 67 | 204 | 2690 | 2 | 16 | 14 | 82 | 2 | 5 | 4 | 10 |
| $H_3$ | 25 | 96 | 15 | 4 | 22 | 111 | 559 | 3 | 16 | 26 | 125 | 2 | 8 | 12 | 40 |
| $H_4$ | 25 | 72 | 11 | 3 | 10 | 21 | 107 | 3 | 10 | 10 | 66 | 3 | 8 | 6 | 21 |
| $H_5$ | 85 | 294 | 13 | 3 | 13 | 65 | 237 | 3 | 10 | 14 | 76 | 2 | 3 | 2 | 4 |
| $H_6$ | 157 | 546 | 14 | 2 | 12 | 66 | 374 | 2 | 7 | 9 | 22 | 2 | 3 | 2 | 4 |

By way of examples, we quote the final presentations of $H_1$ and $H_2$ obtained from Tietze:

$$H_1 = \langle a,b \mid a^4 = (ab^{-1})^3 = abab^{-2}a^{-1}b^{-2} = (ab^{-2})^3 = (a^2b^{-3})^3 = (aba^{-1}b^{-1})^4 = (ab^2)^6 = 1 \rangle,$$
$$H_2 = \langle a,b \mid a^4 = (ab)^2 = b^4 = (ab^{-1})^4 = (a^2b^2)^4 = 1 \rangle.$$

When run automatically with all simplification techniques, the Tietze program will produce, in general, presentations of length somewhere between those quoted under III and IV although, for $H_5$, the automatic run produces the "best possible" result,

$$H_5 = \langle a, b \mid a^2 = b^2 = (ab)^{13} = 1 \rangle.$$

For certain groups, considerably better presentations can be obtained in an interactive run by selecting the order in which the redundant generators are eliminated. Examples studied in [**2**] typify this, where a presentation for $H(n) = \langle a, b^2 \rangle$ of index $2n+3$ in

$$Y(n) = \langle a, b \mid (abab^{-1})^n = ba^{-1}bab^{-1}a, ab^2a^{-1}ba^2b^{-1} = 1 \rangle$$

is found for $n = 5, \pm 7, \pm 8, \pm 10$; this presentation being used again as input to Reidemeister-Schreier and Tietze programs to find $|Y(n)|$ in each case. In Table 2, we give two examples of the performance of Tietze for $n = -8$ and $n = 8$.

*Table 2*

| | III | | | | IV | | | |
|---|---|---|---|---|---|---|---|---|
| | *a* | *b* | *c* | *d* | *a* | *b* | *c* | *d* |
| $H(-8)$ | 2 | 14 | 131 | 807 | 2 | 2 | 27 | 34 |
| $H(8)$ | 4 | 22 | 110 | 881 | 2 | 4 | 38 | 99 |

## References


1. Arrell, D. G. and Robertson, E. F. (1983). A Modified Todd-Coxeter Algorithm, (these Proceedings).
2. Campbell, C. M. and Robertson, E. F. Some problems in group presentations, *J. Korean Math. Soc.* (to appear).
3. Havas, G. (1974). "A Reidemeister-Schreier program", Proc. Second Internat. Conf. Theory of Groups (Canberra, 1973), Lecture Notes in Mathematics, Vol. **372**, 347–356. Springer-Verlag, Berlin.
4. Havas, G. and Richardson, J. S. (1983). Groups of exponent five and class four, *Comm. in Alg.* **11**, 287–304.
5. Havas, G. and Robertson, E. F. (1983). Two Groups which Act on Cubic Graphs, (these Proceedings).
6. Smit, G. De V. (1982). A comparison of three string matching algorithms, *Software Practice and Experience* **12**, 57–66.

# An Aspect of the Nilpotent Quotient Algorithm

M. R. VAUGHAN-LEE

## 1 Introduction

One of the major factors which have made computer calculation of finite $p$-groups practical using the nilpotent quotient algorithm, has been the discovery of relatively small "test sets" for determining whether a power-commutator presentation is consistent, and for determining the exponent of the group defined by the presentation.

In this paper we show how the test sets used up to now can be significantly reduced. We refer the reader to [1] for a detailed description of the nilpotent quotient algorithm. We let $G$ be a finite $d$-generator $p$-group of order at most $p^n$, and we let $P$ be a presentation for $G$ consisting of generators $a_1, a_2, \ldots, a_n$ and relations

$$a_i^P = a_{i+1}^{\alpha(i,i+1)} a_{i+2}^{\alpha(i,i+2)} \ldots a_n^{\alpha(i,n)}$$

$(0 \leqslant \alpha(i,j) < p)$ for $1 \leqslant i \leqslant n$,

$$a_i a_j = a_j a_i a_{i+1}^{\alpha(i,j,i+1)} a_{i+2}^{\alpha(i,j,i+2)} \ldots a_n^{\alpha(i,j,n)}$$

$(0 \leqslant \alpha(i,j,k) < p)$ for $1 \leqslant j < i \leqslant n$.

We assume that if $d < r \leqslant n$ then $a_r$ has a unique definition which is a relation of the form

$$a_i^P = a_{i+1}^{\alpha(i+1)} \ldots a_{r-1}^{\alpha(r-1)} a_r$$

or of the form

$$a_i a_j = a_j a_i a_{i+1}^{\alpha(i+1)} \ldots a_{r-1}^{\alpha(r-1)} a_r.$$

(Relations which are not definitions may also be of this form, of course. The definition of $a_r$ may be used to express $a_r$ in terms of $a_1, a_2, \ldots, a_{r-1}$.) We also assume that if $a_i a_j$ is the left hand side of a definition then $j \leqslant d$. It should be noted that every $d$-generator $p$-group of order $p^n$ has a presentation of this form, and that the presentations generated by the nilpotent quotient algorithm satisfy these conditions.

COMPUTATIONAL GROUP THEORY
ISBN 0-12-066270-1

## 2 Consistency Checking

The presentation $P$ is said to be consistent if $G$ has order $p^n$. A criterion for consistency can be obtained as follows. We let $W$ be the set of normal words on the generators $a_1, a_2, \ldots, a_n$; that is the set of words of the form

$$a_1{}^{\alpha(1)} a_2{}^{\alpha(2)} \ldots a_n{}^{\alpha(n)}. \qquad (0 \leqslant \alpha(i) < p).$$

Note that $W$ has order $p^n$. Any word on the generators can be transformed into a normal word using the collection process. The input to the collection process is a word $w$. If $w$ is normal, then the process stops. Otherwise $w$ contains at least one minimal non-normal subword. The minimal non-normal words are precisely the left hand sides of the relations of $P$. In the collection process a minimal non-normal subword of $w$, $u$ say, is replaced by the right hand side of the corresponding relation. We refer to this as "collecting $u$". This transforms $w$ into a new word $w'$. The process is then repeated with $w'$ in the place of $w$. The process always terminates in a normal word after a finite number of steps. If $w$ contains more than one minimal non-normal subword, we assume that there is a rule for determining which one to collect so that the process is well defined. Typical rules are "collect the rightmost minimal non-normal subword", or "collect the leftmost minimal non-normal subword". Alternatively, we could partially order the set of minimal non-normal words and collect the rightmost (or leftmost) subword among those which are minimal with respect to this partial order.

We define the product $u \,.\, v$ of two elements $u, v$ in $W$ to be the result of collecting the word $uv$ into normal form. This product turns $W$ into a groupoid of order $p^n$. If $W$ is a group, then $W \cong G$, $G$ has order $p^n$ and $P$ is consistent. Wamsley [**2**] showed that the following associativity conditions are sufficient to ensure that $W$ is a group:

$$(a_i \,.\, a_j) \,.\, a_k = a_i \,.\, (a_j \,.\, a_k) \quad \text{for} \quad 1 \leqslant k < j < i \leqslant n;$$

$$(a_j^{p-1} \,.\, a_j) \,.\, a_k = a_j^{p-1} \,.\, (a_j \,.\, a_k) \quad \text{for} \quad 1 \leqslant k < j \leqslant n;$$

$$(a_i \,.\, a_j) \,.\, a_j^{p-1} = a_i \,.\, (a_j \,.\, a_j^{p-1}) \quad \text{for} \quad 1 \leqslant j < i \leqslant n;$$

$$(a_i \,.\, a_i^{p-1}) \,.\, a_i = a_i \,.\, (a_i^{p-1} \,.\, a_i) \quad \text{for} \quad 1 \leqslant i \leqslant n.$$

(Weight considerations imply that some of these conditions are automatically satisfied. But I will ignore this aspect since it is dealt with thoroughly in [**1**] and [**2**].)

We show that $W$ is a group if the above conditions are satisfied for $k \leqslant d$. Given a consistent power-commutator presentation of B(4,4), for example, this reduces the number of associativity conditions from 168 237 to 82 062. (I have taken weight into account in obtaining these figures.)

Parts of the proof rely on a detailed analysis of the collection process. The output from the collection process is a finite sequence $w_1, w_2, \ldots, w_r$ of words on the generators, terminating in a normal word $w_r$. If $P$ is consistent and $W$ is a group, then we can identify a word $a_i a_j \ldots a_k$ with the group product $a_i . a_j . \ldots . a_k$, and then $w_1, w_2, \ldots, w_r$ are different expressions for the same element of $W$. In the case of some collection processes, the same is true even when $P$ is not consistent. For example, suppose that in the collection process we always collect the rightmost minimal non-normal subword. If we identify a word $a_i a_j \ldots a_x a_y a_z$ with the groupoid product $a_i . (a_j . ( \ldots . (a_x . (a_y . a_z)) \ldots ))$ in $W$, then $w_1, w_2, \ldots, w_r$ are different expressions for the same element of $W$. However, for some collection processes there may be no relevant correspondence between words and groupoid products, and so it is important to distinguish between them. On the other hand we will be making an inductive hypothesis that for some $k > 0$, the subgroupoid $W_{k+1}$ consisting of normal words in the generators $a_{k+1}, a_{k+2}, \ldots, a_n$ is a group. When this hypothesis is in force, it will be helpful to identify words (and subwords) on these generators with elements of $W_{k+1}$.

First we will prove that Wamsley's associativity conditions (without the restriction $k \leqslant d$) imply that $W$ is a group. The proof given here is different from that given in [**2**]. Then we will prove that the associativity conditions with $k > d$ are redundant.

For $r = 1, 2, \ldots, n$ we let $W_r$ be the subgroupoid consisting of normal words of the form

$$a_r^{\alpha(r)} a_{r+1}^{\alpha(r+1)} \ldots a_n^{\alpha(n)}.$$

We assume that all Wamsley's associativity conditions hold, and we use induction to show that $W = W_1, W_2, \ldots, W_n$ are all groups. Clearly $W_n$ is a cyclic group of order $p$. We suppose that $W_{k+1}$ is a group for some $1 \leqslant k < n$, and we prove that $W_k$ is a group.

We define a map, $\theta_k : W_{k+1} \to W_{k+1}$ by setting $w\theta_k := u$, where $a_k u$ is the normal word obtained from collecting $wa_k$. We show that $\theta_k$ is an automorphism of $W_{k+1}$ as follows.

First we show that, if $a_i a_j \ldots a_r a_s$ is a normal word in $W_{k+1}$, then

$$(a_i a_j \ldots a_r a_s)\theta_k = a_i\theta_k . a_j\theta_k . \ldots . a_r\theta_k . a_s\theta_k.$$

To calculate $(a_i a_j \ldots a_r a_s)\theta_k$, we must apply the collection process to $a_i a_j \ldots a_r a_s a_k$. The first two words of the output are

$$a_i a_j \ldots a_r a_k (a_s\theta_k), \; a_i a_j \ldots a_k (a_r\theta_k)(a_s\theta_k).$$

At this point there is some ambiguity since there may be more than one minimal non-normal subword. There is one minimal non-normal subword involving $a_k$, and there may also be one in $(a_r\theta_k)(a_s\theta_k)$. However, if we identify $(a_r\theta_k)(a_s\theta_k)$ with an element of the group $W_{k+1}$, then collection of any of its subwords does not change its value as an element of $W_{k+1}$. So we can ignore collection of subwords to the right of $a_k$. Also, there is never any minimal non-normal subword to the left of $a_k$ at any stage in the collection process. So when the collection process is completed we obtain a word $a_k u$, where $u$ is a normal word equal to $a_i\theta_k . a_j\theta_k . \dots . a_r\theta_k . a_s\theta_k$ as an element of $W_{k+1}$. So

$$(a_i a_j \dots a_r a_s)\theta_k = u,$$

and this is the required result. Note that for $k+1 \leqslant i \leqslant n$, $a_i\theta_k = a_i$ modulo $W_{i+1}$. This implies that every element of $W_{k+1}$ can be expressed uniquely as a normal word in the elements $a_{k+1}\theta_k, a_{k+2}\theta_k, \dots, a_n\theta_k$, and hence that $\theta_k$ is a permutation of $W_{k+1}$.

Next notice that the associativity condition

$$(a_i . a_j) . a_k = a_i . (a_j . a_k)$$

is equivalent to the condition

$$(a_j a_i a_{i+1}^{\alpha(i,j,i+1)} \dots a_n^{\alpha(i,j,n)})\theta_k = (a_i\theta_k) . (a_j\theta_k),$$

and the associativity condition

$$(a_j^{p-1} . a_j) . a_k = a_j^{p-1} . (a_j . a_k)$$

is equivalent to the condition

$$(a_{j+1}^{\alpha(j,j+1)} \dots a_n^{\alpha(j,n)})\theta_k = (a_j\theta_k)^p.$$

So the permutation $\theta_k$ is induced homomorphically from a map of the generators of $W_{k+1}$ into $W_{k+1}$, and $\theta_k$ preserves the relations satisfied by the generators. This proves that $\theta_k$ is an automorphism of $W_{k+1}$.

Also the associativity condition that

$$(a_i . a_k) . a_k^{p-1} = a_i . (a_k . a_k^{p-1})$$

is equivalent to the condition that

$$a_i(\theta_k)^p = u^{-1} . a_i . u$$

where $u = a_{k+1}^{\alpha(k,k+1)} a_{k+2}^{\alpha(k,k+2)} \dots a_n^{\alpha(k,n)} \in W_{k+1}$. (The word $u$ is the right hand side of the relation whose left hand side is $a_k^p$.) Finally, the associativity condition

$$(a_k . a_k^{p-1}) . a_k = a_k . (a_k^{p-1} . a_k)$$

is equivalent to the condition that

$$u\theta_k = u.$$

So we can form the split extension $H$ of $W_{k+1}$ by $\theta_k$. The element $(\theta_k)^{-p}u$ is central in $H$, and $H/\langle(\theta_k)^{-p}u\rangle$ is isomorphic to $W_k$. This proves that $W_k$ is a group, and so, by induction, that $W$ is a group.

We now show that the associativity conditions with $k > d$ are redundant. As we saw above, they were used to show that $\theta_k$ was an automorphism of $W_{k+1}$. We will show that if $k > d$, then the definitions of $a_{d+1}, a_{d+2}, \ldots, a_k$ can be used to express $\theta_k$ in terms of $\theta_1, \theta_2, \ldots, \theta_d$. Provided we know that $\theta_1, \theta_2, \ldots, \theta_d$ act as automorphisms on $W_{k+1}$, this will imply that $\theta_k$ is also an automorphism.

So we assume that the associativity conditions hold for $k \leqslant d$, and we make the following triple induction hypothesis. First we assume that $W_{k+1}$ is a group for some $1 \leqslant k < n$. Next we assume that $\theta_1$, $\theta_2, \ldots, \theta_{r-1}$ act as automorphisms on $W_{k+1}$ for some $d < r \leqslant k$. (The fact that $\theta_1, \theta_2, \ldots, \theta_d$ act as automorphisms follows, as above, from the assumed associativity conditions.) Finally we assume that $\theta_r$ acts as an automorphism on $W_{s+1}$ for some $k+1 < s+1 \leqslant n$. (Clearly $\theta_r$ acts as the identity automorphism on $W_n$.) We show that this implies that $\theta_r$ acts as an automorphism on $W_s$. By induction on $s$, this implies that $\theta_r$ acts as an automorphism on $W_{k+1}$. By induction on $r$, this implies that $\theta_k$ acts as an automorphism on $W_{k+1}$. As above, this implies that $W_k$ is a group, and hence by induction on $k$, that $W$ is a group.

First we prove three technical lemmas about the collection process.

*Lemma 1.* Let $W_{k+1}$ be a group, and let $\theta_r$ be an automorphism of $W_{k+1}$ for some $r \leqslant k$. If $u$ is a word in the generators of $W_{k+1}$ whose normal form is $v$, and if the collection process is applied to $ua_r$, then we obtain $a_r(v\theta_r)$.

*Proof.* If $u = v$, or if, in the collection of $ua_r$, $u$ is collected into normal form before any subword involving $a_r$ is collected, then the result follows immediately from the definition of $\theta_r$. In general there are two types of minimal non-normal subwords to consider. One type consists of words of the form $a_p a_r$ for some $p > k$, and the other type involves only generators of $W_{k+1}$. Suppose that at an intermediate stage of the collection process we have a word of the form $w_1 a_r w_2$, where $w_1, w_2$ are words in $a_{k+1}, a_{k+2}, \ldots, a_n$. If a sequence of collections of minimal non-normal subwords of the second type is applied to $w_1 a_r w_2$, then we obtain a word of the form $v_1 a_r v_2$ where $v_1 = w_1$ and $v_2 = w_2$ as elements of the group $W_{k+1}$. On the other hand, if a subword of the first type, $a_p a_r$ say, is collected, then it is replaced by $a_r(a_p\theta_r)$. So if the subwords of the form $a_p a_r$ which are collected in the complete collection process are (in sequence) $a_i a_r$, $a_j a_r, \ldots, a_m a_r$, then $a_m . \ldots . a_j . a_i = v$. (Even though $a_m \ldots a_j a_i$ is not necessarily the same word as $v$.) These subwords are replaced successively by $a_r(a_i\theta_r)$, $a_r(a_j\theta_r), \ldots, a_r(a_m\theta_r)$, and so the result at the end of the collection process is $a_r w$, where $w$ is a normal word in $W_{k+1}$, equal to $(a_m\theta_r) . \ldots . (a_j\theta_r) . (a_i\theta_r)$. But $\theta_r$ is an automorphism of $W_{k+1}$ and so

$$w = (a_m\theta_r) . \ldots . (a_j\theta_r) . (a_i\theta_r) = (a_m . \ldots . a_j . a_i)\theta_r = v\theta_r$$

as required.

*Lemma 2.* Let $W_{k+1}$ be a group and let $a_i a_j \ldots a_r$ be a normal word with $i \leqslant j \leqslant \ldots \leqslant r \leqslant k$. Suppose that $\theta_i, \theta_j, \ldots, \theta_r$ act as automorphisms on $W_{k+1}$, and let $u$ be a word in the generators of $W_{k+1}$ whose normal form is $v$. Then, if the collection process is applied to $u a_i a_j \ldots a_r$, we obtain $a_i a_j \ldots a_r(v\theta_i\theta_j \ldots \theta_r)$.

*Proof.* The proof follows the same lines as the proof of Lemma 1. Let $a_p a_i$, $a_q a_i, \ldots, a_s a_i$ be the sequence of minimal non-normal subwords involving $a_i$ which are collected during the complete process. Then $a_s . \ldots . a_q . a_p = v$, and the final result is $a_i w$, where $w$ is the result of collecting

$$(a_s\theta_i) \ldots (a_q\theta_i)(a_p\theta_i)a_j \ldots a_r.$$

(We are not claiming that this word ever occurs as a subword during the collection process. It might happen, for example, that part of $(a_p\theta_i)a_j \ldots a_r$ is collected before $a_q a_i$ is collected.) But $v\theta_i$ is the normal form of $(a_s\theta_i) \ldots (a_q\theta_i)(a_p\theta_i)$ and so, by induction on the length of $a_i a_j \ldots a_r$,

$$(a_s\theta_i) \ldots (a_q\theta_i)(a_p\theta_i)a_j \ldots a_r$$

collects to $a_j \ldots a_r(v\theta_i\theta_j \ldots \theta_r)$.

This completes the proof of Lemma 2.

*Lemma 3.* Let $W_{k+1}$ be a group and let $a_i a_j \ldots a_t a_r$ be a normal word with $i \leqslant j \leqslant \ldots \leqslant t < r \leqslant k$. Suppose that $\theta_i, \theta_j, \ldots, \theta_t$ act as automorphisms of $W_{k+1}$, and that $\theta_r$ acts as an automorphism of $W_{s+1}$ for some $s > k$. Then, if the collection process is applied to $a_s a_i a_j \ldots a_t a_r$, we obtain $a_i a_j \ldots a_t a_r u$, where $u$ is a normal word in $W_{k+1}$ equal to

$$a_s\theta_r . (a_s^{-1} . (a_s\theta_i\theta_j \ldots \theta_t))\theta_r.$$

*Proof.* As in the proof of Lemmas 1 and 2, we see that when $a_s a_i a_j \ldots a_t a_r$ is collected, we obtain $a_i a_j \ldots a_t v$, where $v$ is the result of collecting a word of the form $a_s a_p \ldots a_q a_r$ with $p, \ldots, q > s$, and $a_s . a_p . \ldots . a_q = a_s\theta_i\theta_j \ldots \theta_t$ as an element of $W_{k+1}$. However the collection of $a_s a_p \ldots a_q a_r$ proceeds, the last minimal non-normal subword involving $a_r$ to be collected is always $a_s a_r$. So the result of collecting $a_s a_p \ldots a_q a_r$ is $a_r u$, where

$$u = a_s\theta_r . (a_p \ldots a_q)\theta_r = a_s\theta_r . (a_s^{-1} . (a_s\theta_i\theta_j \ldots \theta_t))\theta_r.$$

This completes the proof of Lemma 3.

Finally we complete our triple induction. The hypothesis is that $W_{k+1}$ is a group for some $1 \leqslant k < n$, that $\theta_1, \theta_2, \ldots, \theta_{r-1}$ act as automorphisms of $W_{k+1}$ for some $d < r \leqslant k$, and that $\theta_r$ acts as an automorphism of $W_{s+1}$ for some $k+1 < s+1 \leqslant n$. We show that $\theta_r$ acts as an automorphism of $W_s$.

Since $r > d$, $a_r$ has a definition of the form

$$a_j a_k = a_k a_j a_{j+1}^{\alpha(j+1)} \dots a_{r-1}^{\alpha(r-1)} a_r \quad \text{(with } k \leqslant d\text{)},$$

or a definition of the form

$$a_j^p = a_{j+1}^{\alpha(j+1)} \dots a_{r-1}^{\alpha(r-1)} a_r.$$

Suppose that the definition of $a_r$ is of the first form. Then we show that $\theta_r$ acts like $(\theta_k \theta_j \theta_{j+1}^{\alpha(j+1)} \dots \theta_{r-1}^{\alpha(r-1)})^{-1} \theta_j \theta_k$ on $W_s$. Since, by induction, $\theta_1, \theta_2, \dots, \theta_{r-1}$ all act as automorphisms on $W_s$, this proves that $\theta_r$ does also. As part of our inductive hypothesis we assume that $\theta_r$ acts like $\phi^{-1}\theta_j\theta_k$ on $W_{s+1}$, where

$$\phi = \theta_k \theta_j \theta_{j+1}^{\alpha(j+1)} \dots \theta_{r-1}^{\alpha(r-1)}.$$

Now, if $a_s^{\beta(s)} \dots a_n^{\beta(n)}$ is a normal word in $W_s$, then

$$(a_s^{\beta(s)} \dots a_n^{\beta(n)})\theta_r = (a_s\theta_r)^{\beta(s)} \dots . (a_n\theta_r)^{\beta(n)}$$

and so it is sufficient to show that

$$a_s\theta_r = a_s\phi^{-1}\theta_j\theta_k.$$

(Since, by induction, $a_i\theta_r = a_i\phi^{-1}\theta_j\theta_k$ for $i > s$.) But $k \leqslant d$, and so

$$(a_s . a_j) . a_k = a_s . (a_j . a_k),$$

since this is one of the original associativity conditions. Now

$$\begin{aligned}(a_s . a_j) . a_k &= (a_j(a_s\theta_j)) . a_k \\ &= a_k a_j a_{j+1}^{\alpha(j+1)} \dots a_{r-1}^{\alpha(r-1)} a_r (a_s\theta_j\theta_k).\end{aligned}$$

Also, by Lemma 3,

$$\begin{aligned}a_s . (a_j . a_k) &= a_s . (a_k a_j a_{j+1}^{\alpha(j+1)} \dots a_{r-1}^{\alpha(r-1)} a_r) \\ &= a_k a_j a_{j+1}^{\alpha(j+1)} \dots a_{r-1}^{\alpha(r-1)} a_r u,\end{aligned}$$

where $u = a_s\theta_r . (a_s^{-1} . a_s\phi)\theta_r$ as an element of $W_s$. So,

$$a_s\theta_j\theta_k = a_s\theta_r . (a_s^{-1} . a_s\phi)\theta_r.$$

But $(a_s^{-1} . a_s\phi) \in W_{s+1}$, and so,

$$\begin{aligned}(a_s^{-1} . a_s\phi)\theta_r &= (a_s^{-1} . a_s\phi)\phi^{-1}\theta_j\theta_k \\ &= a_s^{-1}\phi^{-1}\theta_j\theta_k . a_s\theta_j\theta_k.\end{aligned}$$

It follows that
$$a_s\theta_j\theta_k = a_s\theta_r . a_s^{-1}\phi^{-1}\theta_j\theta_k . a_s\theta_j\theta_k,$$

and hence that
$$a_s\theta_r = a_s\phi^{-1}\theta_j\theta_k$$

as required.

The proof is similar if the definition of $a_r$ is of the form

$$a_j{}^p = a_{j+1}^{\alpha(j+1)} \dots a_{r-1}^{\alpha(r-1)} a_r.$$

This completes the proof that Wamsley's associativity conditions with $k \leqslant d$ imply that $W$ is a group.

## 3 Exponent Checking

Let $G$ be a finite $p$-group, with a consistent power-commutator presentation $P$ as described above. For $i = 1, 2, \dots, n$, we let the weight of $a_i$, $w(a_i)$, be $k$ if $a_i$ lies in the $k$th term of the lower exponent-$p$-central series of $G$, but not in the $(k+1)$th term. If $u = a_i a_j \dots a_k$ is a word in the generators $a_1, a_2, \dots, a_n$, then we define the weight of $u$ to be $w(a_i) + w(a_j) + \dots + w(a_k)$. It was shown in [1] that the exponent of $G$ is the maximum of the orders of normal words of weight at most $c$, where $c$ is the $p$-class of $G$. The same argument shows that the exponent is also the maximum of the orders of words of weight at most $c$ (not necessarily normal), in the generators $a_1, a_2 \dots, a_d$. Of course there are more of these words than there are normal words of weight at most $c$ in $a_1$, $a_2, \dots, a_n$, but it is much easier to eliminate redundant words of this type. Since conjugate elements of $G$ have the same order, it is only necessary to find the orders of one element of $G$ from each conjugacy class which contains a word of weight at most $c$ in the generators $a_1, a_2, \dots, a_d$. It is not feasible to calculate all the conjugates of a word, but $a_i a_j \dots a_k$ is conjugate to $a_j \dots a_k a_i$ and this simple observation cuts down the number of words whose orders have to be calculated by a factor of about $c$. One other elementary observation is that, if the orders of $a_1, a_2, \dots, a_d$ have been computed to be $e(1), e(2), \dots, e(d)$ respectively, then we can ignore all words which have a subword of the form $a_i{}^{e(i)}$, where $1 \leqslant i \leqslant d$, since they are equal in $G$ to words of shorter length. To be precise, we can define a relation $\sim$ on the set of words on $a_1, a_2, \dots, a_d$ by letting

$$a_i a_j \dots a_k \sim a_j \dots a_k a_i,$$

$$a_i{}^{e(i)} a_j a_k \dots a_r \sim a_j a_k \dots a_r.$$

We form the equivalence relation generated by $\sim$ and pick one representative out of each equivalence class containing a word of weight less than or equal to $c$. Then the exponent of $G$ is the maximum of the orders of these representatives. (It is probably easier to program a computer to produce these representatives than to produce the set of normal words of weight at most $c$.) In $B(4,4)$ for example, which has class 10, there are 129 223 of these representatives, as opposed to 376 727 normal words of weight at most 10.

It is possible to extend $\sim$ by letting

$$a_i a_j \dots a_k \sim a_k{}^{e(k)-1} \dots a_j{}^{e(j)-1} a_i{}^{e(i)-1}$$

and there are many other possibilities of a more or less elementary nature. The problem is to balance the possible gain against programming problems, and also against the time taken to eliminate redundant words. There is no net gain if the time taken to eliminate a word exceeds the time it would take to compute its order. However, it seems that exponent testing is the most time-consuming part of the calculation of large $p$-groups, and it probably pays to eliminate as many words as possible. For example, Havas and Newman report in [**1**] that they were unable to complete a calculation of the largest finite quotient of $\langle a, b : a^2 = 1, b^4 = 1, \text{exponent } 8\rangle$. They reported that they found a pre-image of this group with class 26 and order $2^{205}$. They confirmed that all normal words in this pre-image of weight at most 17 have order dividing 8, and conjectured that the group does indeed have exponent 8. However, they were unable to complete the exponent testing of the group because of the time being taken. There are 2970 normal words of weight at most 17 in this group, whereas there are about 52 000 normal words of weight at most 26. However, I have found a set of only 1290 words of weight at most 26 in $a$, $b$, which suffice to determine the exponent of their group. These words are representatives of equivalence classes of words in $a, b$ as described, above. However the relation $\sim$ was extended by adding a few extra conditions for equivalence, all of an elementary nature.

## References

1. Havas, G. and Newman, M. F. (1980). Applications of computers to questions like those of Burnside. *In* "Burnside Groups" (Ed. J. L. Mennicke), Lecture Notes in Mathematics, Vol. **806**, 211–230. Springer-Verlag, Berlin.
2. Wamsley, J. W. (1974). "Computation in nilpotent groups (theory)". Proc. Second Internat. Conf. Theory of Groups (Canberra, 1973), Lecture Notes in Mathematics, Vol. **372**, 691–700. Springer-Verlag, Berlin.

# A Soluble Group Algorithm

C. R. LEEDHAM-GREEN

## 1 Introduction

The appearance of SOGOS, as described elsewhere in these Proceedings, has emphasized the fact that a finite soluble group can be investigated very efficiently if it is described in terms of an AG-system, that is, roughly, a presentation that exhibits a composition series for the group. This raises the question of producing an AG-system for a finite soluble group that has been defined in some other way.

The algorithm defined here takes a finite presentation $\langle X \mid R\rangle$ of a group $G$, and returns an AG-system for $G/G^{\omega}$, where $G^{\omega}$ is the intersection of the terms of the derived series of $G$, provided that this quotient is finite.

Let us call this algorithm the SQA (soluble quotient algorithm) by analogy with the NQA (nilpotent quotient algorithm) which has been so successful in the investigation of groups of prime-power order, and which is incorporated in the SQA.

I am grateful to John Cannon for drawing to my attention the need for such an algorithm.

## 2 AG-Systems

If $G$ is a finite soluble group, an AG-*system* for $G$ is a presentation of $G$ of the form

$$\langle a_1, a_2, \ldots, a_n \mid a_i^{p_i} = w_{ii}, [a_k, a_j] = w_{jk}, 1 \leqslant i \leqslant n, 1 \leqslant j < k \leqslant n\rangle$$

where $p_1, \ldots, p_n$ are primes, and for all $i \leqslant j$, $w_{ij}$ is a word of the form

$$a_{i+1}^{t(i+1)} \ldots a_n^{t(n)},$$

with $0 \leqslant t(k) < p_k$ for all $k$, and where $G$ has order $p_1 \ldots p_n$.

If the restriction on the order of $G$ is lifted, it is clear that $G$ is still a finite soluble group, and that every element of $G$ can be expressed in the form

$$a_1^{t(1)} \ldots a_n^{t(n)}$$

COMPUTATIONAL GROUP THEORY
ISBN 0-12-066270-1

with $0 \leqslant t(k) < p_k$ for all $k$. Let us call this a *canonical form* for the element. The condition that $G$ is of order $p_1 \dots p_n$ is equivalent to the condition that the canonical form of every element is unique. The condition is expressed by the statement that an AG-system is required to be *consistent*.

If some element $g$ of $G$ is expressed in canonical form in the above notation, $t(k) > 0$, and $t(l) = 0$ for all $l < k$, then $k$ is the *leading index* of $g$.

The fact that every finite soluble group $G$ has an AG-system, follows from the fact that $G$ has a composition series with factors of prime order. In fact, $G$ will have AG-systems with further useful properties which we now develop.

*Proposition 1.* With the above notation, $G$ has an AG-system with the property:

(1) If $1 \leqslant i \leqslant j \leqslant n$, and $w_{ij} = a_{i+1}^{t(i+1)} \dots a_n^{t(n)}$ is defined as above, then, for all $k$,

$$p_k = p_j \text{ or } p_k = p_i \text{ or } t(k) = 0.$$

*Proof.* As $G$ is soluble, $G$ has a Sylow system; that is to say a set of Sylow subgroups, one for each prime dividing $G$, such that, if $P$ and $Q$ belong to the system, then $PQ = QP$. It is clearly possible to pick a Sylow system and to choose each $a_i$ to be in a Sylow subgroup of this system. Such an AG-system clearly satisfies property (1).

If $G$ is any group, define $K_\alpha = K_\alpha(G)$ for $\alpha = 0, 1, \dots$ as follows: $K_0 = G$; if $\alpha > 0$ the $K_\alpha$ is the least normal subgroup of $G$, such that $K_{\alpha-1}/K_\alpha$ is nilpotent, provided such a $K_\alpha$ exists.

Clearly $G$ is soluble if and only if, for some $l \geqslant 0$, $K_l = 1$. The least such $l$ is called the *fitting height* of $G$.

If the SQA is adapted to construct the largest soluble $\pi$-quotient of $G$, where $\pi$ is some recursive set of primes, then $K_\alpha$ should be defined so that $K_{\alpha-1}/K_\alpha$ is the largest nilpotent $\pi$-quotient of $K_{\alpha-1}$. The only property that is essential is that $K_{\alpha-1}/K_\alpha$ should be the largest nilpotent quotient of $K_{\alpha-1}/K_{\alpha+1}$ for all $\alpha > 0$.

Clearly, any finite soluble group has an AG-system with the following property:

(2) There are integers $0 = m_1 < m_2 < \dots < m_l = n$ such that if $\alpha \geqslant 0$ then $K_\alpha$ is generated by $(a_j : j > m_\alpha)$.

If $G$ is any group, $p$ is a prime, and $n > 0$, define

$$\lambda_n(G) = \gamma_1(G)^{p^{n-1}} \gamma_2(G)^{p^{n-2}} \dots \gamma_n(G),$$

where $G = \gamma_1(G) \supseteq \gamma_2(G) \dots$ is the lower central series of $G$. We call $\lambda_n(G)$ the $n$th term of the *lower p-central series* of $G$. Then, $\lambda_{n-1}(G)/\lambda_n(G)$ is the largest quotient of $\lambda_{n-1}(G)$ that is central in $G$ and of exponent $p$. The basic properties of this series are discussed in section VIII, §1 of [**3**].

Given an AG-system for a finite soluble group, satisfying condition (2) above, an AG-system for $K_{\alpha-1}/K_\alpha$ is obtained by taking the AG-generators in $K_{\alpha-1}-K_\alpha$ and removing the generators in $K_\alpha$ from the relevant $w_{ij}$. Let us call this the *induced* AG-system for $K_{\alpha-1}/K_\alpha$.

If $P$ is a finite $p$-group, a PC (power-commutator) *system* for $P$ is an AG-system,

$$P=\langle a_1,a_2.\ldots,a_n|a_i^p=w_{ii},[a_k,a_j]=w_{jk},1\leqslant i\leqslant n,1\leqslant j<k\leqslant n\rangle$$

where, for all $i\leqslant j$, $w_{ij}$ is a word of the form

$$a_{j+1}^{t(j+1)}\ldots a_n^{t(n)},$$

with $0\leqslant t(k)<p$ for all $k$, and such that

(A) the central series

$$P\supset\langle a_2,\ldots,a_n\rangle\supset\langle a_3,\ldots,a_n\rangle\supset\ldots\supset\langle a_n\rangle\supset\langle 1\rangle$$

refines the lower $p$-central series, and

(B) the derived group of $P$ is generated by some subset of $(a_i)$.

The Canberra NQA, which was available at the Symposium, constructs PC-systems very efficiently (see also [**5**]).

If $N$ is a finite nilpotent group, an AG-system for $N$ can be constructed by forming a PC-system for each Sylow subgroup of $N$ and requiring the PC-generators for distinct Sylow subgroups to commute. We call such an AG-system an NQ-*system.*

Clearly, any finite soluble group has an AG-system satisfying properties (1) and (2) and also the following property:

(3) If $1\leqslant\alpha\leqslant l$, the induced AG-system on $K_{\alpha-1}/K_\alpha$ is an NQ-system.

The final condition to be imposed on an AG-system is less obvious. We start with the following result.

*Proposition 2.* Let $G$ be a finite group, $N$ be a nilpotent normal subgroup of $G$, $K$ be a unitary ring and $A$ be a Noetherian $KG$-module such that $H_0(N,A)=0$. Then $H^\lambda(G,A)=H_\lambda(G,A)=0$ for all $\lambda\geqslant 0$.

*Proof.* For a generalization of this result, and variations on this theme, see [**4**].

*Proposition 3.* If $K_\alpha$ is abelian, then $K_\alpha$ has a complement in $G$ and any two complements of $K_\alpha$ in $G$ are conjugate.

*Proof.* Let $G_\alpha=G/K_\alpha$ and $N=K_{\alpha-1}/K_\alpha$. As $N$ is the largest nilpotent quotient of $K_{\alpha-1}$, $H_0(N,K_\alpha)=0$. So, by Proposition 2, $H^1(G,K_\alpha)=H^2(G,K_\alpha)=0$.

*Proposition 4.* $G$ has an AG-system with properties (1), (2) and (3), and the additional property:

(4) If $w_{jk} = a_{j+1}^{t(j+1)} \dots a_n^{t(n)}$, in canonical form, where $j \leqslant k \leqslant m_\alpha$, then $t(i) = 0$, whenever $i > m_\alpha$ (i.e. $a_i \in K_\alpha$) and $a_i \notin K_\alpha'$.

*Proof.* In view of condition (B) above this is just a re-wording of Proposition 3.

If $G = \langle X \mid R \rangle$ is a finite presentation of the finite soluble group $G$, the SQA will construct an AG-system satisfying conditions (1) to (4), and will give the image of each element of $X$ in $G$ in canonical form. This output will be called a *special* AG-system for $G$.

## 3 Computational Prerequisites

### 3.1 The NQA (nilpotent quotient algorithm)

Given a finite presentation $\langle X \mid R \rangle$ of a group $G$ and a prime $p$, the NQA will return the largest $p$-quotient $P$ of this group, and the $p$-covering group $Q$ of $P$.

In more detail, a power-commutator system is produced for $G/\lambda_i(G)$ for $i = 2, 3 \dots$, halting when $\lambda_i(G) = \lambda_{i+1}(G)$ or at some pre-assigned value, where $\lambda_i(G)$ is as defined in section 2.

The $p$-covering group $Q$ is the largest extension of $P$ by a central kernel $M$ of exponent $p$ with $M \subseteq \Phi(Q)$, the Frattini subgroup of $Q$. $M$ is functorial in $P$, being naturally isomorphic to $H_2(P, \mathbb{Z}/p)$, and is called the *p-multiplicator* of $P$; any automorphism of $P$ lifts to an automorphism of $Q$, and induces an automorphism of $M$ that is independent of the lifting. $M$ is generated by a set of elements of the form $a^p$ or $[a, b]$.

Let $a, b$ map to $\bar{a}, \bar{b}$ in $P$. To calculate the action of an automorphism $\theta$ of $P$ on such a generator, pick elements $c, d$ of $Q$ which map respectively to $\bar{a}\theta$ and $\bar{b}\theta$ in $P$, and evaluate $c^p$ or $[c, d]$, as the case may be.

The Canberra NQA also has the essential feature of calculating the image in $P$ of the generators $X$; thus it produces a special AG-system for $P$.

### 3.2 The AQA (abelian quotient algorithm)

Given a finite presentation, this returns the abelian quotient of the group defined. This is just a matter of reducing a matrix to Smith normal form. The only efficient program I know for doing this is due to Havas.

### 3.3 Solving a G-module

Given a finite group $G$, and a finite subset $S$ of a free $G$-module $F_M$ of finite rank, this returns the quotient of $F_M$ by the $G$-module generated by $S$. Although a special case of the AQA, this problem is discussed in section 8.

### 3.4 A submodule program

Given a finite (soluble) group $G$, a prime $p$, and a finite KG-module $V$, where $K$ is the Galois field of $p$ elements, this should return the lattice of submodules of $V$, and the quotient of one submodule by another contained within it. As $G$ will be given by a special AG-system, the structure of $V$ can be investigated using the given generators $X$ or the AG-generators of $G$.

### 3.5 Fox derivatives

If $w$ is an element of the free group $F$ on the generators $(x_i : i \in I)$ the Fox derivative $\partial w/\partial x_i$ in $\mathbb{Z}F$ is easily calculated by the rules

$$\partial x_j/\partial x_i = \delta_{ij};\ \partial(uv)/\partial x_i = (\partial u/\partial x_i)v + \partial v/\partial x_i.$$

If $G$ is a finite soluble group, with a special AG-system, generated by $X$, a program is needed to calculate the image in $\mathbb{Z}G$ of the Fox derivatives of an arbitrary word in $X \cup X^{-1}$.

### 3.6 AG-systems

Given an arbitrary word in the AG-generators of a special AG-system, we will need to be able to collect the word into canonical form, as in section 2. In fact the exponents of some of the AG-generators, both in the word to be collected and in the $w_{ij}$, will be indeterminates, so the calculation is done in a pseudo AG-system, as discussed in section 6.

## 4 The General Strategy

Given a finite presentation $G = \langle X \mid R \rangle$, the algorithm will construct special AG-systems for the quotients $G/K_\alpha$, $\alpha = 1, 2, \ldots$, as in section 2. When the finite group $G/K_\alpha$ has been constructed, there are four mutually exclusive possibilities:

(i) $K_{\alpha+1} = K_\alpha$;
(ii) $K_\alpha/K_{\alpha+1}$ is finite but not trivial;
(iii) $K_\alpha$ has an infinite abelian quotient;
(iv) $K_\alpha$ has an infinite residually nilpotent quotient with finite lower central factors.

In cases (i) and (iii) a suitable message is printed, and the algorithm halts. In case (ii) a special AG-system for $G/K_{\alpha+1}$ is constructed. In case (iv) the algorithm continues indefinitely, constructing larger and larger extensions of $G/K_\alpha$ by finite nilpotent kernels. There is probably no algorithm to determine whether one is in case (ii) or case (iv).

If $\pi$ is a finite set of primes, and the SQA is set to calculate the largest soluble $\pi$-quotient of $G$, an increase in efficiency is achieved; though cases (iii) and (iv) above cannot now be distinguished.

Progressing from $G/K_\alpha$ to $G/K_{\alpha+1}$ involves two completely different steps. The first calculates $G/K_\alpha'$, and is discussed in section 5; the second proceeds from $G/K_\alpha'$ to $G/K_{\alpha+1}$, and is discussed in section 6.

## 5 Increasing the Fitting Height

In this section an algorithm is developed that solves the following problem. Let $\langle X \mid R\rangle$ be a finite presentation of the group $G$, and assume that, for some $\alpha \geqslant 0$, $G/K_\alpha$ is finite, and that we have a special AG-system for $G/K_\alpha$. The algorithm determines whether or not $G/K_\alpha'$ is finite, and in the affirmative case constructs a special AG-system for $G/K_\alpha'$.

A modification of the algorithm is discussed in case a set $\pi$ of primes is given so that $G/K_\alpha$ is restricted to being a $\pi$-group.

We first need the following general result.

*Proposition 5.* Let $G$ be any group, and let

$$1 \longrightarrow S \longrightarrow F \xrightarrow{\delta} G \longrightarrow 1$$

be exact, when $F$ is free on $(x_i : i \in I)$. Then there is an exact sequence,

$$0 \longrightarrow S/S' \xrightarrow{\theta} M_F \xrightarrow{\phi} IG \longrightarrow 0,$$

where $M_F$ is the free $G$-module on a set $(e_i : i \in I)$; $IG$ is the augmentation ideal of $G$,

$$r\theta = \textstyle\sum_i e_i\, \partial r/\partial x_i,$$

where $\partial r/\partial x_i$ is the $i$th Fox derivative of $r$, and

$$e_i\phi = 1 - x_i\delta.$$

*Proof.* This is well known, and arises from the beginning of the Gruenberg resolution $M_F \to \mathbb{Z}G \to \mathbb{Z}$, see §3 of [**1**].

Now apply Proposition 5, using $G/K_\alpha$ in place of $G$. Let us write $G_\alpha$ for $G/K_\alpha$. The surjection $\delta : F \to G_\alpha$ is defined by the special AG-system for $G_\alpha$. Let $TS'$ be the sub-$G_\alpha$-module of $S/S'$ generated by the relators $R$ of $G$. Clearly $K_\alpha/K_\alpha'$ is isomorphic, as $G_\alpha$-module, to $S/T$. It is now easy to construct $G/K_\alpha'$.

*Algorithm 1* The input and output are as described at the beginning of this section, except that the calculation of the image of $X$ in $G/K_\alpha'$ is postponed for a moment.

If $X = (x_i : i \in I)$, and $R = (r_j : j \in J)$, solve the $G$-module $M$, with generators $(e_i : i \in I)$ and relators $(\sum e_i \partial r_j / \partial x_i : j \in J)$.

$d$: = the torsion-free rank of $M$; (so $d \geqslant rk_{\mathbb{Z}} IG = |G| - 1$)
if $d > |G| - 1$ then halt; ($K_\alpha/K_\alpha'$ is infinite)
else
 $T(M)$: = torsion module of $M$; (the maximum finite submodule)
 construct an AG-system for $G/K_\alpha'$ as the split extension of $T(M)$ by $G_\alpha$ (Proposition 3.)

If $T(M)$ is calculated as the direct sum of its primary components, the AG-system will satisfy condition (1) of a special AG-system. Conditions (2) and (3) are automatic.

It remains to calculate the image of $X$ in $G/K_\alpha'$. This could be done using the above theory as follows. The exact sequence

$$0 \longrightarrow S/T \longrightarrow M_F/T \longrightarrow IG_\alpha \longrightarrow 0$$

splits, as $H^2(G_\alpha, S/T) = 0$. Construct an explicit splitting. This gives rise to a homomorphism $\rho$ from $M_F/T$ onto $S/T$. For each $x$ in $X$ take the image of $x$ in $G/K_\alpha'$, which is the split extension of $K_\alpha/K_\alpha'$ by $G_\alpha$, to be $(x\rho, x\delta)$. A more efficient approach may be as follows.

For each $x$ in $X$, the exponents of the AG-generators in $K_\alpha$ appearing in the canonical expression for the image of $x$ in $G/K_\alpha'$ must be determined. Regard them at first as indeterminates. Impose the relations in $R$. The only further condition on the indeterminates is that the image of $X$ in $G/K_\alpha'$ should generate $G/K_\alpha'$. It is sufficient to ensure that it generates $K_\alpha/K_\alpha'$ as its image in $G/K_\alpha$ generates $G/K_\alpha$. This leads to the following algorithm.

*Algorithm 2.* The input and output are described above.

For each $x_i$ in $X$, form the image of $x_i$ in $G/K_\alpha'$ by taking its image in $G/K_\alpha$, and adjoining $a_j^{t(i,j)}$ for each AG-generator $a_j$ in $K_\alpha$, where the $t(i,j)$ are indeterminates, with $0 \leqslant t(i,j) < p_j$;

for each $r$ in $R$, evaluate $r$ in $G/K_\alpha'$ as a word in the $x_i$, and equate the resulting element of $K_\alpha/K_\alpha'$ to the identity;
use this set of linear equations to eliminate redundant $t(i,j)$;
$U$: = the image of $X$ in $G/K_\alpha'$;
$V$: = the empty subset of $K/K_\alpha'$;
for each $u_1$ in $U$ do
 for each $u_2$ in $U$ do
  {if $u_1 = u_2$ then
   {$i$: = leading index of $u_1$; $v$: = $u_1^{p_i}$}

```
        else v: = [u_1, u_2];
        if v ∉ U then
          if v ∉ K_α then
            U: = U join (v)
          else
            {V: = V join (v);
             if, for some values of the indeterminates, V generates K_α/K_α' as
             G-module, then return these values and halt
            }
    }
```

If a set of primes $\pi$ is given so that we are only interested in the $\pi$-component of $K_\alpha/K_\alpha'$, various obvious modifications to Algorithm 1 can be made. For example, if $\pi$ is finite, work modulo each prime in $\pi$ separately (or modulo the product of some or all of the primes in $\pi$). If working modulo $q$, calculate $M_F/(T+qM_F)$ and obtain $S/(T+qS)$ as the kernel of the homomorphism of $M_F/(T+qM_F)$ onto $IG_\alpha/q(IG_\alpha)$ induced by $\phi$.

In this case, instead of $K_\alpha/K_\alpha'$ we obtain the largest homomorphic image of $K_\alpha$ that is the direct product of elementary abelian $\pi$-groups, and do not discover if $K_\alpha/K_\alpha'$ is infinite. Of course $K_\alpha/K_\alpha'$ is finite if, and only if, for some prime $p$, it has no non-trivial elementary abelian $p$-factor, and is infinite if it does have a non-trivial elementary-abelian $p$-factor, but cannot have any $p$-torsion; for example if $K_{\alpha-1}/K_\alpha$ is a $p$-group.

A final observation. If $\alpha = 0$, so $K_\alpha = G$ and $G_\alpha$ is trivial, Algorithm 1 reduces, as it should, to calculating the group obtained by abelianizing the relations $R$. This is because the Fox derivative of a word evaluated in an abelian group is the exponent sum of the given generator in that word.

For a clear account of the main ideas of this section, and an interesting bibliography, see [**2**].

## 6 Consistency and Collection

Let $G$ have a presentation

$$\langle a_1, a_2, \ldots, a_n | a_i^{p_i} = w_{ii}, [a_k, a_j] = w_{jk} \rangle,$$

as in section 2. This is an AG-system if it is consistent, that is, if $G$ has order $\prod_1^n p_i$. We start by obtaining necessary and sufficient conditions for this to happen.

Let $v_{jk} = a_k w_{jk}$, so that $a_k^{a_j} = v_{jk}$. Strictly speaking, $v_{jk}$ is a word in canonical form obtained from $a_k w_{jk}$ by collection, as described below.

Let $w$ be an element of $G$ expressed as a word in the $a_i$ and their inverses; $w$

can be written in canonical form by repeated cancellation and performing the following operations:

(i) Replace the subword $a_k a_j$ $(j < k)$ by $a_j a_k w_{jk}$;

(ii) If $t < 0$, replace the subword $a_i^t$ by $a_i^{t+p_i} w_{ii}^{-1}$;

(iii) If $t \geqslant p_i$, replace the subword $a_i^t$ by $a_i^{t-p_i} w_{ii}$.

The collecting process is not always unique; in general there will be a choice of operations to perform. The presentation is consistent if, and only if, the result of collecting an arbitrary word is independent of the strategy adopted for collection. Write $w_1 \equiv w_2$ if there is a way of collecting $w_1$ and $w_2$ which reduces them to the same canonical form. It is not clear that this is an equivalence relation.

*Proposition 6.* (cf. [**6**]). The above presentation is consistent if, and only if, the following conditions are satisfied:

(i) If $i < j$, let $u(i,j)$ be obtained from $w_{jj}$ by replacing $a_t$ by $v_{it}$ for all $t$. Then $u(i,j) \equiv v_{ij} \dots v_{ij}$, the concatenation of $p_j$ copies of $v_{ij}$.

(ii) If $i < j < k$, let $u(i,j,k)$ be obtained from $v_{jk}$ by replacing $a_t$ by $v_{it}$ for all $t$. Then, $u(i,j,k) \equiv v_{ij}^{-1} v_{ik} v_{ij}$.

(iii) If $i < j$ and $1 \leqslant m \leqslant p_i$ define $U(i,j,m)$ by $U(i,j,1) = a_j$; if $m > 1$, then $U(i,j,m)$ is obtained from $U(i,j,m-1)$ by replacing $a_t$ by $v_{it}$ for all $t$. Then $U(i,j,p_i) \equiv w_{ii}^{-1} a_j w_{ii}$

*Proof.* Assume that we have a consistent AG-system, and use $\wedge$ to denote exponentiation.

The words in condition (i) represent $(a_j \wedge p_j) \wedge a_i$ and $(a_j \wedge a_i) \wedge p_j$ respectively; as these are equal the words must collect to the same (unique) canonical form.

The words in condition (ii) represent $(a_k \wedge a_j) \wedge a_i$ and $(a_k \wedge a_i) \wedge (a_j \wedge a_i)$ respectively, and these are equal.

The words in condition (iii) represent $a_j \wedge (a_i \wedge p_i)$.

Assuming that the conditions are satisfied we now prove, by descent on $i$, that

$$\langle a_i, \dots, a_n | a_j^{p_j} = w_{jj}, [a_k, a_j] = w_{jk} \rangle$$

is consistent.

As $w_{nn}$ is the empty word this is trivial if $i = n$. Assume that $i < n$, and that $\langle a_{i+1}, \dots, a_n \rangle$ is consistent. To check that $\langle a_i, \dots, a_n \rangle$ is consistent, it is sufficient to prove that $a_j \mapsto v_{ij}$, $j = i+1, \dots, n$, defines an automorphism $\beta$ of $\langle a_{i+1}, \dots, a_n \rangle$, and that $\beta^{p_i}$ is conjugation by $w_{ii}$; for then $\langle a_i, \dots, a_n \rangle$ is the quotient of the split extension of $\langle a_{i+1},\ldots,a_n\rangle$ by an infinite cycle $a_i$ acting as $\beta$, divided out by the central subgroup generated by $a_i^{p_i}w_{ii}^{-1}$.

Conditions (i) and (ii) ensure that $a_j \mapsto v_{ij}$ defines an automorphism of $\langle a_{i+1},\ldots,a_n\rangle$, and by condition (iii) the $p_i$th power of this automorphism is induced by $w_{ii}$.

Note that the words to be collected lie in $\langle a_{i+1},\ldots,a_n\rangle$, which is consistent by assumption, so that the strategy used for collection is immaterial.

The problem that will face us in the SQA will be slightly different, in that we will have a presentation

$$G = \langle a_1, a_2, \ldots, a_{n+d} \mid a_i^{p_i} = w_{ii}, [a_k, a_j] = w_{jk}, 1 \leqslant i \leqslant n+d,$$
$$1 \leqslant j < k \leqslant n+d\rangle,$$

a set $X$ of generators of $G$, a set $R$ of relators in $X$; and each $x$ in $X$ will be given in canonical form. However, some of the exponents of the $a_i$ in the $w_{jk}$, and in the expression for $x$ in $X$, will be indeterminate. The question of whether the relators become trivial in $G$ is left open.

Specifically there will be an integer $m_\alpha$, $1 \leqslant m_\alpha < n$, and the following conditions will be satisfied:

(i) $$\langle a_{m_\alpha+1}, \ldots, a_{n+d} \mid a_i^{p_i} = w_{ii}, [a_k, a_j] = w_{jk}, m_\alpha < i \leqslant n+d,$$
$$m_\alpha < j < k \leqslant n+d\rangle$$

is a PC-system for a group $Q$, the extension of a $p$-group $P$ by its $p$-multiplicator $M = \langle a_i : i > n\rangle$. In particular, the $w_{ij}$, $m_\alpha < i \leqslant j$, are all completely determined. More generally, $M$ may be a quotient of the $p$-multiplicator of $P$.

(ii) Let $\bar{M}$ be the image of $M$ in $G$. Then $\bar{M}$ is a normal subgroup of $G$, and the special AG-system for $G/\bar{M}$ obtained by ignoring the $a_i$ with $i > n$ is consistent. In particular, $a_i$ can only have an indeterminate exponent in some $w_{jk}$, or in the canonical expression for some $x$, if $i > n$. Note that $\bar{M}$ will be isomorphic to $M$ if, and only if, the indeterminates take values that give a consistent AG-system.

(iii) If $n < k$, and $j \leqslant k$, then $w_{jk}$ is completely determined. (Of course $w_{jk}$ is the empty word if $n < j \leqslant k$ as $M$ is elementary abelian). This turns $M$ into a $G$-module, in view of (iv) below.

(iv) If $m_\alpha < j < k$ then:

if $w_{jj}$ lies in $M$, then (i) of Proposition 6 is satisfied for all $i$;
if $w_{jk}$ lies in $M$, then (ii) of Proposition 6 is satisfied for all $i$.

Note that condition (iv) states that the action of $G$ on $M$ is induced by the action of $G$ on $P$.

We call the above a *pseudo* AG-system.

The question now arises of finding values for the indeterminates that will give rise to a special AG-system, if necessary after dividing out some submodule $N$ of $M$.

To this end, we construct a matrix $\mathbf{C}$ with $d$ columns ($d = \dim M$), whose entries are polynomials of degree one over GF($p$) in the indeterminate coefficients, as follows. For each relation $w_1 \equiv w_2$ of Proposition 6, that is not automatically satisfied, collect the two sides. They will differ by an element, $a_{n+1}^{t(1)} \ldots a_{n+d}^{t(d)}$ of $M$, where each $t(i)$ is a polynomial of degree one over GF($p$) in the indeterminate coefficients. Then $t(1), \ldots, t(d)$ is a row of $\mathbf{C}$. The order in which the rows of $\mathbf{C}$ appear is immaterial; in fact $\mathbf{C}$ can be replaced by any row-equivalent matrix.

Another $d$-rowed matrix $\mathbf{D}$ with entries of the same type is obtained as follows. Evaluate the words in $R$; each of these words will again produce an element of $M$ of the above type, though in this case indeterminates in the expression for the elements of $X$ will also appear. $\mathbf{D}$ is obtained from $\mathbf{C}$ by adjoining new rows consisting of the vectors formed in this way. Again $\mathbf{D}$ can be replaced by any row-equivalent matrix.

The matrices $\mathbf{C}$ and $\mathbf{D}$ will be called respectively the *consistency* matrix and the *relation-consistency* matrix of the given pseudo AG-system.

Now let $N$ be a $G$-submodule of $M$, and let $\mathbf{A}$ be the matrix of the projection of $M$ onto $M/N$. Dividing out by $N$ gives rise, in a natural way, to a pseudo AG-system with consistency matrix $\mathbf{CA}$ and relation-consistency matrix $\mathbf{DA}$. Let us say that $N$ is *co-consistent* if the indeterminates can be chosen so that $\mathbf{CA}$ is zero, and is *co-relation-consistent* if they can be chosen so that $\mathbf{DA}$ is zero.

Let $\bar{N}$ denote the image of $N$ in $G$. Clearly $N$ is co-consistent if, and only if, the indeterminate exponents can be chosen so that, after dividing out by $\bar{N}$, a consistent AG-system is obtained; and $N$ is co-relation-consistent if, and only if, they can be chosen so that, in addition, the relators vanish in $G$, so that a special AG-system is obtained. With a slight abuse of notation, we denote the group constructed in this way by $G/N$.

The question arises of whether there are unique co-consistent and co-relation-consistent modules. In the former case I suspect that the answer is negative, though we shall prove an affirmative answer in the latter case.

Even if there is a unique minimal co-consistent submodule $N$ of $M$, different values of the indeterminates may give rise to non-isomorphic groups $G$.

*Example.* Let $G/\bar{M}$ be $S_4$, the symmetric group on 4 letters, and $P$ be the normal 4-subgroup of $S_4$. Then the 2-multiplicator $M$ of $P$ is the direct sum of irreducible submodules $N_1$ and $N_2$ of dimensions 1 and 2 respectively, and $N_2$ is the unique minimal co-consistent submodule of $M$, but $G$ can be either of the two non-isomorphic covering groups of $S_4$.

## 7 Constructing the Nilpotent Sections

For purposes of notational convenience, assume throughout this section that $K_{\alpha+1}$ is trivial. We continue to write $G_\alpha$ for $G/K_\alpha$.

An algorithm is developed which, given a special AG-system for $G/K_\alpha'$, or for $G/\phi(K_\alpha)$, produces a special AG-system for $G$ provided that $G$ is finite. The algorithm is valid when $K_\alpha$ is any nilpotent normal subgroup of $G$.

Given a special AG-system for $G$ modulo each Sylow complement of $K_\alpha$, it is a triviality to construct a special AG-system for $G$. We therefore assume that $K_\alpha$ is a $p$-group for some prime $p$.

As it is also a triviality to construct a special AG-system for $G/\phi(K_\alpha)$ given one for $G/K_\alpha'$, we assume most of the time that we have one of the former.

The algorithm constructs a special AG-system for $G/\lambda_\beta(K_\alpha)$, for $\beta = 3, 4, \ldots$. If it finds $\lambda_n(K_\alpha) = \lambda_{n+1}(K_\alpha)$ for some $n$, which implies that $\lambda_n(K_\alpha)$ is trivial, the algorithm halts. The case $\beta = 2$ is assumed to be given. Let

$$\langle a_1, a_2, \ldots, a_n | a_i^{p_i} = w_{ii}, [a_k, a_j] = w_{jk}, 1 \leqslant i \leqslant n, 1 \leqslant j < k \leqslant n\rangle$$
$$x = a_1^{t(x,1)} \ldots a_n^{t(x,n)}, x \in X,$$

be a special AG-system for $G/\lambda_\beta(K_\alpha)$. Now construct a pseudo AG-system

$$\langle a_1, a_2, \ldots, a_{n+d} | a_i^{p_i} = w_{ii}', [a_k, a_j] = w_{jk}'\rangle$$
$$x = a_1^{t(x,1)} \ldots a_{n+d}^{t(x,n+d)}, x \in X,$$

where

$$\langle a_{m_\alpha+1}, \ldots, a_{n+d} \,|\, a_i^{p_i} = w_{ii}', [a_k, a_j] = w_{jk}',$$
$$m_\alpha < i \leqslant n+d, m_\alpha \leqslant j < k \leqslant n+d\rangle$$

is a PC-system for the $p$-covering group of

$$P = \langle a_{m+1}, \ldots, a_n \,|\, a_i^{p_i} = w_{ii}, [a_k, a_j] = w_{jk}\rangle.$$

The exponent of $a_u$ in $w_{jk}'$, $j \leqslant k$, is the exponent of $a_u$ in $w_{jk}$ if $u \leqslant n$, and is an indeterminate if $u > n$, except if $j > m_\alpha$, when it is determined by the structure of $Q$, or if $k > n$, when it is determined (as in section 6) by the action of $G$ on $M$, or if $p_j \neq p \neq p_k$, when it is zero.

Clearly, a special AG-system for $G/\lambda_{\beta+1}(K_\alpha)$ is obtained by dividing out a suitable minimal co-relation-consistent submodule of $M$ and giving suitable values to the indeterminates. We shall need a uniqueness theorem; but first the following well known result.

*Lemma 1.* Let $Q$ be the $p$-covering group and $M$ be the $p$-multiplicator of the $p$-group $P$. If $N_1$ and $N_2$ are subgroups of $M$ such that $Q/N_1$ is isomorphic to $Q/N_2$ via an isomorphism $\theta$ inducing the identity on $P$, then $N_1 = N_2$.

*Proof.* Let $z$ be an element of $M$, say,

$$z = \prod_j g_j^p \prod_k [h_k, y_k],$$

where $g_j, h_k$ and $y_k$ are elements of $Q$ for all $j, k$. Let $g_j N_1 \theta = \bar{g}_j N_2$, $h_k N_1 \theta = \bar{h}_k N_2$ and $y_k N_1 \theta = \bar{y}_k N_2$ for all $j, k$. As $g_j \equiv \bar{g}_j \bmod M$, $g_j^p = \bar{g}_j^p$, similarly $[h_k, y_k] = [\bar{h}_k, \bar{y}_k]$. So $zN_1\theta = zN_2$, and $N_1 = N_2$.

We can now prove our uniqueness theorem.

*Theorem.* The above pseudo AG-system gives rise to a unique minimal co-relation-consistent submodule $N$ of $M$, and if values for the indeterminates are chosen in any way to produce a special AG-system for $G/N$, as in section 6, then $G/N$ is isomorphic to $G/\lambda_{\beta+1}(K_\alpha)$.

*Proof.* Clearly, $G/\lambda_{\beta+1}(K_\alpha)$ is isomorphic to $G/N$ for some co-relation-consistent module $N$, and for some choice of the indeterminates. If $N_0$ is also co-relation-consistent, and a special AG-system for $G/N_0$ is constructed, then $G/N_0$ is a homomorphic image of $G/\lambda_{\beta+1}(K_\alpha)$, and hence is isomorphic to $G/\lambda_{\beta+1}(K_\alpha)$ if $N_0 = N$, however the indeterminates were chosen. If $N_0 \neq N$, then $G/N_0$ is isomorphic to $G/N$, for some submodule $N_1$ of $M$ containing $N_0$, and some choice of the indeterminates, and the isomorphism can be shown to induce the identity on $G/M$. It follows from Lemma 1 that $N_0 = N_1$, so $N_0 \supseteq N$ as required.

It is easy to see, by a similar argument, that if $N_0$ is any co-relation-consistent submodule of $M$ then $G/N_0$ is uniquely determined up to isomorphism, which partially justifies our abuse of notation.

If $N$ is any submodule of $M$, let $\mathbf{A}_N$ denote the matrix of the projection of $M$ onto $N$. In computational terms, our problem is to find the unique minimum submodule $N$ of $M$ such that the linear equations $\mathbf{DA}_N = \mathbf{0}$ are consistent, and to produce a solution, where $\mathbf{D}$ is defined in section 6. If, for some $N$, the equations $\mathbf{DA}_N = \mathbf{0}$ are consistent, it is easy to find the general solution in terms of an expression for some "bound" indeterminates in terms of other "free" indeterminates. The bound indeterminates can then be eliminated from $\mathbf{D}$. Iterating this procedure gives rise to linear expressions for an increasing set of bound indeterminates in terms of a decreasing set of free indeterminates. This gives rise to the following simple algorithm:

*Algorithm 3.*

```
N: = M;

repeat
   if N has no maximal submodule N_0 such that DA_{N_0} = 0 is
   consistent then
      {give arbitrary (e.g. zero) values to the free indeterminates;
       evaluate the bound indeterminates;
       terminate
      }
```

```
else
  {find one such submodule N_0;
   solve DA_{N_0} = 0,

   eliminate the bound indeterminates from D;
   N: = N_0
  }
until the program terminates.
```

This simple algorithm can be improved as follows. Find the Jacobson radical $J(M)$ of $M$, and split $M/J(M)$ into the direct sum $M_1 \oplus \ldots \oplus M_r$ of blocks of direct sums of isomorphic irreducibles, with the irreducible direct summands of distinct blocks being non-isomorphic. Assume for the time being that $J(M) = 0$. Then $N = N_1 \oplus \ldots \oplus N_r$, where each $N_j$ can be calculated by working modulo the complement of $M_j$. This reduces the problem to the case in which $M$ is the direct sum of isomorphic irreducibles. Now suppose that it has been shown that the maximal submodule $K_0$ of $M$ does not contain $N$, but that $K_1$ does, so we are looking at submodules of $K_1$. Clearly, the maximal submodule $K_0 \cap K_1$ does not contain $N$ and need not be checked. Elaboration of this principle should improve efficiency.

Turning to the case in which $M$ is not semi-simple, the above technique can be used to calculate $N + J(M)$. Define $N^{(0)} = M, N^{(j)} = N + J(N^{(j-1)})$ for $j > 0$. Having calculated $N^{(j)}$, it is easy to use the above ideas to calculate $N^{(j+1)}$; clearly, $N$ can only be contained in maximal submodules of $N^{(j)}$ that do not contain $J(N^{(j-1)})$.

If $N^{(j+1)} = N^{(j)}$ then $N^{(j)} = N$, and we have finished.

As a final observation, suppose that $P$ is abelian. In this case $Q/M_0$ will be abelian for some submodules $M_0$ of $M$. In this case, by Proposition 4, we can take $w'_{jk} = w_{jk}$ whenever $j \leqslant k \leqslant n$. It will be possible to deal immediately with $M_0$, if $K_\alpha/K_\alpha'$ rather than $K_\alpha/\phi(K_\alpha)$, was calculated when the fitting height was increased.

## 8 Conclusion

The weak point in this algorithm is the passage in section 5 from $G_\alpha = G/K_\alpha$ to $G/K_\alpha'$. As this requires more than $|G|^3$ operations, we are restricted to groups that are nilpotent-by-small. While minor improvements can easily be made, it seems clear that the only way to make serious improvements is to perform the matrix manipulation over $\mathbb{Z}(G_\alpha)$ rather than over $\mathbb{Z}$.

To illustrate this point, we turn to Wamsley's paper [**6**]. His soluble quotient algorithm consists of going straight down the derived series. He obtains a presentation for $G^{(n-1)}/G^{(n)}$, as a module over $G/G^{(n-1)}$, and whose relations correspond to the defining relations of $G$ together with the consistency relations for $G/G^{(n)}$, of section 6.

Following Wamsley, consider the group

$$G = \langle a, b, c \mid [b, a] = b^3, [c, b] = c^3, [c, a] = a^3 \rangle.$$

This group, according to [**6**], is of order $3^9 \times 7^2$, has derived length 3 and fitting height 2, and is super-soluble.

Using our algorithm, one would first calculate $G/G'$, which is obviously elementary abelian of order $3^3$, then $G_1 = G/K_1$, which is of order $3^9$, then $G/K_2$, which is of order $3^9 \times 7^2$ and hence is $G$ itself. One would then check, as in section 7, that this is the largest quotient of $G$ of fitting class 2, and then that this is the largest soluble quotient of $G$ as in section 5.

Rather than carry out this procedure, we illustrate the technique of section 5 by calculating $G/G''$. We will however cheat by using module operations and a limited amount of intelligence, rather than just reducing the appropriate matrix blindly to Smith normal form.

Calculating the Fox derivatives of our relators gives the following matrix:

$$\mathbf{A} = \begin{bmatrix} 1-b & a-1-(1+b+b^2) & 0 \\ 0 & 1-c & b-1-(1+c+c^2) \\ 1-c-(1+a+a^2) & 0 & a-1 \end{bmatrix},$$

whose columns correspond to the generators $e_1$, $e_2$, $e_3$ of $M_F$, which in turn correspond to $a$, $b$ and $c$ respectively. The entries of this matrix lie in $\mathbb{Z}(G/G')$.

Dividing out the submodule $M$ generated by $f_1 = e_1(1+a+a^2)$, $f_2 = e_2(1+b+b^2)$ and $f_3 = e_3(1+c+c^2)$, we get a module defined by the matrix

$$\mathbf{B} = \begin{bmatrix} 1-b & a-1 & 0 \\ 0 & 1-c & b-1 \\ 1-c & 0 & a-1 \end{bmatrix}.$$

As $(1-a)(1+a+a^2) = (1-b)(1+b+b^2) = (1-c)(1+c+c^2) = 0$, the submodule we have divided out by lies in the kernel of the map $\phi: M_F \to I(G/G')$.

Denote the rows of $\mathbf{B}$ by $r_1$, $r_2$ and $r_3$. Then $r_1$ is annihilated by $1+a+a^2$ and by $1+b+b^2$, so all consequences of $r_1$ are $\mathbb{Z}$-linear combinations of

$$r_1 a^x b^y c^z, 0 \leqslant x, y, z \leqslant 2, x, y \leqslant 1.$$

This gives 12 relators. Similarly the second row produces 12 relators. Now note that

$$r_1(1-c) + r_2(1-a) + r_3(b-1) = 0,$$

so we need only take

$$r_3 a^x c^z, \ 0 \leqslant x, z \leqslant 1$$

to obtain 4 more relators; and every consequence of the relators is a $\mathbb{Z}$-linear combination of the 28 relators we have produced. But $IG$ is of rank 26 as an abelian group, and $28+26 = 54$, which is the $\mathbb{Z}$-rank of $M_F$. So $M$ is the kernel of $\phi$, and the relations between the above rows, together with the relations $f_1(1-a) = f_2(1-b) = f_3(1-c) = 0$, which are defining relations for the submodule of $M_F$ generated by $(f_1, f_2, f_3)$, give rise to defining relations for $G'/G''$. (Concealed in this construction there lies an efficient free resolution of $\mathbb{Z}$ over $G/G'$ up to dimension 3.) Thus, replacing $r_1$ in the equation $r_1(1+b+b^2) = 0$ by the first row of $\mathbf{A}$ gives (if we replace $f_1, f_2, f_3$ by their images in $G'/G''$),

$$f_2(a-4) = 0,$$

since $f_2 b = f_2$, Hence, $f_2 a = 4f_2$, and $r_1(1+a+a^2) = 0$ gives

$$f_1(1-b) = 21f_2.$$

Similarly, we get the relations

$$f_1(c+2) = f_3(b-4) = 0, f_2(1-c) = 21f_3, f_2(a-1) = 3f_1.$$

Also, the relation $r_1(1-c)+r_2(1-a)+r_3(b-1) = 0$ gives

$$f_1(b-1)+f_2(1-c)+f_3(1-a) = 0. \qquad (*)$$

It is easy to sée that the $t \times t$ circulant whose first row is $1, -n$, followed by $t-2$ zeros, reduces in Smith normal form to a diagonal matrix consisting of $t-1$ unit entries with $1-n^t$ as the last entry. The relations $f_1(c+2) = f_2(a-4) = f_3(b-4) = 0$, thus imply that $f_1$, $f_2$ and $f_3$ have orders dividing 9, 63 and 63 respectively. It is now easy to obtain the relation

$$-3f_1+63f_2+63f_3 = 0,$$

which is now equivalent to $(*)$.

Thus $G'/G''$ is defined, as a $G/G'$-module, by the following relations:

$$\begin{bmatrix} -3 & 21 & 21 \\ 0 & 63 & 0 \\ 0 & 0 & 63 \\ a-1 & 0 & 0 \\ 0 & b-1 & 0 \\ 0 & 0 & c-1 \\ c+2 & 0 & 0 \\ 0 & a-4 & 0 \\ 0 & 0 & b-4 \\ b-1 & 21 & 0 \\ 0 & c-1 & 21 \\ -3 & 0 & a-1 \end{bmatrix}.$$

It is easy to see that these relations are consistent, in that the abelian group generated by the rows is a $G$-module; thus $G'/G''$ has order $3 \times 63 \times 63$. It is now a triviality to write down an AG-system for $G/G''$.

Wamsley proceeds to calculate $G''/G'''$, which turns out to be a central 3-cycle. We could obtain this 3-cycle by the procedure described in section 7. As the calculation of $G'/G''$ only requires the reduction of a rather sparse $81 \times 81$ matrix to Smith normal form, we could obtain $G$ mechanically with a slight amendment of the procedures described here. Of course the question of proving that the group obtained is the whole of $G$, and not just a proper quotient, is a different matter.

## References

1. Gruenberg, K. W. (1970). Cohomological Topics in Group Theory. Lecture Notes in Mathematics, Vol. **143**, Springer-Verlag, Berlin.
2. Howie, J. R. and Johnson, D. L. (1981). An algorithm for the second derived factor group. *In* "Groups St. Andrews 1981" (Eds. C. M. Campbell and E. F. Robertson). L.M.S. Lecture Notes **71**, Cambridge University Press.
3. Huppert, B. and Blackburn, N. (1982). Finite Groups II. *In* "Grundlehren der Mathematischen Wissenschaften", Vol. **242**, Springer-Verlag, Berlin.
4. Robinson, D. (1976). The vanishing of certain homology and cohomology groups, *J. Pure and Appl. Alg.* **7**, 145–167.
5. Vaughan-Lee, M. R. (1983). An aspect of the Nilpotent Quotient Algorithm, (these Proceedings).
6. Wamsley, J. W. (1977). Computing soluble groups. Conference on the Theory of Groups (Canberra, 1975), Lecture Notes in Mathematics, Vol. **573**. Springer-Verlag, Berlin.

# Finite Groups

# Algorithms for Finite Soluble Groups and the SOGOS System

R. LAUE, J. NEUBÜSER AND U. SCHOENWAELDER

## 1 Introduction and Basic Facts

At the heart of the matter in C. Sims' techniques for the algorithmic investigation of permutation groups [**23**], [**24**], [**25**], is the successive approximation of the result modulo the members of a chain of stabilizers. In contrast to this most successful algorithmic application of a classical notion, so far rather little use has been made in computational group theory of the possibility of approximating the solution of problems such as finding the centralizer, normalizer, join or intersection of subgroups, or determining all maximal abelian subgroups, all normal subgroups or all subgroups, by solving them first in a homomorphic image, a technique which is so basic in algebraic argument. The reason for this may be that, at first sight, this seems to need simultaneous handling of data for different groups and of mappings between them and hence a considerable organization and overhead.

However, we will show in sections 2–5 of this paper that, at least for a finite soluble group $G$, this seeming difficulty can be overcome by working with special generating systems defined in the following way.

Let $G = G_0 > G_1 > \ldots > G_n = 1$ be a subnormal series of $G \neq 1$ with cyclic factors, i.e. $G_i \lhd G_{i-1}$ for all $i = 1, \ldots, n$ and $G_{i-1} = \langle G_i, g_i \rangle$, say, then the sequence $(g_1, \ldots, g_n)$ has been called an AG-system [**10**] for $G$. (If needed, the unit group will be assigned the empty sequence as its AG-system.) Let $e_i$ be the order of $G_{i-1}/G_i$—one may of course assume without loss of generality that $e_i$ is a prime and we shall do so for most of the paper and call such a system a PAG-system—then in terms of the AG-system the group $G$ has a presentation

(1.1)
$$g_i^{e_i} = w_{ii}(g_{i+1}, \ldots, g_n) \quad \text{for} \quad 1 \leqslant i \leqslant n$$
$$[g_i, g_j] = w_{ij}(g_{j+1}, \ldots, g_n) \quad \text{for} \quad 1 \leqslant j < i \leqslant n$$

and each element of $G$ can be uniquely expressed in the form

(1.2)
$$g_1^{v_1} \ldots g_n^{v_n} \quad \text{with} \quad 0 \leqslant v_i < e_i,$$

COMPUTATIONAL GROUP THEORY
ISBN 0-12-066270-1

i.e. $\exp: g_1^{v_1} \ldots g_n^{v_n} \mapsto (v_1, \ldots, v_n)$ defines a bijective mapping $G \to \mathbb{Z}_{e_1} \times \ldots \times \mathbb{Z}_{e_n}$ which is uniquely determined by $(g_1, \ldots, g_n)$.

Multiplication of the elements can be accomplished by a collection process using the relations (1.1) (c.f. [**8**] for a survey of collection methods).

Such multiplication was first implemented for 2-groups, [**16**], and soon after for arbitrary finite soluble groups [**14**], [**15**], [**10**]. It is now incorporated [**6**] in the CAYLEY system [**2**]. It has become even more applicable since the NQ-algorithm [**19**] allows a $p$-step central series to be found (and hence a presentation of the form (1.1) from an arbitrary finite presentation of a finite $p$-group.

Early programs also used the fact that (1.2) defines a numbering of the elements of $G$; therefore, a list of these need not be stored and this improves the efficiency of programs for the determination of subgroup lattices [**16**], [**17**]. A bottleneck of these programs, pointed out in the latter paper, was the description of subgroups by their characteristic function on the set of all cyclic subgroups of prime power order in $G$.

The next major step was a proposal of M. F. Newman (private communication, 1977) to fix a PAG-system, $(g_1, \ldots, g_n)$ of $G$, and to describe each subgroup $1 < U \leqslant G$ by a generating set of elements in the form (1.2), which is modelled after an echelonized basis for a row-vector space. Such a system will correspond to a composition series of $U$ obtained by eliminating repetitions from the series of the subgroups $G_i \cap U, i = 0, \ldots, n$. To define it formally, let us introduce some notation. For an element $g = g_1^{v_1} \ldots g_n^{v_n} \neq 1$, we shall denote the exponent $v_i$ of $g_i$ also by $v_i(g)$. If $v_i = 0$ for $i = 1, \ldots, k-1$ and $v_k \neq 0$, we call $v_k$ the leading exponent and $k$ the weight of $g$ and denote them by $v_k = \lambda(g)$, $k = w(g)$. We now formulate as follows.

(1.3) *Definition.* Let $1 < U \leqslant G$ and $(g_1, \ldots, g_n)$ a PAG-system of $G$. A generating system $(u_1, \ldots, u_r)$ of $U$ is called the canonical generating system (CGS) of $U$ with respect to $(g_1, \ldots, g_n)$, iff:

(i) $(u_1, \ldots, u_r)$ is a PAG-system for $U$;
(ii) $w(u_i) > w(u_j)$ for $i > j$;
(iii) $\lambda(u_i) = 1$ for $i = 1, \ldots, r$;
(iv) $v_{w(u_i)}(u_j) = 0$ for $i \neq j$.

We see that this definition requires that the exponent vectors $\exp(u_i)$ form a row-echelonized matrix of a form like

$$\begin{bmatrix} \exp u_1 \\ \exp u_2 \\ \exp u_3 \end{bmatrix} = \begin{bmatrix} 0 & 1 & * & * & 0 & * & * & 0 & * \\ 0 & 0 & 0 & 0 & 1 & * & * & 0 & * \\ 0 & 0 & 0 & 0 & 0 & 0 & 0 & 1 & * \end{bmatrix}$$

It is clear that this data structure allows easy passing to factor groups $G/G_i$ when $G_i \lhd G$ (command: FACTOR GROUP).

The two most important statements about the CGS of a subgroup $1 < U \leqslant G$ are (i) that it is indeed uniquely defined with respect to the fixed PAG-system $(g_1, \ldots, g_n)$ and (ii) that there is a "non-commutative Gauß algorithm" (NCGA) which will construct it from any finite generating system of $U$. We shall give details of this algorithm in section 2. Here we point out that it immediately provides us with methods for solving a number of further problems:

(1.4) to form the CGS of $\langle U, g \rangle$ from that of $U$ and an element $g \in G$ (command: JOIN);

(1.5) hence, to test containment $g \in U$ (command: TEST ELEMENT);

(1.6) to form the CGS of $\langle U, V \rangle$ from those of subgroups $U$ and $V$ of $G$ (command: JOIN);

(1.7) hence, to test $U \leqslant V$ (command: TEST SUBGROUP);

(1.8) to form the CGS of $U^V$ from those of $U$ and $V$ (command: NORMAL CLOSURE);

(1.9) hence to test $V \leqslant N_G(U)$ and in particular $U \lhd G$ (command: TEST NORMAL);

(1.10) to form the CGS of $[U, V]$ and hence to determine derived series and lower central series of $G$ (commands: COMMUTATOR, DERIVED SUBGROUP, DERIVED SERIES, LOWER CENTRAL SERIES);

(1.11) to construct a subnormal series from $G$ "towards" a subgroup $U$ by forming $H_0 = G$ and $H_i = U^{H_{i-1}}$ for $i \geqslant 1$ until the series becomes stationary. This will be a subnormal series from $G$ to $U$, iff $U \lhd\lhd G$ (command: SUBNORMAL CHAIN).

Already these applications of the NCGA indicate that, with the use of their canonical generating systems, the ability to handle subgroups $U$ of a finite soluble group $G$ no longer critically depends on the order of $G$, but rather depends on the length of its AG-system (or, to be more realistic, on some function of the indices $e_i$ that grows much slower than their product). It is promising, therefore, to look for more algorithms which make use of this data structure.

A first example of such an algorithm was a method for the determination of conjugacy classes of elements in $p$-groups [7]. This method proceeds inductively via increasing factor groups of a $p$-step central series, starting at each step from the result of the previous one (cf. also section 4). This algorithm may thus be considered as yet another application of the same kind of "homomorphism principle" that underlies the strategy of binary search or, in number theoretic programs, the application of Hensel's lemma [**13**]. In our case, the domain on which a group $G$ operates is mapped ($G$-)homomorphically; the problem—here the determination of orbits—is supposed to be solved in that homomorphic image domain and the result is then lifted back to the original domain.

There is, however, yet another kind of "homomorphism principle" which is perhaps more peculiar to our situation. We may start from a normal subgroup $N$ of the acting group $G$, determine the orbits of $N$ and let the factor group $G/N$ act on the set of these orbits of $N$.

We shall see that many of the algorithms for finite soluble groups to be described in sections 3–6 may be understood as instances of one or both of these principles, which are quite generally known to reduce the computing requirements about logarithmically.

For some of the algorithms, "finite soluble" might in fact have been replaced by "polycyclic"; however, we shall nowhere follow this up. On the other hand, some algorithms need stronger prerequisites and are introduced only for $p$-groups.

Since it is a main working horse for many further algorithms, we describe the NCGA rather explicitly in section 2; however technical details given here will not be needed in the subsequent sections.

Sections 3–5 will contain the mathematical description of further algorithms that have so far been implemented; section 6 an outlook on some possibilities for generalizations; while in section 7 we survey the design and present state of the SOGOS system which combines the so far implemented parts of this project. The paper closes with examples of the use of SOGOS.

SOGOS was implemented by F.-L. Chomse [**4**], E. Oppelt [**20**] and U. Paul [**21**] as part of their work for their diploma theses; further students are presently starting to work on its extension; we would like to thank all of them for the enthusiasm and dedication with which they have worked on this project. We would also like to thank M. F. Newman and C. R. Leedham-Green, who have contributed most valuable ideas to its development and V. Felsch for helpful discussions and advice in the implementation.

## 2 The "Non-Commutative Gauß Algorithm"

As mentioned in the introduction, the AG-system that is made the basis of the "canonical generating systems" of all subgroups $U \neq 1$ has to be a PAG-system, i.e. the subgroups $G = G_0 > G_1 > \ldots > G_n = 1$ are supposed to form a composition series of $G$. Therefore, after elimination of duplicates, the subgroups $U \cap G_i$ form a composition series

$$U = U_0 > U_1 > \ldots > U_r = 1 \text{ of } U.$$

Let $(u_1, \ldots, u_r)$ be the CGS of $U$; then $U_{i-1} = \langle U_i, u_i \rangle$. These facts have some immediate consequences:

(2.1) for each $u \in U$, there exists a unique $u_i \in \{u_1, \ldots, u_r\}$ with $w(u) = w(u_i)$. $u \in \langle u_i, u_{i+1}, \ldots, u_r \rangle$;

(2.2) if $u, v \in U$ and $w(u) = w(v) = w(u_i)$, then $uU_i = vU_i$ iff $\lambda(u) = \lambda(v)$;

(2.3) the CGS of $U$ is uniquely defined with respect to the fixed PAG-system $(g_1, \ldots, g_n)$ of $G$;

(2.4) the order of $U$ is equal to

$$\prod_{i=1}^{r} e_{w(u_i)}.$$

These statements are also the background for the "non-commutative Gauß algorithm" mentioned in the introduction. A key step in this algorithm is the procedure to "*insert*" an element $u \in G$ into a subsequence of a canonical generating system of a subgroup.

(2.5) Let $K = (u_1, u_2, \ldots, u_s)$ be a sequence of elements $u_i \neq 1$ of $G$ with $w(u_1) < w(u_2) < \ldots < w(u_s)$ and let $u$ be a further element of $G$.

(1) If there exists $u_i \in K$ with $w(u_i) = w(u)$, find $k \in \mathbb{N}$ such that $w(uu_i^{-k}) > w(u)$, replace $u$ by $uu_i^{-k}$ and go to (1);

(2) if $u \neq 1$, either there exists $i \in \{1, \ldots, s\}$ such that $w(u_i) < w(u) < w(u_{i+1})$ or $w(u_s) < w(u)$; extend $K$ by putting $u$ in between $u_i$ and $u_{i+1}$, or behind $u_s$, respectively;

(3) return to the main program.

This procedure is used at three points in the NCGA which constructs a canonical generating system $K$ for a subgroup $U \neq 1$ given by some finite generating set $M$. This algorithm proceeds by "inserting", one after the other, the elements from $M$ into $K$ (steps 1, 2). After each such "insertion", it extends $K$ to a PAG-system of the subgroup generated by the elements so far "inserted" into $K$, by repeated formation of powers and commutators (step 3). In this process an element of $K$ gets "marked" when all its commutators with all other "marked" elements in $K$ have been formed and "inserted". Finally the "norming" conditions (iii) and (iv) of definition 1.3 are enforced (step 5). We now give a formal description.

*Non-commutative Gauß algorithm.* Let $M$ be a set of elements of $G$, containing at least one non-identity element.

(1) Initialize $K$ as the empty sequence;

(2) choose $u \in M$ and remove it from $M$, "insert" $u$ into $K$;

(3) if all elements in $K$ are marked go to (4), else
choose an unmarked element $v$ in $K$;
"insert" $v^{e_{w(v)}}$ into $K$;
for all marked elements $x$ in $K$ "insert"
$[x, v]$ into $K$;
mark $v$;
go to (3);
(4) if $M \neq \emptyset$, go to (2);
(5) for each $u \in K$ with $\lambda(u) \neq 1$ find $m \in \mathbb{N}$ such
that $\lambda(u^m) = 1$ and replace $u$ by $u^m$;
for each $u \in K$ and each $v \in K$ with $w(v) > w(u)$
and $v_{w(v)}(u) \neq 0$ replace $u$ by $uv^{-v_{w(v)}(u)}$;
(6) stop.

The choices to be made in step 2 and 3 may be specified in various ways. We remark that the efficiency of this algorithm can be (and is, in SOGOS) improved, e.g. by starting from the CGS of a subgroup of $\langle M \rangle$ if such is known, and by avoiding the formation of commutators of powers of the same element.

## 3 Orbit Algorithm for Soluble Groups and Applications

Standard problems to be solved for a group $G$ acting on a set $\Omega$ are:

(i) to give a set of representatives for the orbits;
(ii) to compute the stabilizer of one point $\omega$;
(iii) to compute the orbit of $\omega$;
(iv) to decide whether two points lie in the same orbit.

We use the homomorphism principle in the two aspects mentioned in section 1 to find effective algorithmic solutions. The first algorithm is due to C. R. Leedham-Green and was already applied in [**1**]. It is based on the observation that an orbit $\Delta$ of a normal subgroup $N$ of $G$ is a block of $G$ on $\Omega$, i.e. $\Delta \cap \Delta^g \neq \phi$ for some $g \in G$ implies $\Delta = \Delta^g$.

If $G = \langle N, g \rangle$ and $|G/N| = p$ is a prime number, then for any $\omega \in \Delta$, either $G = N\,\mathrm{stab}_G(\omega)$ and $|\mathrm{stab}_G(\omega)/\mathrm{stab}_N(\omega)| = p$, or $N = N\,\mathrm{stab}_G(\omega)$ and $\mathrm{stab}_G(\omega) = \mathrm{stab}_N(\omega)$. Which case occurs, can be decided by testing whether $\omega^g \in \Delta$. If $\omega^g \in \Delta$, then $\omega^g = \omega^x$ for some $x \in N$ and $s = gx^{-1} \in \mathrm{stab}_G(\omega)$. Since $s \notin N$, we have $G = N\,\mathrm{stab}_G(\omega)$ and $\mathrm{stab}_G(\omega) = \langle \mathrm{stab}_N(\omega), s \rangle$. If $\omega^g \notin \Delta$ then $\Delta \cap \Delta^g = \phi$ and $\mathrm{stab}_G(\omega) \leqslant \mathrm{stab}_G(\Delta) = N$. Thus the orbit of $\omega$ under $G$ is

$$\Delta \,\dot\cup\, \Delta^g \ldots \dot\cup\, \Delta^{g^{p-1}}.$$

It is easy to see that stepping up a composition series $1 = G_n < G_{n-1} < \ldots < G_0 = G$, where in each step $G_{i-1}$ and $G_i$ play the role of $G$ and $N$ respectively in the above discussion, we obtain the following algorithm.

(3.1) *Orbit algorithm for soluble groups.* Let $G$ be a soluble group with PAG-system $(g_1, \ldots, g_n)$ acting on a set $\Omega$, and let $\omega \in \Omega$ be given. Then the orbit $\Delta = \omega^G$, a PAG-system $S$ of $\mathrm{stab}_G(\omega)$, and a set $R$ to be discussed below, are obtained as follows.

Initialize $\Delta = \{\omega\}$, $S$ as the empty sequence, $R = \phi$.

Let $i$ run from $n$ to 1.

If $\omega^{g_i} \notin \Delta$, then adjoin to $\Delta$ the elements of $\Delta^{g_i}, \Delta^{g_i^2}, \ldots, \Delta^{g_i^{e_i - 1}}$ and put $g_i$ into $R$.

If $\omega^{g_i} \in \Delta$, say $\omega^{g_i} = \omega^x$, then append $g_i x^{-1}$ to $S$.

(3.2) *Remark.* It is not necessary to store the, possibly, rather big set $\Delta$ because, at each step, we can run through the elements $\delta \in \Delta$ (without any repetition or searching) and compare $\omega^{g_i}$ with $\delta$. For this reconstruction of $\Delta$, one can use the small set $R$, since $\Delta = \omega^T$, where

$$T = \{g_n{}^{x(n)v_n} \ldots g_1{}^{x(1)v_1} \mid \begin{array}{lll} x(i) = 0 & \text{if} & g_i \notin R, \\ x(i) = 1 & \text{if} & g_i \in R, 0 \leqslant v_i < e_i \end{array}\}.$$

Of course this saving of space is bought at the cost of time.

An immediate application of 3.1 is an algorithm for the computation of the normalizer of a subgroup (command: NORMALIZER).

This in turn allows the computation of a Sylow $p$-subgroup for a given prime divisor $p$ of $|G|$ (command: SYLOW SUBGROUP). We start with the largest $i$ such that $e_i = p$. If $g_i$ has order $p^x k$, $p \nmid k$, then $U = \langle g_i{}^k \rangle$ is a nontrivial $p$-subgroup. Now we compute a PAG-system of $N_G(U)$ which runs through $U$. In the same way we find in $N_G(U)/U$ a $p$-subgroup $U_1/U$ and proceed as before with $U_1$ instead of $U$. The iteration stops with a Sylow $p$-subgroup of $G$.

In algorithm 3.1, the computation of the whole orbit seriously limits its applicability. Fortunately, in many situations, orbit sizes can be reduced by the other aspect of the homomorphism principle.

(3.3) *Reduction of orbit size.* Let a group $G$ act on two sets $\Omega_0, \Omega_1$ and let $\alpha: \Omega_0 \to \Omega_1$ be compatible with these actions, i.e. $\omega^{\alpha g} = \omega^{g\alpha}$ for all $\omega \in \Omega_0$ and $g \in G$. Then,

(i) $\mathrm{stab}_G(\omega) \leqslant \mathrm{stab}_G(\omega^\alpha)$ for each $\omega \in \Omega_0$.

(ii) For $\omega_1, \omega_2 \in \Omega_0$ there exists some $g \in G$ s.t. $\omega_1{}^g = \omega_2$ if and only if
there exists some $y \in G$ s.t. $\omega_1{}^{\alpha y} = \omega_2{}^{\alpha}$ and
there exists some $x \in \mathrm{stab}_G(\omega_2{}^{\alpha})$ s.t. $\omega_2 = \omega_1{}^{yx}$.

(iii) A full set $\Gamma$ of representatives for the $G$-orbits on $\Omega_0$ is obtained by first computing a full set $\Pi$ of representatives for the $G$-orbits on $\Omega_1$ and $\mathrm{stab}_G(\pi)$ for each $\pi \in \Pi$ and then computing full sets $\Pi(\pi)$ of representatives for the $\mathrm{stab}_G(\pi)$-orbits on

$$\pi^{\alpha^{-1}} = \{\omega \in \Omega_0 \mid \omega_1{}^{\alpha} = \pi\}.$$

Then $\Gamma = \dot{\bigcup}_{\pi \in \Pi} \Pi(\pi)$.

The proof is straightforward. It is important to recognize that the orbit of $\omega$ under $G$ is thus described by

$$|G : \mathrm{stab}_G(\omega^{\alpha})| + |\mathrm{stab}_G(\omega^{\alpha}) : \mathrm{stab}_G(\omega)|$$

group elements, so if

$$|G : \mathrm{stab}_G(\omega_2{}^{\alpha})| = a \text{ and } |\mathrm{stab}_G(\omega_2{}^{\alpha}) : \mathrm{stab}_G(\omega_2)| = b$$

in part (ii), we only have to inspect $a+b$ points to decide whether $\omega_1$ and $\omega_2$ lie in the same orbit instead of $ab$ points.

A first application of this principle is an algorithm which computes the intersection of two subgroups $U$, $V$ of a soluble group $G$ (command: MEET). We consider the operation of $V$ on the "space" $G/U$ of all right cosets of $U$ in $G$ given by multiplication from the right. Since the stabilizer $\mathrm{stab}_V(U)$ of the point $U$ under this action is equal to $U \cap V$, we can apply 3.1. Of course the orbit length $|UV|/|U|$ will in general be too large for a direct application of 3.1. But if we have a chain of subgroups $G = U_0 > U_1 > \ldots > U_k = U$, then for $i = 1, \ldots, k$ we have mappings $\alpha_i : G/U_i \to G/U_{i-1}$ between the coset spaces, which are compatible with the action of $V$. Therefore, we can apply 3.3, which means that we first compute

$$V_1 = \mathrm{stab}_V(U_1) = U_1 \cap V$$

and then, $$V_i = \mathrm{stab}_{V_{i-1}}(U_i) = U_i \cap V \text{ for } i = 2, \ldots, k.$$

Thus we apply 3.1, $k$ times, with orbit lengths of at most $\max_i |U_{i-1} : U_i|$. Of course the principle 3.3 does not need solubility of $G$, but in soluble groups we have a good chance of finding a chain of subgroups $U_i$, where the indices $|U_{i-1} : U_i|$ are fairly small. The implemented algorithm computes first the chain 1.11 and then refines the obtained factors by means of the given PAG-system of $G$. If $U$ is subnormal in $G$, then all resulting indices will be prime numbers. Of course the approach is symmetric in $U$ and $V$ so that the algorithm can choose, as $U$, that subgroup for which the indices of the chain are smallest.

The same idea can, of course, be used to compute representatives of the double cosets $UgV = \cup \{Ugv \mid v \in V\}$ and $\mathrm{stab}_V(Ug) = U^g \cap V$. This has been described for a special purpose in [**12**].

## 4 Applications of the Orbit Algorithm to *p*-groups

Whereas, in the preceding section, the reduction of orbit size was applied to the operation of right-multiplication on right cosets, we here use the principle for the conjugation operation. In $p$-groups, some composition series are already chief series. Therefore, for this section, we assume that $G$ is a $p$-group given by a PAG-system corresponding to a chief series $G = G_0 > G_1 > \ldots > G_n = 1$. Then we can use the natural homomorphisms $\alpha_i: G/G_i \rightarrow G/G_{i-1}$ for the reduction of orbit size.

Let us first recall, briefly, the algorithm for the computation of representatives for the conjugacy classes of elements in a $p$-group [7]. As a starting point we may choose the $G$-set $G/G_1$. Then each element represents a conjugacy class and its stabilizer is the whole group. In the iteration step we know, already, a full set of representatives $G_{i-1}r$ of the conjugacy classes of $G/G_{i-1}$ together with a PAG-system of each stabilizer $\mathrm{stab}_G(G_{i-1}r)$. We now have to find representatives in $G/G_i$ and their stabilizers. As in 3.3 (iii), we let $S = \mathrm{stab}_G(G_{i-1}r)$ act on the full pre-image of $G_{i-1}r$ in $G/G_i$. Since $|G_{i-1}/G_i| = p$, this coset consists of $p$ elements. Now the $p$-group $S$ has orbits of length $p$ or 1 on this coset. Thus the orbit algorithm 3.1, yields the new representatives and PAG-systems of the stabilizers in effectively only one step. This allows a particularly easy and efficient implementation (command: CLASSES).

This algorithm can easily be modified to find the centralizer of a given element or subgroup (command: CENTRALIZER). Then the following idea of Leedham-Green allows the computation of the normalizer of a subgroup much faster than in the general situation 3.1 (command: P NORMALIZER).

*Algorithm for the computation of a normalizer.* Let $1 < U < G$ and $U_i = U \cap G_i$ for $0 \leqslant i \leqslant n$. Let $C_0 = G$ and compute recursively $C_{i+1}$ from $C_i$ by

$$C_{i+1}/U_{n-i} = C_{C_i/U_{n-i}}(U_{n-i}u), \text{ where } \langle U_{n-i}, u\rangle = U_{n-i-1},$$
$$\text{for } i = 0, 1, \ldots, n-1.$$

Then $C_n = N_G(U)$.

*Proof.* It is clear that $C_n$ leaves $U_{n-n} = U$ invariant. On the other hand $N_G(U)$ normalizes $U$ and each $G_i$, hence each $U_i$. Since $|U_{i-1}/U_i|$ is either trivial or $p$, the $p$-group $N_G(U)$ centralizes $U_{i-1}/U_i$ for each $i$. Thus, $N_G(U)$ is contained in each $C_i$.

For the rest of this section we consider conjugacy classes of subgroups of our $p$-group $G$. The approach is again based on 3.3, but we need some more theoretical background.

*Lemma 1.* Let $N \trianglelefteq G$, $|N| = p$, and $N \leqslant V \leqslant G$. If $V = \langle N, v_1, \ldots, v_d \rangle$, where $V/N$ has minimal number of generators $d$ then either $N \leqslant \Phi(V)$ or $U = \langle v_1, \ldots, v_d \rangle$ is a complement of $N$ in $V$. In the first case $\{v_1, \ldots, v_d\}$ is a minimal generating set of $V$ and $V$ is the only pre-image of $V/N$ in $G$. In the second case (Fig. 1) if $N = \langle x \rangle$, a minimal set of generators of $V$ is given by $\{x, v_1, \ldots, v_d\}$. Then $V$ and all subgroups $U_f = \langle v_1 f(v_1), \ldots, v_d f(v_d) \rangle$, where $f \in \mathrm{Hom}(U, N)$, are the pre-images of $V/N$ in $G$. They are given by minimal sets of generators.

*Proof.* If $U = V$ then $N \leqslant \Phi(V)$ (p. 275 of [**9**]). Then each maximal subgroup of $V$ contains $N$ so that $V$ is the only pre-image of $V/N$ in $G$.

If $U \neq V$ then $NU = V$ and $N \cap U < N$, hence $N \cap U = 1$. Of course then $U \cong V/N$ and $U$ has minimal number of generators $d$. Thus, $V = N \times U$ has $\{x, v_1, \ldots, v_d\}$ as a minimal generating set. If $L$ is any proper subgroup of $V$ which is a pre-image of $V/N$, then $LN = V$ and $L \cap N = 1$. Therefore also $L$ is a complement of $N$ in $V$. Since $L \cong V/N$, also $|L/\Phi(L)| = p^d$. Now $\Phi(V) \leqslant L$ and $L/\Phi(V)$ is elementary abelian of order $p^d$. Since $L/\Phi(L)$ is the largest elementary abelian factor group of $L$, this implies $\Phi(L) = \Phi(V)$. Thus, the complements of $N$ in $V$ are just the complete pre-images of the complements

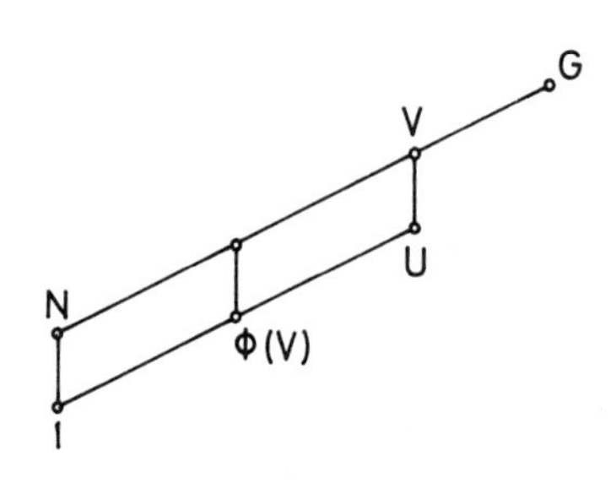


*Fig. 1*

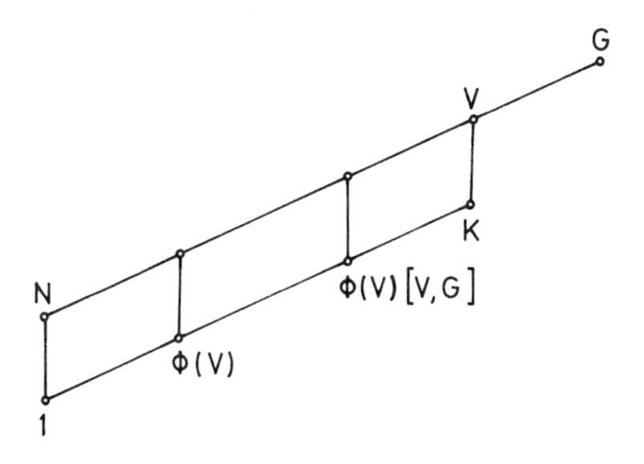


*Fig. 2*

of $N\Phi(V)/\Phi(V)$ in $V/\Phi(V)$, which can be interpreted as a vector space of dimension $d+1$ over $GF(p)$. This yields the description above.

Now it is easy to compute all subgroups of the $p$-group $G$. One steps down the chief series of the $G_i$ and finds, by Lemma 1, all pre-images of the subgroups of $G/G_{i-1}$ in $G/G_i$ in each step. However, since the number of subgroups increases rapidly, we apply such an approach only to the computation of all normal subgroups. Here a modification is needed to single out the normal subgroups in Lemma 1 (command: NORMAL SUBGROUPS).

If $U$ in Lemma 1 is normal in $G$, then also $V = NU$ is normal in $G$. Therefore, we only need to consider the normal subgroups of $G/N$, which we assume to be computed before. Now if $V = N \times U$ then $U \lhd G$ iff $V/U$ is a central factor of $G$. This is the case iff $\Phi(V)[V, G] \leqslant U$. Thus, one only has to take those complements $K$ of $N$ in $V$ which contain $\Phi(V)[V, G]$ (Fig. 2). It is a simple matter of linear algebra to find minimal generating systems of these complements.

The (non-normal) subgroups of $G$ are computed only up to conjugation (command: REPRESENTATIVES). We apply 3.3 by stepping down the chief series and in each step we could use 3.1. Fortunately, in this situation, we can avoid computing orbits and find representatives directly again by means of linear algebra in each step.

Assume that a set of representatives for the conjugacy classes of subgroups of $G/G_{i-1}$ is given by a minimal generating system for each representative and that we have a PAG-system for the stabilizers with respect to the conjugation operation in $G$. Then by 3.3, for each representative $V/G_{i-1}$ we have to let the stabilizer $N_G(V)$ act on the set of all pre-images of $V/G_{i-1}$ in $G/G_i$. So we assume, without loss of generality, that $V/G_{i-1} \lhd G/G_{i-1}$ and also $\Phi(V/G_i) = 1$.

To simplify our notation, we consider the situation that a $p$-group $G$ acts on a vector space $V$ over $GF(p)$ of dimension $d+1$ and that $G$ fixes a 1-dimensional subspace $N$ of $V$. We want to find representatives of the orbits of $G$ on the set of all $p^d$ complements of $N$ in $V$ and a PAG-system for each stabilizer in $G$. This problem in turn will be solved by applying iteratively the following result for the special case $[V, G] = N$, which is also contained in a more general version in [**10**].

*Lemma 2.* Suppose $[V, G] = N$ and let $U_1, U_2$ be complements of $N$ in $V$. Then $U_1{}^g = U_2$ for some $g \in G$ iff $U_1 \cap C_V(G) = U_2 \cap C_V(G)$, where $C_V(G)$ is the set of all vectors in $V$ fixed by each $g \in G$. All orbits of $G$ on the set of all complements have the same length

$$|V:C_V(G)| = |G:C_G(V)|.$$

*Proof.* Let $U$ be a complement of $N$ in $V$ (Fig. 3). If $s \in G$ stabilizes $U$, then $s$ acts trivially on $U$, since $s$ acts trivially on $V/N$. But then $s$ acts trivially on $V = \langle N, U \rangle$. Therefore, $\mathrm{stab}_G(U) = C_G(V)$ and all orbits have length $|G:C_G(V)|$. It is clear that if $U_1{}^g = U_2$ then $(C_V(G) \cap U_1)^g = C_V(G) \cap U_2$. But $C_V(G)$ is centralized by $g$ so that $C_V(G) \cap U_1 = C_V(G) \cap U_2$. Thus only those $|V:C_V(G)|$ complements of $N$ in $V$ can lie in one orbit which have the same intersection with $C_V(G)$. So it only remains to be shown that $|G:C_G(V)| = |V:C_V(G)|$. Without loss of generality, we may assume $C_G(V) = 1$.

Each $g \in G$ corresponds to a homomorphism $f_g: V \to N$ where $f_g(v) = -v + v^g \in [V, G] = N$. The map $g \to f_g$ embeds $G$ into $\mathrm{Hom}(V, N)$. If $v \in C_V(G)$, then $f_g(v) = -v + v = 0$ and hence $C_V(G) \leqslant \ker f_g$ for all $g \in G$. Thus, actually, $G$ is isomorphic to a subgroup of $\mathrm{Hom}(V/C_V(G), N) \cong V/C_V(G)$; hence $G$ is elementary abelian and $|G| \leqslant |V:C_V(G)|$.

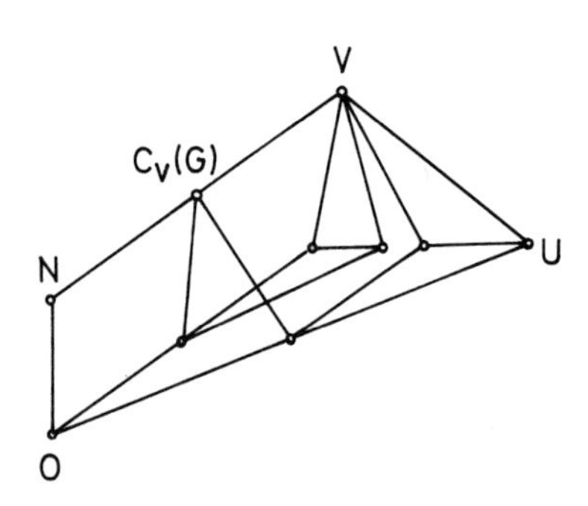


*Fig. 3*

On the other hand, if $|G| = p^r$ and $G = \langle g_1, \ldots, g_r \rangle$, then

$$C_V(G) = \bigcap_{i=1}^{r} C_V(g_i) = \bigcap_{i=1}^{r} \ker f_{g_i} \quad \text{and} \quad V/\ker f_{g_i} \cong \operatorname{im} f_{g_i} = N.$$

Thus $|V/C_V(G)| \leqslant p^r = |G|$.

Returning to the problem stated just before Lemma 2, we now consider the chain $V_0 = V, V_i = [V_{i-1}, G]N$, $V_{k-1} > V_k = N$. Then each $V_i$ is invariant under $G$, and $G$ acts on the set $\Omega_i$ of all complements of $N$ in $V_i$. If $U \in \Omega_{i-1}$, then $\alpha_i(U) = U \cap V_i$ defines a mapping $\alpha_i : \Omega_{i-1} \to \Omega_i$, which is compatible with the action of $G$. We can apply 3.3 with $\alpha = \alpha_i$ for $i = k-1, \ldots, 0$ to obtain representatives for the orbits of $G$ on $\Omega_0$ and the corresponding stabilizers.

To illustrate the algorithm, we consider the last step. We assume that we have already computed a full set $\Pi$ of representatives for the orbits of $G$ on the set $\Omega_1$ of complements of $N$ in $V_1 = N[V, G]$, and for each $K \in \Pi$ we have a PAG-system of $\mathrm{stab}_G(K) = N_G(K)$ (Fig. 4). Let $K \in \Pi$ and $S = \mathrm{stab}_G(K)$. Then we have to find representatives for the orbits of $S$ on the complete pre-image of $K$ under $\alpha_1$, which is the set of all complements of $N$ in $V$ whose intersection with $V_1$ is $K$. Clearly, we only have to consider the factor $V/K$ and the action of $S$ on the set of all complements of $V_1/K$ in $V/K$. Now $[V, S] \leqslant V_1$, so that either $[V, S] \leqslant K$ and all complements of $V_1/K$ in $V/K$ lie in different orbits and $S$ is the stabilizer of each of these complements, $[V/K, S] = V_1/K$ and we can apply Lemma 2 to find representatives for the orbits of $S$ on these complements. Then $C_S(V/K)$ is the stabilizer of each representative.

We have described an algorithm which computes, for a subgroup $V/G_{i-1}$ given by a minimal set of generators and its stabilizer $\mathrm{stab}_G(V/G_{i-1}) = N_G(V)$ given by a PAG-system, a complete set of representatives for the conjugacy classes of subgroups $U/G_i$ for which $G_{i-1}U/G_i$ is conjugate to $V/G_i$. The representatives are given by a minimal generating system and in addition, their normalizers are given by PAG-systems. It is important to notice that the orbit computation is avoided by using some linear algebra.

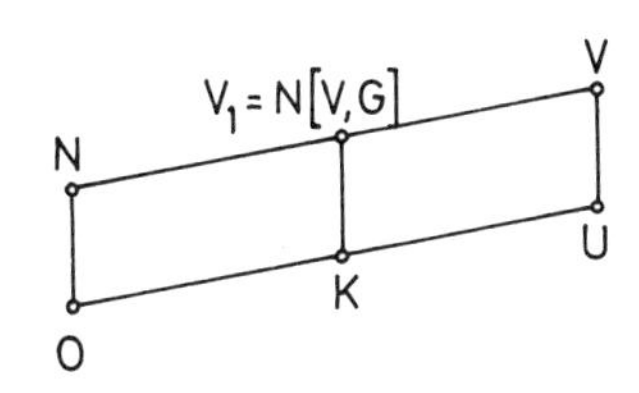


*Fig. 4*

Actually determining $C_V(G)$ when applying Lemma 2 means solving a system of linear equations. Thus, only the number of representatives which has to be stored, controls the problem size that can be tackled. The orbit lengths may be rather big, depending on the actual prime $p$.

Usually, for large groups, one does not really want representatives for all conjugacy classes of subgroups. Fortunately, the above algorithm allows one to compute rather locally within a group. First, one can compute representatives of only those conjugacy classes of subgroups which are contained in a specified normal subgroup. Second, one can compute supplements (command: SUPPLEMENTS) or complements (command: COMPLEMENTS) of one normal subgroup $N$ in another normal subgroup $M$ of $G$ up to conjugation within $G$.

Quite generally, the algorithms can be modified to compute only subgroups with some homomorphism invariant property, such as being abelian or normal (commands: ABELIAN REPRESENTATIVES, ABELIAN COMPLEMENTS, NORMAL COMPLEMENTS etc.).

Since for an abelian normal subgroup $N$ of $G$, $|H^1(G/N, N)|$ is the number of classes of complements in the split extension, the order of $H^1(G/N, N)$ can be computed. We include, in section 8, examples with $N$ abelian of type $(p^2, p)$ and $(p^3, p^2, p)$ and $G/N$ a Sylow $p$-subgroup of $\operatorname{Aut} N$. While the orders of these cohomology groups are, of course, a power of $p$, we also give an example showing that for non-abelian $N$, the number of conjugacy classes of complements need not be a power of $p$.

## 5 Further Algorithms for $p$-Groups

As in section 4, we assume that the $p$-group $G$ is given by a PAG-system $(g_1, \ldots, g_n)$ which corresponds to a chief series $G = G_0 > \ldots > G_n = 1$. Then $\langle g_i \rangle G_i/G_i$ is a central subgroup of $G/G_i$ of order $p$. We use this fact in the following two algorithms.

### 5.1 Maximal abelian subgroups of a $p$-group

Suppose we are given a list of all maximal abelian subgroups of $G/G_{i-1}$ ordered by the size of the subgroups, where each subgroup is given by its CGS with respect to the PAG-system $(g_1 G_{i-1}, \ldots, g_{i-1} G_{i-1})$ of $G/G_{i-1}$. We seek to construct a similar list for $G/G_i$. To simplify notation we assume that $G_i = 1$ and set $X = G_{i-1}$, $x = g_i$, i.e. $X = \langle x \rangle$.

Every maximal abelian subgroup of $G$ is contained in the complete pre-image $A$ under $G \to G/X$ of some maximal abelian subgroup of $G/X$.

If $A' \neq 1$, what are the maximal abelian subgroups of $A$? The subgroup $A' = X$ has order $p$. Therefore $[a^p, b] = [a, b]^p = 1$ for all $a, b$ in $A$, i.e.

$a^p \in Z(A)$, and $\bar{A} = A/Z(A)$ is elementary abelian. As a vector space over $GF(p)$ it carries a non-degenerate alternating bilinear form $f$ defined by

$$f(\bar{u}, \bar{v}) = d \quad \text{if} \quad [u, v] = x^d, 0 \leqslant d < p.$$

The maximal abelian subgroups of $A$ are exactly the complete pre-images under $A \to \bar{A}$ of the maximal isotropic subspaces of the symplectic space $(\bar{A}, f)$.

As $\bar{A}$ is an orthogonal sum of hyperbolic planes, we construct a PAG-system $(a_1, \ldots, a_r)$ of $A$ such that

$$a_r = x,$$

$$\langle a_{s+1}, \ldots, a_r \rangle = Z(A) \text{ and}$$

$$\langle \bar{a}_1, \bar{a}_2 \rangle + \langle \bar{a}_3, \bar{a}_4 \rangle + \ldots + \langle \bar{a}_{s-1}, \bar{a}_s \rangle = \bar{A}$$

with $f(\bar{a}_1, \bar{a}_1) = f(\bar{a}_2, \bar{a}_2) = 0$, $f(\bar{a}_1, \bar{a}_2) = 1$ etc. as follows. From the given PAG-system for $A/X$ we have a PAG-system $(h_1, \ldots, h_r)$ for $A$ with $h_r = x$. We then find non-commuting generators $h_i, h_j$ to produce a hyperbolic plane $\langle \bar{a}_1, \bar{a}_2 \rangle$ and adjust the remaining generators $a_k$ to generate modulo $Z(A)$ the orthogonal complement of $\langle \bar{a}_1, \bar{a}_2 \rangle$ in $\bar{A}$.

In fact, pick $i$ maximal and then $j < i$ maximal such that $[h_j^b, h_i] = x$ for some integer $b$. Set $a_1 = h_j^b$, $a_2 = h_i$. For $k \neq i, j$ set

$$h_k' = h_k \quad \text{if} \quad [h_k, a_1] = 1 \quad \text{and}$$

$$h_k' = h_k^{-1} a_2^{p-m} \quad \text{if} \quad [h_k, a_1] = x^m \neq 1.$$

Then $[h_k', a_1] = 1$. Similarly set

$$h_k'' = h_k' \quad \text{if} \quad [h_k', a_2] = 1 \quad \text{and}$$

$$h_k'' = h_k' a_1^{p-m} \quad \text{if} \quad [h_k', a_2] = x^n \neq 1.$$

Then $[h_k'', a_2] = 1$ and still $[h_k'', a_1] = 1$. Iteration will complete the decomposition of $\bar{A}$.

The maximal isotropic subspaces of $\bar{A}$ have dimension $t$ where $s = 2t$. Therefore the maximal abelian subgroups of $A$ have order $p^{r-t}$. There will be

$$I = \prod_{i=1}^{t} (p^i + 1)$$

of them. For $p = 2$ these numbers are listed in Table 1.

At this point, we may use a preprocessed library which contains all maximal isotropic subspaces of a non-degenerate symplectic space of dimension $s = 2t$ over $\mathrm{GF}(p)$ in terms of the hyperbolic-plane generators of the symplectic space. This will yield a list of the $I$ maximal abelian subgroups of $A$ in terms of the PAG-system $(a_1, \ldots, a_r)$ of $A$.

*Table 1*

| $s=2t$ | 2 | 4 | 6 | 8 | 10 | 12 |
|---|---|---|---|---|---|---|
| $t$ | 1 | 2 | 3 | 4 | 5 | 6 |
| $I$ | 3 | 15 | 135 | 2295 | 75 735 | 4 922 775 |

The maximal abelian subgroups of $G$ are among the maximal abelian subgroups of the various groups $A$ obtained from maximal abelian subgroups of $G/X$. However, some of these candidates may be contained in others. For the following containment test, we take the groups $A$ with candidates of large order first. Only now a set of candidates will be read from the library and their generators will be transformed into a CGS with respect to the given PAG-system for $G$. If $Y$ is in the list of subgroups of $G$ already known to be maximal abelian and if $K$ is the current candidate, then $|Y| \geqslant |K|$ where, if $|Y| > |K|$, then the CGS of $Y$ contains at least one generator $g$ such that no canonical generator of $K$ has weight $w(g)$. In such a case, take a canonical generator $g$ of $Y$ not among those of $K$. If $g$ does not commute with all canonical generators of $K$, then $Y \not\geqslant K$ as $Y$ is abelian. If $g$ commutes with all generators of $K$, then $K$ is not maximal and must be discarded. In this way a list of all maximal abelian subgroups of $G$ is obtained.

Such an algorithm has been implemented for the prime $p=2$ (SOGOS subsystem SYMPY). For that purpose a library of maximal isotropic subspaces of the non-degenerate symplectic space

$$V = \langle b_1, b_2 \rangle + \ldots + \langle b_{s-1}, b_s \rangle$$

of each dimension $s \leqslant 10$ was produced as follows. The cases $s=2$ and 4 are obvious. For $s \geqslant 6$ pick $a \neq 0$ in $V$ and adjust the basis elements $b_i$ to obtain a decomposition

$$V = \langle a, c_2 \rangle + \ldots + \langle c_{s-1}, c_s \rangle.$$

The maximal isotropic subspaces of the $(s-2)$-dimensional space

$$V' = \langle c_3, c_4 \rangle + \ldots + \langle c_{s-1}, c_s \rangle$$

are known by induction. We add the vector $a$ to each of the bases of those spaces and obtain bases for all maximal isotropic subspaces of $V$ *which contain a*. They may be transformed into canonical generating systems with respect to $(b_1, \ldots, b_s)$. We choose another element $a'$ and compute all maximal isotropic subspaces which contain $a'$, and discard those which contain the element $a$. We now have all maximal isotropic subspaces which contain $a$ or $a'$ without duplicates. We may continue using an interactive program for $s=6$,

8 and 10 until we obtain the correct number $I$ of subspaces. The 10-dimensional case was completed in a five hour terminal session using 764 seconds CPU-time on a CYBER 175.

Incorporated into SYMPY is the possibility of computing all subgroups of $G$ which are maximal among the abelian subgroups of exponent at most $p^k$, for fixed $k$. In this case, the candidates $K$ of the general algorithm are replaced by

$$\Omega_k(K) = \{g \in K \mid g^{p^k} = 1\}.$$

As $|\Omega_k(K)|$ does not depend on $|A|$ and $|Z(A)|$ alone, an *a priori* ordering of the candidates $\Omega_k(K)$ by their order is no longer possible. Therefore the containment test must now take care of the possibility $\Omega_k(Y) < \Omega_k(K)$ also.

As an example (section 8) we take a Sylow 2-subgroup of order $2^{10}$ of the semidirect product V(4,2)GL(4,2) which is isomorphic to the Sylow 2-subgroups of the simple groups GL(5,2), $M_{24}$ and He.

### 5.2 Intersection of subgroups of a *p*-group

We use again the fact that a $p$-group $G$ has a PAG-system $(g_1, \ldots, g_n)$ which arises from a chief series ($p$-step central series) $G = G_0 > G_1 > \ldots > G_n = 1$. For $p$-groups the following intersection algorithm suggested by Leedham-Green (private communication, 1981) proves to be more efficient than the general intersection algorithm (MEET) described at the end of section 3. We offer an example in section 8 which uses Chomse's implementation (command: P MEET) of Leedham-Green's algorithm.

For a subgroup $H$ of $G$ and $0 \leqslant i \leqslant n$ let $H_i$ denote the subgroup of all elements of $H$ of weight $\geqslant i+1$. Then $H_i = H \cap G_i$.

In order to compute $U \cap V$ from canonical generating systems for subgroups $U$ and $V$ of $G$, we wish to construct a PAG-system for each member of the sequence

$$1 = U_n \cap V_n, \ldots, U_{i+1} \cap V_{i+1}, U_i \cap V_i, \ldots, U_0 \cap V_0 = U \cap V.$$

Suppose we have already constructed a PAG-system for $U_{i+1} \cap V_{i+1}$. There are four possibilities:

(1) $U_i = U_{i+1}$ and $V_i = V_{i+1}$;
(2) $U_i > U_{i+1}$ and $V_i = V_{i+1}$;
(3) $U_i = U_{i+1}$ and $V_i > V_{i+1}$;
(4) $U_i > U_{i+1}$ and $V_i > V_{i+1}$.

In case (1), $U_i \cap V_i = U_{i+1} \cap V_{i+1}$. In general $|U_i \cap V_i|$ is linked with the size of the product $U_i V_i$ which need not be a subgroup of $G$ by the order formula

$$|U_i||V_i| = |U_i \cap V_i||U_i V_i|.$$

Similarly

$$|U_{i+1}||V_{i+1}| = |U_{i+1} \cap V_{i+1}||U_{i+1}V_{i+1}|.$$

In case (2), we have $|U_i| = p|U_{i+1}|$, $|V_i| = |V_{i+1}|$, and $|U_iV_i| = p|U_{i+1}V_i| = p|U_{i+1}V_{i+1}|$, therefore again

$$U_i \cap V_i = U_{i+1} \cap V_{i+1}.$$

Case (3) is similar.

However, in case (4) we have to be more careful. Let $u \in U_i$, $v \in V_i$ be the element of weight $i$ in the CGS for $U$, $V$ respectively. As in case (2), $|U_iV_{i+1}| = p|U_{i+1}V_{i+1}|$ and $U_i \cap V_{i+1} = U_{i+1} \cap V_{i+1}$. Then observe that

(4a) $\quad |U_iV_i| = p|U_iV_{i+1}| \qquad$ or $\quad$ (4b) $\quad |U_iV_i| = |U_iV_{i+1}|$.

In case (4a), $v \notin U_iV_{i+1}$. The order formulae for the products $U_iV_i$ and $U_iV_{i+1}$ show that $U_i \cap V_i = U_i \cap V_{i+1} = U_{i+1} \cap V_{i+1}$. In the second case (4b), we have $v \in U_iV_{i+1}$, $v = h_0k_0$, say, with $h_0$ of weight $i$ in $U_i$ and $k_0 \in V_{i+1}$. Therefore $h_0 \in U_i \cap V_i$. From the order formulae we get

$$|U_i \cap V_i| = p|U_i \cap V_{i+1}| = p|U_{i+1} \cap V_{i+1}|.$$

We could add $h_0$ to the PAG-system of $U_{i+1} \cap V_{i+1}$ to obtain a PAG-system of $U_i \cap V_i$, while in all previous cases we retain the PAG-system of $U_{i+1} \cap V_{i+1}$.

So we need for case (4), a procedure to decide whether $v \notin U_iV_{i+1}$ or $v \in U_iV_{i+1}$. It turns out that it suffices for this decision to have a list of elements $x$ of $U_{i+1}V_{i+1}$, one for each weight occurring in $U_{i+1}V_{i+1}$, along with a factorization $x = yz$, $y \in U$, $z \in V$ for $x$. Such a list of triples $(x, y, z)$ will be called a *system of representatives for the product* $U_{i+1}V_{i+1}$. In our inductive approach we now start with a PAG-system for $U_{i+1} \cap V_{i+1}$ and a system $R_{i+1}$ of representatives for the product $U_{i+1}V_{i+1}$ and seek to construct a PAG-system for $U_i \cap V_i$ along with a system $R_i$ of representatives for the product $U_iV_i$.

In case (1), we keep $R_i = R_{i+1}$. In case (2), we add $(u, u, 1)$ to $R_{i+1}$. The new system $R_i$ is a system of representatives for the product $U_iV_{i+1} = U_iV_i$ as all new elements lie in cosets of the form $u'V_{i+1}$ with $u' \in U_i \backslash U_{i+1}$ and have weight $i$. Case (3) is similar.

In case (4), first add $(u, u, 1)$ to $R_{i+1}$ to obtain a system $R$ of representatives for the product $U_iV_{i+1}$, as in case (2). Then look at $v$ of weight $i$. Let $j$ denote the weight of $w = u^{-1}.v.1$. Then $i < j$ and

$$v \in U_iV_{i+1} \quad \text{iff} \quad w \in U_iV_{i+1}.$$

Next see whether $R$ contains a triple $(x, y, z)$ where $x$ has weight $j$. If $(x, y, z)$

exists in $R$, then from the leading factors $g_j^a$ of $w$ and $g_j^b$ of $x$, we determine $c$ with $a+bc \equiv 0 \bmod p$ and set $w' = y^c w z^c$. Then

$$w' \equiv wx^c \equiv g_j^a g_j^{bc} \equiv 1 \bmod G_j$$

as $w$ and $g_j$ lie in the centre of $G \bmod G_j$. This shows that $w'$ has larger weight than $w$. Moreover

$$w \in U_i V_{i+1} \quad \text{iff} \quad w' \in U_i V_{i+1}.$$

We may now treat $w'$ in the same way as $w$ and so on until we arrive at

(i) $w^+ = 1$, or
(ii) an element $w^+$ whose weight $r$ does not occur among the weights of elements of $U_i V_{i+1}$.

By construction, $w^+$ has the form $w^+ = (\ldots y^c u^{-1}) v (z^c \ldots) = hvk$ with $h = \ldots y^c u^{-1} \in U_i$, $k \in V_{i+1}$. So,

$$v \in U_i V_{i+1} \quad \text{iff} \quad w^+ \in U_i V_{i+1} \quad \text{iff} \quad w^+ = 1.$$

Case (i) with $w^+ = 1$, which corresponds to case (4b) in the preliminary discussion above, leads to the element $h = \ldots y^c u^{-1}$ of weight $i$ in $U_i \cap V_i$, which will be added (in place of $h_0$) to the PAG-system of $U_{i+1} \cap V_{i+1}$ to yield a PAG-system for $U_i \cap V_i$. Then $R_i = R$ is a system of representatives for the product $U_i V_i = U_i V_{i+1}$.

In case (ii) = (4a), the PAG-system of $U_{i+1} \cap V_{i+1}$ is retained for $U_i \cap V_i$ and we add the triple $(w^+, h, vk)$ with the new weight $r$ to $R$. It remains to show that this new system $R_i$ is a system of representatives for the product $U_i V_i$.

To prove this property of $R_i$, suppose there is an element $w^- \in U_i V_i$ with a further new weight $s$ not represented in $R_i$. In particular, $s \neq i$. Clearly $w^-$ is not in $U_i V_{i+1}$. Therefore $w^- = (u^{-a} u_1 v_1) v^a$ with $u_1 \in U_{i+1}$, $v_1 \in V_{i+1}$, $a \not\equiv 0 \bmod p$. We compare $w^-$ with $w^+$:

$$\begin{aligned} w^- &= (u^{-a} u_1)(v_1 v^a) = (u_2 u^{-a})(v^a v_2) = \\ &= (u_3 h^a)(vk)^a v_3 = u_3 (w^+)^a v_3 \end{aligned}$$

with suitable elements $u_2, u_3 \in U_{i+1}$, $v_2, v_3 \in V_{i+1}$. Then $w^- \equiv (u_3 v_3) g_r^{ba} \bmod G_r$ where $g_r^b \neq 1$ is the leading factor of $w^+$. Here $ba \not\equiv 0 \bmod p$. Therefore, $g_r^{ba}$ has weight $r$. The element $u_3 v_3$ of $U_{i+1} V_{i+1}$ has a weight $t \neq r$. Therefore, the weight $s$ of $w^-$ is the minimum of $t$ and $r$; hence $s = t$ or $s = r$. But then $s$ is not a further new weight, a contradiction.

We conclude this section with a practical and two theoretical remarks to this description and explanation of Leedham-Green's algorithm. If, in case (4), $m$ denotes an integer such that $R$ contains elements of all weights $l$ for $m \leqslant l \leqslant n$, then instead of waiting for $w^+ = 1$ in case (i), one may stop already with a $w^*$ of weight $\geqslant m$. One may choose $m$ independent of $i$ as the least integer $m$ such that each of $U$ and $V$ contains canonical generators of all weights $l$ with $m \leqslant l \leqslant n$. Second, one observes that in the special case where $UV$ is a subgroup of $G$, the elements $x$ of a system of representatives for the product $UV$ form a PAG-system for the subgroup $UV$. Finally, there is a close analogy between this algorithm and a method for calculating sums and intersections in vector spaces and modules, which we learned from Prof. H. Zassenhaus [**28**].

## 6 Further Plans for Soluble Groups

In this section we discuss some ideas which have not yet been implemented.

The algorithms discussed in section 4 rely very much on $G$ being a $p$-group. Corresponding algorithms for soluble groups will therefore need other tools but can follow a similar strategy.

Representatives for the conjugacy classes of elements can be computed by the soluble orbit algorithm 3.1. Of course, one should reduce the orbit sizes by means of 3.3. For that purpose one needs a chain of normal subgroups $G = H_0 > H_1 > \ldots > H_s = 1$. Then the natural homomorphism $\alpha_i : G/H_i \to G/H_{i-1}$ can be used to apply 3.3. Since the size of the orbits in each step is bounded by $|H_{i-1}/H_i|$, it would be desirable that the normal subgroups $H_i$ form a chief series. Then for a $p$-group $G$, we obtain just the algorithm described at the beginning of section 4.

For the computation of a set of representatives for the conjugacy classes of subgroups, we propose an algorithm which also proceeds along a chief series of $G$. To describe the iteration step, we assume that $N$ is a minimal normal subgroup of $G$ and we know a representative $V/N$ of each conjugacy class of subgroups of $G/N$. Moreover, for each representative $V/N$ we also know $N_{G/N}(V/N)$. We assume that both groups $V/N$ and $N_{G/N}(V/N)$ are given by a CGS. From that we obtain a CGS for $V$ and $N_G(V)$.

We have to compute representatives of those conjugacy classes of subgroups of $G$, which have some representative $U$ with $NU = V$, and the corresponding normalizers. We proceed along a chain $N = N_0 > N_1 > \ldots > N_s = 1$, where $N_{i-1}/N_i$ is a chief factor of $V$. Suppose we have already computed up to conjugation in $N_G(V)$ all subgroups $S$, such that $NS = V$ and $N_{i-1} \leqslant S$. Then we have to find those supplements $U$ of $N$ in $V$ for which $N_i \leqslant U$ but $N_{i-1} \not\leqslant U$ and, if there are any, find out whether such a $U$ is conjugate to one of the already stored representatives (which can happen, since the $N_i$ need not be invariant under $N_G(V)$). Obviously, this general strategy yields the desired representatives.

Let us discuss some details for the proposed algorithm. First, we must be

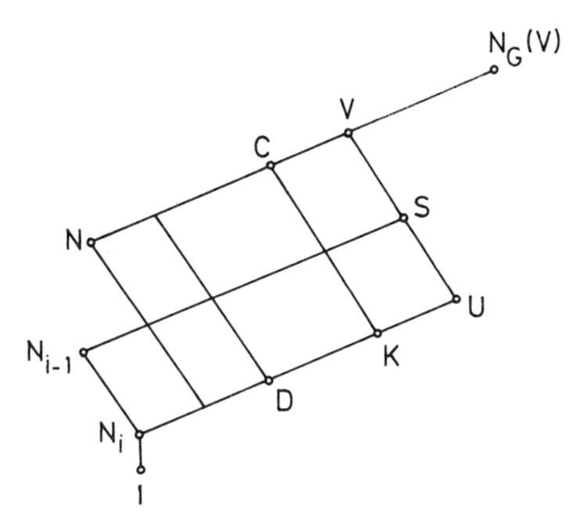


*Fig. 5*

able to find a chief series of a given soluble group. Such a series is needed once for $G$ and in the iteration step for each $V$ considered. By forming $p$th powers and commutator subgroups, it is easy to construct a series with elementary abelian factors. Then one could apply R. Parker's algorithm [these Proceedings] [**25**], for example, and [**21**], to refine this series to a chief series.

Second, we have to find, in the above terminology, the supplements $U$. Suppose some $U$ exists. Then (cf. Fig. 5) $N_i \leqslant N_{i-1} \cap U < N_{i-1}$ and $N_{i-1} \cap U$ is normal in the abelian group $N$ and in $U$. Since $NU = V$ and $N_{i-1}/N_i$ is a chief factor of $V$, $N_i = N_{i-1} \cap U$. $S = N_{i-1}U$ is a supplement of $N$ in $V$ that contains $N_{i-1}$. Since $NS = V$ and $N \leqslant C = C_V(N_{i-1}/N_i)$, $S$ acts on $N_{i-1}/N_i$ irreducibly. Then $U/N_i$ is a complement of the minimal normal subgroup $N_{i-1}/N_i$ in $S/N_i$, hence a maximal subgroup of $S/N_i$. Let $K = \mathrm{core}_S(U)$. Then $S/K$ is a primitive soluble group with a unique minimal normal subgroup and just one conjugacy class of complements (cf. pp. 159–160 of [**9**]).

Since $N_{i-1}K/K_s \cong N_{i-1}/N_i$, the unique minimal normal subgroup of $S/K$ is $N_{i-1}K/K$ and $U/K$ is one of its complements. Thus $U$ is already determined by $K$ up to conjugation in $S$. In addition

$$N_{i-1}K = C_S(N_{i-1}K/K) = C_S(N_{i-1}/N_i).$$

To find our subgroup $U$, we can run through a set of representatives for the orbits of $V$ on the set of those supplements $S$ of $N$ in $V$ that contain $N_{i-1}$. Then for each $S$ we form $C \cap S = C_S(N_{i-1}/N_i)$ and $D = (C \cap S)^p[C \cap S,$

$C \cap S]N_i$ where $p$ divides $|N|$. If a maximal subgroup $U$ of $S$ exists that is a supplement of $N$ in $V$ and contains $N_i$, then $D$ is contained in $\text{core}_S(U)$. Thus, we look for all complements $K/D$ of $N_{i-1}D/D$ in $C \cap S/D$ which are normal in $S/D$. Since $C \cap S/D$ is elementary abelian, these complements can be described by one fixed normal complement $K/D$ and $\text{Hom}_S(K/D, N_{i-1}/N_i)$, similar to section 4. For each $S$ we apply the following algorithm to find one complement $L = U/K$ of $M = N_{i-1}K/K$ in $S/K = H$.

*Algorithm for the computation of a complement L of the unique minimal normal subgroup M of a non-cyclic soluble primitive group H.*

(i) Compute $A < H$ such that $A/M$ is an elementary abelian normal subgroup of $H/M$.
(ii) Compute a Sylow $q$-subgroup $Q$ of $A$, where $q$ divides $|A/M|$.
(iii) Compute $L = N_H(Q)$.

For these candidates for new conjugacy classes of supplements of $N$ in $V$, we have to decide whether they are conjugate to one of the already stored representatives under $N_G(V)$. The mappings $\alpha: U \rightarrow \text{core}_V(U)$ and $\beta: \text{core}_V(U) \rightarrow U \cap N$ are compatible with conjugation in $N_G(V)$ and thus can be used for a reduction of orbit size.

Like in the case of $p$-groups, this algorithm can also be used for a local analysis of the given soluble group. Unfortunately we cannot simplify orbit computations as much as in section 4, but we can reduce the orbit sizes and thus make larger structures accessible. We imagine that choosing a chief series with isomorphic chief factors collected together can be used to improve the described algorithm.

## 7 The SOGOS System

Most algorithms described in the previous sections have been implemented as part of the system SOGOS (**so**luble **g**roups' **o**perating **s**ystem). SOGOS is, with minor exceptions, written in FORTRAN 4, ANSI 66 standard; the program text presently consists of about 40 000 FORTRAN cards.

Four previously existing program packages have been incorporated into SOGOS.

(i) For storage allocation, packing of information etc. J. Cannon's STACKHANDLER [**3**], which fulfills the same task in CAYLEY [**2**] and CAS [**18**].
(ii) The basic printing routines have been taken over from CAYLEY.
(iii) The Canberra implementation NQ 2.0 of the Nilpotent Quotient Algorithm [**19**] has been incorporated in order to produce an AG-system for

a $p$-group which is given to the system by an arbitrary finite presentation (command: NILPOTENT QUOTIENT).

(iv) Element multiplication by collection is done by preprocessor-generated routines. The program package AG82, which has been implemented by V. Felsch for this purpose, is an improved version of AG76 which is described in [**6**]. One of the improvements is the addition of a general consistency test for AG-systems [**27**], [**19**] (command: CHECK CONSISTENCY).

As far as possible, SOGOS was made compatible with CAYLEY. So information about groups is internally stored in a CAYLEY-compatible "group table". Exchange of data with CAYLEY is possible via files through special commands (SAVE GROUP, GET GROUP). The format of relations for the input is the same as for CAYLEY.

SOGOS has its own command language SOGOL (**so**luble **g**roups' **o**perating **l**anguage). SOGOL commands that invoke the execution of one of the algorithms described in the paper, have been quoted with that algorithm. It remains to mention that, of course, there are a number of administrative commands, e.g. for entering and leaving the system, for input and output of data, for setting up—as in CAYLEY—a group table in which record of the information available for this group is kept, for changing parameters etc. For a subgroup or a factor group of a given group, a PAG-system and defining relations can be determined and made the basis for the further investigation of this subgroup or factor group (commands: GROUP and FACTOR GROUP).

At present, SOGOL follows the general design philosophy that SOGOS is mainly a system for interactive use of commands corresponding to the algorithms described in this paper. So, while one can define identifiers and evaluate arithmetic expressions as well as expressions for group elements, that can involve multiplications, powering, conjugation and forming commutators, there are at present no means of writing programs of one's own, using the commands of SOGOL in loops or conditional statements. Experience gathered in the use of SOGOS has shown that, in spite of the general philosophy, this is a drawback that should be (and is being) removed.

The implementation of SOGOS and the simple structure of its command language allows rather easy addition of new algorithms, an advantage that we want to keep in further extensions. The portability of SOGOS depends mainly on the adaption of STACKHANDLER to the new host machine. Once STACKHANDLER is running on it, only very little changes have to be made. In fact, an implementation of SOGOS on an ICL2980 at Queen Mary College London, on which STACKHANDLER was available, took only a day or two.

The SOGOS system is available on tape. It is expected that at the time this paper is printed, a more complete version of the present preliminary user manual [**5**] will be available with the system.

## 8 Examples

In parts I–III, the groups considered are split extensions $G = [M]A$ of a group $M$ of $p$-power order by a $p$-Sylow subgroup $A$ of the automorphism group Aut $M$.

I) SOGOS computes representatives for the conjugacy classes of complements of $M$ in $G$. (Note that the normal subgroup $M$ is referred to by a number, 4 in (1)–(3) and 12 in (4), which it obtained in a preceding computation.)

(1) $M$ abelian of type $(p^2, p)$.

Input

```
GROUP :         ´[ M ] A - 1´;   P = 5;
GENERATORS : XP, Y, X, A, B, C;
RELATIONS :  XP↑P, Y↑P, X↑P = XP, ( X, A ) = XP, A↑P, ( X, B ) = Y,
             B↑P, ( Y, C ) = XP, ( B, C ) = A, C↑P;
ORDER ;  COMPLEMENTS, NORMAL = 4, DROP;
```

Output

```
GROUP [ M ] A - 1           ORDER : 15,625 = 5↑6

FACTOR GROUP 3 COMPLETED *     1 REPRESENTATIVE(S) STORED
FACTOR GROUP 4 COMPLETED *    25 REPRESENTATIVE(S) STORED
FACTOR GROUP 5 COMPLETED *   125 REPRESENTATIVE(S) STORED
GIVEN FACTOR COMPLETED * 25 REPRESENTATIVE(S) STORED
```

(2) $M$ non abelian of order $p^3$ and exponent $p$.

Input

```
GROUP :         ´[M ] A - 2´;    P = 5;
GENERATORS : Z, Y, X, C, B, A;
RELATIONS :  Z↑P, Y↑P, ( Y, X ) = Z, X↑P, ( X, C ) = Z, C↑P,
             ( X, B ) = Y, B↑P, ( Y, A ) = Z, ( B, A ) = C, A↑P;
ORDER;   COMPLEMENTS, NORMAL = 4, DROP;
```

Output

```
GROUP [ M ]  A - 2          ORDER : 15,625 = 5↑6

FACTOR GROUP 3 COMPLETED *      1 REPRESENTATIVE(S) STORED
FACTOR GROUP 4 COMPLETED *     25 REPRESENTATIVE(S) STORED
FACTOR GROUP 5 COMPLETED *    125 REPRESENTATIVE(S) STORED
GIVEN FACTOR COMPLETED * 50 REPRESENTATIVE(S) STORED
```

(3) $M$ non abelian of order $p^3$ and exponent $p^2$.

Input

```
GROUP :        '[ M ] A - 3';   P = 5;
GENERATORS : XP, Y, X, C, B, A;
RELATIONS :   XP↑P, Y↑P, ( Y, X ) = XP, X↑P = XP, ( X, C ) = XP, C↑P,
              ( X, B ) = Y, B↑P, ( Y, A ) = XP, ( B, A ) = C, A↑P;
ORDER;   COMPLEMENTS, NORMAL = 4, DROP;
```

Output

```
GROUP [ M ] A - 3              ORDER : 15,625 = 5↑6

FACTOR GROUP 3 COMPLETED *       1 REPRESENTATIVE(S) STORED
FACTOR GROUP 4 COMPLETED *      25 REPRESENTATIVE(S) STORED
FACTOR GROUP 5 COMPLETED *     125 REPRESENTATIVE(S) STORED
GIVEN FACTOR COMPLETED * 25 REPRESENTATIVE(S) STORED
```

(4) $M$ abelian of type $(p^3, p^2, p)$

Input

```
GROUP :        '[ M ] A - 4';    P = 3;
GENERATORS : XP2, YP, XP,  Z, Y, X, AP, CP, DP, A, B, E, W1, C,
              D,  F, W2;
RELATIONS :   XP2↑P, YP↑P, XP↑P = XP2, Z↑P, ( XP, A ) = XP2,
              ( Z, W1 ) = XP2, ( XP, C ) = YP, ( YP, D ) = XP2,
              ( Z, W2 ) = YP, Y↑P = YP, X↑P = XP, ( X, AP ) = XP2,
              AP↑P, ( X, CP ) = YP, CP↑P, ( Y, DP ) = XP2,
              ( X, A ) = XP,  ( Y, B ) = YP, ( X, E ) = Z,
              ( X, C ) = Y, ( Y, D ) = XP, ( CP, D ) = AP,
              ( Y, F ) = Z, DP↑P, A↑P = AP, B↑P, E↑P, ( E, W1 ) = AP,
              ( DP, C ) = AP↑-1, ( A, C ) = CP, ( B, C ) = CP↑-1,
              ( A, D ) = DP↑-1, ( B, D ) = DP, ( E, W2 ) = CP, W1↑P,
              C↑P = CP, ( C, D ) = B↑-1 * A * DP↑-1 * CP * AP,

              D↑P = DP, ( W1, F ) = DP↑-1, ( C, F ) = E,
              ( D, W2 ) = W1↑-1, F↑P, ( F, W2 ) = B, W2 ↑ P;
ORDER;   COMPLEMENTS, NORMAL = 12, DROP;
```

Output

```
GROUP [ M ] A - 4              ORDER : 129,140,163 = 3↑17

FACTOR GROUP 11 COMPLETED *       1 REPRESENTATIVE(S) STORED
FACTOR GROUP 12 COMPLETED *      81 REPRESENTATIVE(S) STORED
FACTOR GROUP 13 COMPLETED *     729 REPRESENTATIVE(S) STORED
FACTOR GROUP 14 COMPLETED *     243 REPRESENTATIVE(S) STORED
FACTOR GROUP 15 COMPLETED *    6561 REPRESENTATIVE(S) STORED
FACTOR GROUP 16 COMPLETED *    6561 REPRESENTATIVE(S) STORED
GIVEN FACTOR COMPLETED * 81 REPRESENTATIVE(S) STORED
```

II) For $M$ elementary abelian of order $p^6$, SOGOS computes the upper and lower central series of $G = [M]A$. Two subgroups $S.27$ and $S.28$ are computed to demonstrate the run time differences of the algorithms P MEET and MEET. Similarly the different algorithms P NORMALIZER and NORMALIZER are compared.

(5) $M$ abelian of type $(p, p, p, p, p, p)$.

Input

```
GROUP :        '[ M ] A - 5';   P = 5;
GENERATORS : V1, V2, V3, V4, V5, V6, A1, A2, A3, A4, A5, A6, A7,
             A8, A9, A10, A11, A12, A13, A14, A15;
RELATIONS :  V1↑P, V2↑P, V3↑P, V4↑P, V5↑P, V6↑P, A1↑P, A2↑P, A3↑P,
             A4↑P, A5↑P, A6↑P, A7↑P, A8↑P, A9↑P, A10↑P, A11↑P, A12↑P,
             A13↑P, A14↑P, A15↑P, ( V5, A2 ) = V1, ( V4, A4 ) = V1,
             ( V5, A5 ) = V2, ( V3, A7 ) = V1, ( V4, A8 ) = V2,
             ( V5, A9 ) = V3, ( V2, A11 ) = V1, ( V3, A12 ) = V2,
             ( V4, A13 ) = V3, ( V5, A14 ) = V4, ( V6, A1 ) = V1,
             ( V6, A3 ) = V2, ( V6, A6 ) = V3, ( V6, A10 ) = V4,
             ( A4, A10 ) = A1↑-1, ( A3, A11 ) = A1, ( A4, A14 ) = A2↑-1,
             ( V6, A15 ) = V5, ( A2, A15 ) = A1↑-1, ( A6, A7 ) = A1,
             ( A7, A9 ) = A2↑-1, ( A8, A10 ) = A3↑-1, ( A5, A11 ) = A2,
             ( A8, A11 ) = A4, ( A6, A12 ) = A3, ( A9, A12 ) = A5,
             ( A7, A13 ) = A4↑-1, ( A8, A14 ) A5↑-1,
             ( A5, A15 ) = A3↑-1, ( A9, A15 ) = A6↑-1,
             ( A11, A12 ) = A7↑-1, ( A10, A13 ) = A6,
             ( A12, A13 ) = A8↑-1, ( A13, A14 ) = A9↑-1,
             ( A14, A15 ) = A10↑-1;
ORDER; LOWER CENTRAL SERIES; UPPER CENTRAL SERIES;
SGR1 = CENTRALIZER ( V6 ); SGR2 = TRANSFORM ( SGR1, A15 );
P MEET, SGR1, SGR2; TIME; MEET; SGR1, SGR2; TIME;
P NORMALIZER, SGR1; TIME; NORMALIZER, SGR1; TIME;
```

Output

```
LOWER CENTRAL SERIES OF S.1
===========================


S.1      ORDER : 5↑21
S.23     ORDER : 30,517,578,125 = 5↑15
S.24     ORDER : 9,765,625 = 5↑10
S.25     ORDER : 15,625 = 5↑6
S.26     ORDER : 125 = 5↑3
S.21     ORDER : 5
S.22     ORDER : 1


UPPER CENTRAL SERIES OF S.1
===========================


S.22     ORDER : 1
S.21     ORDER : 5
S.26     ORDER : 125 = 5↑3
S.25     ORDER : 15,625 = 5↑6
S.24     ORDER : 9,765,625 = 5↑10
S.23     ORDER : 30,517,578,125 = 5↑15
S.1      ORDER : 5↑21


S.27 IS THE CENTRALIZER OF ELEMENT V6 IN S.1
S.27     ORDER : 152,587,890,625 = 5↑16


S.28 IS THE CONJUGATE SUBGROUP OF S.27
WITH TRANSFORMING ELEMENT A10
S.28     ORDER : 152,587,890,625 = 5↑16


S.29 IS THE INTERSECTION OF S.27 AND S.28
S.29     ORDER : 1,220,703,125 = 5↑13
```

```
 TIME USED FOR LAST COMMAND :      0.058 CPU-SECONDS
 TIME USED FOR WHOLE RUN :        50.509 CPU-SECONDS


S.29 IS THE INTERSECTION OF S.27 AND S.28
S.29      ORDER : 1,220,703,125 = 5↑13


 TIME USED FOR LAST COMMAND :      8.307 CPU-SECONDS
 TIME USED FOR WHOLE RUN :        58.820 CPU-SECONDS


S.32 IS THE NORMALIZER OF S.27
S.32      ORDER : 762,939,453,125 = 5↑17


 TIME USED FOR LAST COMMAND :      1.138 CPU-SECONDS
 TIME USED FOR WHOLE RUN :        59.963 CPU-SECONDS


S.32 IS NORMALIZER OF S.27
S.32      ORDER : 762,939,453,125 = 5↑17


 TIME USED FOR LAST COMMAND :     34.831 CPU-SECONDS
 TIME USED FOR WHOLE RUN :        94.799 CPU-SECONDS
```

III) SOGOS computes the maximal abelian subgroups and the maximal elementary abelian subgroups of a Sylow 2-subgroup of $M_{24}$.

(6) $M$ abelian of type (2,2,2,2).

Input

```
GROUP     :     '[ M ] A - 6';
GENERATORS : V1, V2, V3, V4, A1, A2, A3, A4, A5, A6;
RELATIONS  : V1↑2, V2↑2, V3↑2, V4↑2, A1↑2, A2↑2, A3↑2, A4↑2, A5↑2,A6↑2,
             ( V2, A6 ) = V1, ( V3, A3 ) = V1, ( V3, A5 ) = V2,
             ( V4, A1 ) = V1, ( V4, A2 ) = V2, ( V4, A4 ) = V3,
             ( A2, A6 ) = A1, ( A3, A4 ) = A1, ( A4, A5 ) = A2,
             ( A5, A6 ) = A3;
ORDER; DELETE, RELATIONS;
SNOLD = START ( 2 );
SNNEW = STUFE ( 2, 10, SNOLD );
PANORAMA, SNNEW;
```

Output

```
GROUP [ M ] A - 6          ORDER : 1,024 = 2↑10

G R U P P E   :   [ M ] A - 6 STUFE : 10

O R D N U N G   :   2 HOCH 6
1. ISOMORPHIETYP MIT 2 VERTRETERN   TYP : ( 1 , 1 , 1 , 1 , 1 , 1 )

O R D N U N G : 2 HOCH 5
2. ISOMORPHIETYP MIT    4 VERTRETERN TYP : ( 1 , 1 , 1 , 1 , 1 )
3. ISOMORPHIETYP MIT   21 VERTRETERN TYP : ( 2 , 1 , 1 , 1 )
```

```
O R D N U N G : 2 HOCH 4
4. ISOMORPHIETYP MIT  28 VERTRETERN TYP : ( 1 , 1 , 1 , 1 )
5. ISOMORPHIETYP MIT 248 VERTRETERN TYP : ( 2 , 1 , 1 )
6. ISOMORPHIETYP MIT   8 VERTRETERN TYP : ( 3 , 1 )
7. ISOMORPHIETYP MIT  84 VERTRETERN TYP : ( 2 , 2 )

GESAMTANZAHL DER UNTERGRUPPEN 395

G R U P P E   :  [ M ] A - 6  STUFE : 10

O R D N U N G :  2 HOCH 6
1. ISOMORPHIETYP MIT   2 VERTRETERN TYP : ( 1 , 1 , 1 , 1 , 1 , 1 )

O R D N U N G :  2 HOCH 5
2. ISOMORPHIETYP MIT   4 VERTRETERN TYP : ( 1 , 1 , 1 , 1 , 1 )

O R D N U N G :  2 HOCH 4
3. ISOMORPHIETYP MIT  28 VERTRETERN TYP : ( 1 , 1 , 1 , 1 )

GESAMTANZAHL DER UNTERGRUPPEN 34
```

IV) With the aid of SOGOS it is shown that for each of the orders $2^a$ for $8 \leqslant a \leqslant 11$, there are exactly three isomorphism types of groups with exponent 4 and minimal number of generators 2.

To find the isomorphism types of factor groups $B(4,2)/N$ of order $2^8$–$2^{11}$, we let an automorphism $\alpha$ of order 3 act on the set of all normal subgroups of order 2–$2^4$ of $B(4,2)$. SOGOS computes these normal subgroups and it is easily seen that they are all contained in $S.8 = Z_2(B(4,2))$. Moreover they lie in $S.52 = Z(B(4,2))$ or contain $S.11 = B(4,2)''$. Both factors $S.8/S.11$ and $S.52$ are elementary abelian of order $2^3$. Now, any automorphism of order 3 of $V(3,2)$ fixes just one hyperplane and one 1-dimensional subspace and has orbits of length 3 on the set of the other non trivial subspaces. All orbits can be deduced from the orbits on the 1-dimensional subspaces. The subgroups $S.12$, $S.75$, $S.74$ of $B(4,2)''$ form one orbit. $S.76$ is left invariant and $S.77$, $S.78$, $S.79$ form the last orbit of $\alpha$ on the set of all minimal normal subgroups. SOGOS shows that $B(4,2)/S.76$ and $B(4,2)/S.79$ are not isomorphic. Thus, the orbits of Aut $B(4,2)$ are the orbits of $\alpha$ in this case. It is easily seen that, then, $S.11$, $S.72$ and $S.73$ represent the orbits of Aut $B(4,2)$ on the subgroups of order 4 of $S.52$. Similarly SOGOS shows that $B(4,2)/S.51$ and $B(4,2)/S.55$ are not isomorphic and we conclude that, together with $B(4,2)/S.52$, they represent the isomorphism types of factor groups of order $2^9$. Then, $S.48$, $S.49$, and $S.56$ represent the orbits of Aut $B(4,2)$ on the normal subgroups of order 16 of $B(4,2)$.

Input

```
GROUP : B42;
GENERATORS : A, B;
RELATIONS :  A↑4, B↑4, ( A * B )↑4, ( A↑2 * B )↑4, ( A↑3 * B )↑4,
             ( A↑2 * B↑2 )↑4, ( A * B * A * B↑3 )↑4,
             ( A↑3 * B↑3 * A * B )↑4, ( A↑2 * B↑2 )↑4,
             ( A↑3 * B↑2 )↑4, ( A↑3 * B↑3 )↑4, ( A↑2 * B↑3 )↑4;
´B ( 4, 2 )´ = NILPOTENT QUOTIENT ( GROUP = B42, PRIME = 2, AG );
GROUP : ´B ( 4, 2 )´;   NORMAL SUBGROUPS;
TRANSFER, REPRESENTATIVES;  DISPLAY, SUBGROUPS, GENERATORS, MAXORD=16;
```

## Output

```
GROUP B ( 4, 2 ) COMPLETED * 2 - CLASS : 5 * ORDER : 2 ↑ 12

GIVEN FACTOR COMPLETED * 81 REPRESENTATIVE(S) STORED

STORED SUBGROUPS OF GROUP B ( 4, 2 )
====================================

ORDER : 1

S.13 = < IDENTITY >

ORDER : 2

S.12 = < G.1 >
S.74 = < G.2 G.1 >
S.75 = < G.2 >
S.76 = < G.5 G.4 G.3 G.2 G.1 >
S.77 = < G.5 G.4 G.3 G.2 >
S.78 = < G.5 G.4 G.3 G.1 >
S.79 = < G.5 G.4 G.3 >

ORDER : 4 = 2↑2

S.11 = <G.2, G.1>
S.72 = <G.5 G.4 G.3 G.2, G.1>
S.73 = <G.5 G.4 G.3, G.1>
S.80 = <G.5 G.4 G.3 G.1, G.2 G.1>
S.81 = <G.5 G.4 G.3, G.2 G.1>
S.82 = <G.5 G.4 G.3 G.1, G.2>
S.83 = <G.5 G.4 G.3, G.2>

ORDER : 8 = 2↑3

S.10 = <G.3, G.2, G.1>
S.50 = <G.4 G.3, G.2, G.1>
S.51 = <G.4, G.2, G.1>
S.52 = <G.5 G.4 G.3, G.2, G.1>
S.53 = <G.5 G.4, G.2, G.1>
S.54 = <G.5 G.3, G.2, G.1>
S.55 = <G.5, G.2, G.1>

ORDER : 16 = 2↑4

S.9  = <G.4, G.3, G.2, G.1>
S.48 = <G.5 G.4, G.3, G.2, G.1>
S.49 = <G.5, G.3, G.2, G.1>
S.56 = <G.5 G.3, G.4 G.3, G.2, G.1>
S.57 = <G.5, G.4 G.3, G.2, G.1>
S.58 = <G.5 G.3, G.4, G.2, G.1>
S.59 = <G.5, G.4, G.2, G.1>
```

## Input

```
FAC76 = FACTOR GROUP ( GROUP = 'B ( 4, 2 )', MODUL = S.76 );
FAC79 = FACTOR GROUP ( GROUP = 'B ( 4, 2 )', MODUL = S.79 );
FAC51 = FACTOR GROUP ( GROUP = 'B ( 4, 2 )', MODUL = S.51 );
FAC55 = FACTOR GROUP ( GROUP = 'B ( 4, 2 )', MODUL = S.55 );
GROUP : FAC76;   ORDER;  REPRESENTATIVES;
GROUP : FAC79;   ORDER;  REPRESENTATIVES;
GROUP : FAC51;   ORDER;  REPRESENTATIVES;
GROUP : FAC55;   ORDER;  REPRESENTATIVES;
```

Output

```
GROUP FAC79                      ORDER : 2

GIVEN FACTOR COMPLETED * 2442 REPRESENTATIVE(S) STORED

GROUP FAC76                      ORDER : 2

GIVEN FACTOR COMPLETED * 3981 REPRESENTATIVE(S) STORED

GROUP FAC51                      ORDER : 8 = 2↑3

GIVEN FACTOR COMPLETED * 1071 REPRESENTATIVE(S) STORED

GROUP FAC55                      ORDER : 8 = 2↑3

GIVEN FACTOR COMPLETED * 647 REPRESENTATIVE(S) STORED
```

## References


1. Ascione, J. A., Havas, G. and Leedham-Green, C. R. (1977). A computer aided classification of certain groups of prime power order, *Bull. Austral. Math. Soc.* **17**, 257–274, 317–320, (microfiche supplement).
2. Cannon, J. J. (1982), (Preprint). A language for group theory. University of Sydney.
3. Cannon, J. J., Gallagher, R. and McAllister, K. (1972, revised 1974). Stackhandler: A language extension for low level set processing. Programming and implementation manual. Technical Report No. 5, Computer-Aided Mathematics Project. University of Sydney.
4. Chomse, F.-L. (1982). SOGOS II. Methoden zum Rechnen in nilpotenten Gruppen, Diplomarbeit, RWTH Aachen.
5. Chomse, F.-L., Oppelt, E. and Paul, U. (1982). SOGOL Handbuch (vorläufige fassung). Lehrstuhl D für Mathematik, RWTH Aachen.
6. Felsch, V. (1976). A machine independent implementation of a collection algorithm for the multiplication of group elements. *In* "SYMSAC '76. Proc. of the 1976 ACM Symposium on Symbolic and Algebraic Computation" (Ed. R. D. Jenks), 159–166. Assoc. Comput. Mach., New York.
7. Felsch, V. and Neubüser, J. (1979). An algorithm for the computation of conjugacy classes and centralizers in *p*-groups. Lecture Notes in Computer Science, Vol. **72**, 452–465. Springer-Verlag, Berlin.
8. Havas, G. and Nicholson, T. (1976). Collection. *In* "SYMSAC '76. Proc. of the 1976 ACM Symposium on Symbolic and Algebraic Computation" (Ed. R. D. Jenks), 9–14. Assoc. Comput. Mach., New York.
9. Huppert, B. (1967). "Endliche Gruppen I". Springer-Verlag, Berlin.
10. Jürgensen, H. (1970). Calculation with the elements of a finite group given by generators and defining relations. *In* "Computational Problems in Abstract Algebra" (Ed. J. Leech), 47–57. Pergamon Press, Oxford.
11. Laue, R. (1978). On normal p-subgroups with large centers which cannot be contained in the Frattini subgroup, *Israel J. Math.* **29**, 155–166.
12. Laue, R. (1982). Computing double coset representatives for the generation of solvable groups. Lecture Notes in Computer Science, Vol. **144**, 65–70. Springer-Verlag, Berlin.

13. Lauer, M. (1982). Computing by homomorphic images. *Computing*, Suppl. 4, 139–168.
14. Lindenberg, W. (1962). Über eine Darstellung von Gruppenelementen in digitalen Rechenautomaten, *Numerische Mathematik* **4**, 151–153.
15. Lindenberg, W. (1963). "Die Struktur eines Übersetzungsprogramms zur Multiplikation von Gruppenelementen in digitalen Rechenautomaten", *Mitt. Rhein.-Westfäl. Inst. Instrum. Math. Bonn* **2**, 1–38.
16. Neubüser, J. (1961). Bestimmung der Untergruppenverbände endlicher $p$-Gruppen auf einer programmgesteuerten elektronischen Dualmaschine, *Numerische Mathematik* **3**, 271–278.
17. Neubüser, J. (1971). Computing moderately large groups: some methods and applications, *SIAM-AMS Proc.* **4**, 183–190.
18. Neubüser, J., Pahlings, H. and Plesken, W. (1983). CAS; Design and use of a system for the handling of characters of finite groups, (these Proceedings).
19. Newman, M. F. (1976). Calculating presentations for certain kinds of quotient groups. *In* "SYMSAC '76. Proc. of the 1976 ACM Symposium on Symbolic and Algebraic Computation" (Ed. R. D. Jenks), 2–8. Assoc. Comput. Mach., New York.
20. Oppelt, E. (1981). SOGOS, ein Programmsystem zur Handhabung von Untergruppen auflösbarer Gruppen, Diplomarbeit, RWTH Aachen.
21. Paul, U. (1981). Algorithmische Berechnung der maximalen abelschen Untergruppen endlicher $p$-Gruppen, Diplomarbeit, RWTH Aachen.
22. Ruland, D. (1982). Konstruktive Reduktion modularer Darstellungen, Diplomarbeit, RWTH Aachen.
23. Sims, C. C. (1970). Computational methods in the study of permutation groups. *In* "Computational Problems in Abstract Algebra" (Ed. J. Leech), 169–183. Pergamon Press, Oxford.
24. Sims, C. C. (1971a). Determining the conjugacy classes of a permutation group, *SIAM-AMS Proc.* **4**, 191–195.
25. Sims, C. C. (1971b). Computation with permutation groups. *In* "Proc. of the Second Symposium on Symbolic and Algebraic Manipulation" (Ed. S. R. Petrick), 23–28. Assoc. Comput. Mach., New York.
26. Thackray, J. G. (1979). Reduction of modules in non-zero characteristics. Lecture Notes, Amer. Math. Soc. Summer Inst. on Finite Group Theory, University of California, Santa Cruz.
27. Wamsley, J. W. (1974). Computation in nilpotent groups (theory). Proc. Second Internat. Conf. Theory of Groups (Canberra, 1973), Lecture Notes in Mathematics, Vol. **372**, 691–700. Springer-Verlag, Berlin.
28. Zassenhaus, H. (1966). The sum-intersection method. Manuscript, Ohio State University, Columbus.

# An Interactive Program for Computing Subgroups

VOLKMAR FELSCH AND GÜNTER SANDLÖBES

## 1 Introduction

In the late fifties, J. Neubüser designed and implemented an algorithm for computing all soluble subgroups of moderately large groups [**10**], [**11**]. Since then his method, usually referred to as cyclic extension method, has been further developed and extended in several re-implementations [**6**], [**5**], [**16**], [**1**]. A new version, which is being written at Aachen, is now almost finished. It is already partially available in the last release of the group theory program system CAYLEY [**3**], [**4**], where it replaced the 1974 implementation of H. Steinmann [**16**], and it will be fully available in the next release. Moreover, it is planned to make this new version available as a stand-alone program. It is the purpose of this paper to report on two major improvements of the new program upon its predecessors, namely its capability of computing non-soluble subgroups as well as soluble ones, and its interactive facilities.

## 2 Computing Non-soluble Subgroups

The 1963 implementation [**6**] is the only one of the programs mentioned above, which provided an algorithm for finding all subgroups of a given group. It worked quite satisfactorily for the small groups which could be handled by the then existing computers, but the underlying method is too clumsy to be extended to larger groups.

A different, merely combinatorial algorithm [**7**] turned out to be fairly inefficient.

In the following we will describe a new approach. Let us first recall a few known facts which we will need.

### 2.1 The cyclic extension method

The computation of the soluble subgroups of a given group $G$ starts from an initial set $S$ consisting of just the unit subgroup. Then, for each subgroup (or,

COMPUTATIONAL GROUP THEORY
ISBN 0-12-066270-1

in fact, for each conjugacy class representative) $H$ in $S$, all its different cyclic extensions $\langle H, g\rangle$ in $G$ of prime index over $H$ are determined and, together with their conjugates, added to $S$. In any implementation of this process known to us, the subgroups are represented by certain bit strings, namely by their characteristic functions on the set of all cyclic subgroups of prime power order of $G$. It is an essential consequence of this representation that the algorithm can decide by a few cheap bit string operations, whether some cyclic extension $\langle H, g\rangle$ of prime index over $H$ will, or will not, yield a subgroup already contained in $S$, before actually performing the time-consuming construction itself. This makes the algorithm efficient.

## 2.2 Perfect subgroups

Obviously, the same algorithm can be applied to find all subgroups of $G$ including the non-soluble ones, if only the set $S$ is initialized to contain, besides the unit subgroup, at least one representative of each conjugacy class of perfect subgroups of $G$. Therefore, we have implemented an appropriate method to determine such a set of subgroups. It works as follows.

As a first step, the soluble residuum $R$ of $G$ is computed. If $R$ is trivial, then $G$ is soluble and nothing has to be done. Otherwise, $R$ is perfect, and all perfect subgroups of $G$ are contained in $R$. In this case, $R$ will be checked against a catalogue of (isomorphism types of) perfect groups which is an integrated part of the program. It is the aim of the procedure to find an isomorphism $\psi$ from some suitable catalogue group $P$ onto $R$, and then to map an associated list of perfect subgroups of $P$ via $\psi$ into $R$ and hence into $G$.

Any two groups in the catalogue are distinguished by some appropriately chosen invariants (e.g. order, number of conjugacy classes of certain length containing elements of certain order, etc.) which may be easily checked for $R$. So the set of catalogue groups can immediately be reduced to at most one candidate $P$ possibly isomorphic to $R$.

If there is such a candidate $P = \langle h_1, \ldots, h_n\rangle$, then a backtrack search procedure will be applied which is similar to certain algorithms that have been developed to compute automorphisms [**13**], [**9**]. It is used to investigate suitable $n$-tuples $g_1, \ldots, g_n$ of elements of $R$ for being images of $h_1, \ldots, h_n$ under some isomorphism from $P$ to $R$. For this purpose, two sets of words in the generators of $P$ are provided by the catalogue, namely

(1) a defining set of relators $r_j(h_i)$ for $P$, and
(2) a set of "anti-relators" $t_j(h_i)$, i.e. a set of words representing just one non-trivial element of each minimal normal subgroup of $P$.

If $r_j(g_i) = 1$ in $R$ for all $j$, then there exists a homomorphism $\psi$ mapping each $h_i$ onto the corresponding $g_i$. Moreover, if $t_j(g_i) \neq 1$ in $R$ for all $j$ (that is why we call the words $t_j$ anti-relators), then $\psi$ is injective and hence, as $R$ is finite, an isomorphism.

Finally, there is a third list of words in the catalogue. It contains

(3) for each class of conjugate perfect subgroups of $P$, a set of generators $s_j(h_i)$ of some representative subgroup.

As soon as some $n$-tuple $g_1, \ldots, g_n$ has successfully passed the two tests above, this list is used to obtain generators $s_j(g_i)$ for at least one representative of each conjugacy class of perfect subgroups of $G$.

## 2.3 Initiation of a perfect group for the catalogue

If no isomorphism has been found by the procedure described in section 2.2, the catalogue does not yet contain an isomorphic image of $R$. Hence the user has to extend the lists appropriately and then to restart his computation.

The job of adding a new perfect group $P$ to the catalogue includes, on the one side, the determination of a presentation and anti-relators, and the choice of characterizing invariants for $P$, while on the other side it comprises the computation of all (conjugacy classes of) perfect subgroups of $P$. Whereas existing CAYLEY routines and some simple hand manipulations suffice to do the first of these tasks, we had to implement a new procedure for the second one. Its basic idea is to reduce the problem of finding the perfect subgroups of $P$ to the problem of finding perfect subgroups of smaller groups which, by induction, may be assumed to be already covered by the catalogue.

More precisely, the procedure makes use of the fact that any perfect proper subgroup $Q$ of $P$ is either simple, or $Q$ possesses some minimal normal proper subgroup $M$. If $M$ is normal in $P$, then $Q$ is the pre-image of one of the perfect subgroups of the factor group $P/M$ which can be obtained from the catalogue. If, on the other hand, $M$ is not normal in $P$, then $Q$ is among the perfect subgroups of the normalizer $N_P(M)$ which again can be obtained from the catalogue.

Therefore, the procedure starts with the determination of all classes of characteristically simple proper subgroups of $P$, where it uses the cyclic extension method to construct the elementary abelian ones, and a slightly generalized version of the algorithm described in section 2.2 to find the non-abelian ones by searching for monomorphisms from characteristically simple catalogue groups into $P$. Then it runs through a set of class representatives and determines the perfect subgroups of the relevant factor groups or normalizers as sketched above. This process yields, up to conjugation, all perfect proper subgroups of $P$.

A few words have to be added concerning our induction hypothesis. It is our intention that the catalogue should contain all perfect groups up to some order. The determination of all perfect groups of order less than $10^4$ [**14**], [**15**], was a preparatory step in this direction. In general, however, omissions will not do any harm. If the process fails for some $P/M$ or $N_P(M)$ because its soluble residuum is not yet contained in the catalogue, then the user first has to add that group to the catalogue and then to restart the initiation of $P$. The only exception is the assumption that the catalogue indeed covers all non-abelian characteristically simple proper subgroups of $P$, but this can be guaranteed by the present knowledge of finite simple groups.

## 3 Interactive Facilities and Space Saving Options

Like its predecessors, the new implementation of the subgroup lattice algorithm provides means for automatically computing the subgroups of a given group $G$ of suitable size. This is done by the following default procedure.

### Step 1

*Initialization.* Initialize the characteristic functions of the subgroups by fixing a numbering of the cyclic subgroups of prime power order. If $G$ is not soluble, then determine class representatives for the perfect subgroups (using the method described in section 2.2) and save them in a temporary hoard list. Then initialize the actual set $S$ of "known" subgroups by the unit subgroup, the whole group, and the centre.

### Step 2

*Perfect subgroups.* If the hoard is empty, then go to step 3. Otherwise extract a perfect subgroup from the hoard. If it is not yet contained in $S$, then add it to $S$ and close $S$ under conjugation, normalizers, and centralizers. Repeat this step until the hoard is exhausted.

### Step 3

*Cyclic extensions.* Run through all class representatives in $S$ and compute for each of them, all cyclic extensions of prime index which are not yet contained in $S$. Add them to $S$, and in each case close $S$ under conjugation, normalizers, and centralizers.

### Step 4

*Final operations.* Sort the (classes of) subgroups in $S$ in some prescribed way by their orders. Then print the "subgroup lattice" of $G$, i.e. a listing of all subgroups which, in particular, includes for each subgroup a list of its maximal subgroups.

This automatic approach has several disadvantages. A user who only wants to look at a few particular subgroups is forced to compute all subgroups and hence to waste a possibly significant amount of space and time before he can start his investigation. Moreover, the program will fail to assemble all subgroups of $G$ because of insufficient space in many cases where the available workspace would suffice to store those subgroups which are of real interest. But even in the case of a successful run, the user may find himself hopelessly overtasked by the problem of tracing certain subgroup relations through thousands of subgroups listed in a heavy paper output.

Therefore, in addition to the default procedure, the new implementation offers a range of optional commands which enable the user to build up and manipulate partial or complete subgroup lattices interactively. In particular, the user can ask explicitly for certain subgroups such as subgroups from the hoard, cyclic extensions of given subgroups, normalizers, centralizers, intersections, joins, Sylow subgroups, or subgroups generated by given elements. These subgroups will be constructed or just read off from existing lists, then they will be identified in $S$, or added to $S$ if necessary. In each case, the respective command admits a lot of optional parameters which may be used to specify further conditions such as maximal number of subgroups to be added, order restrictions, group-theoretic properties, or embedding properties (e.g. prescribing a supergroup which has to contain the new subgroups). Further commands allow the user to sort the subgroups by their orders, to determine their maximal subgroups or their minimal supergroups with respect to the current set of known subgroups, to display more or less explicit information on prescribed classes, or, if all subgroups are known, to compute Burnside's "table of marks" [**12**]. All these commands may be interchanged arbitrarily provided only that the initialization step has been called just once.

Moreover, there are options which provide the user with a certain control of the space requirements. In particular, he can decide to save space at the cost of increasing computing time by not storing the characteristic functions of those subgroups which are not chosen as representatives of their conjugacy classes. In this case, instead of using the standard version, he may choose a version in which the corresponding bit strings are saved on disk files, or another one in which they are just deleted and recomputed by conjugation whenever needed. An interactive command allows switching from one of these versions to another at any state of the computation.

These space saving options significantly increase the range of groups which can be handled in a given workspace. We used them, for instance, just as a test example for the capability of the program, to compute all 48 337 subgroups of the alternating group $A_8$ of order 20 160. Though this group possesses 4117 cyclic subgroups of prime power order, and hence each characteristic function occupies 69 machine words (on a Cyber 175 with 60 bit words), we were able to compute its elements and subgroups and to save a lot of information on them, in a total workspace of 34 000 machine words only.

The program and its predecessors have frequently been used e.g. to construct examples and counterexamples for conjectures, to help in the investigation of Galois groups, and in complete classifications of various classes of groups of smaller order [**2**], [**8**], as well as in teaching group theory.

## Acknowledgements

We thank Klaus Lux for his help in implementing the perfect subgroups procedure. The program has been developed and used on the Cyber 175 of the "Rechenzentrum der RWTH Aachen". We are grateful for the facilities made available to us.

## References

1. Bohmann, B. (1978). Ein Verfahren zur Bestimmung von Untergruppenverbänden, das die Benutzung von Sekundärspeicher zuläßt, und seine Implementation, Diplomarbeit, RWTH Aachen.
2. Brown, H., Bülow, R., Neubüser, J., Wondratschek, H. and Zassenhaus, H. (1978). "Crystallographic groups of four-dimensional space". Wiley, New York.
3. Cannon, J. J. (1974). A general purpose group theory program, Proc. Second Internat. Conf. Theory of Groups (Canberra, 1973), Lecture Notes in Mathematics, Vol. **372**, 204–217. Springer-Verlag, Berlin.
4. Cannon, J. J. (1982). (Preprint). A Language for Group Theory. University of Sydney.
5. Dreyer, P. (1970). Ein Programm zur Berechnung der auflösbaren Untergruppen von Permutationsgruppen, Diplomarbeit, University of Kiel.
6. Felsch, V. and Neubüser, J. (1963). Ein Programm zur Berechnung des Untergruppenverbandes einer endlichen Gruppe, *Mitt. Rhein.-Westfäl. Inst. Instrum. Math. Bonn* **2**, 39–74.
7. Gerhards, L. and Lindenberg, W. (1965). Ein Verfahren zur Berechnung des vollständigen Untergruppenverbandes endlicher Gruppen auf Dualmaschinen, *Numerische Mathematik* **7**, 1–10.
8. Laue, R. (1982). Zur Konstruktion und Klassifikation endlicher auflösbarer Gruppen, *Bayreuther Mathematische Schriften* **9**.
9. Leon, J. S. (1983). Computing automorphism groups of combinatorial objects, (these Proceedings).
10. Neubüser, J. (1960). Untersuchungen des Untergruppenverbandes endlicher Gruppen auf einer programmgesteuerten elektronischen Dualmaschine, *Numerische Mathematik* **2**, 280–292.
11. Neubüser, J. (1971). Computing moderately large groups: Some methods and applications, *SIAM-AMS Proc.* **4**, 183–190.
12. Plesken, W. (1982). Counting with groups and rings, *J. Reine Angew. Math.* **334**, 40–68.
13. Robertz, H. (1976). Eine Methode zur Berechnung der Automorphismengruppe einer endlichen Gruppe, Diplomarbeit, RWTH Aachen.

14. Sandlöbes, G. (1977). Perfekte Gruppen bis zur Ordnung $10^4$, *Schriften zur Informatik und Angew. Math.* **36**, RWTH Aachen.
15. Sandlöbes, G. (1981). Perfect groups of order less than $10^4$, *Comm. in Alg.* **9**, 477–490.
16. Steinmann, H. (1974). Ein transportables Programm zur Bestimmung des Untergruppenverbandes von endlichen auflösbaren Gruppen, Diplomarbeit, RWTH Aachen.

# An Introduction to the Group Theory Language, Cayley

JOHN J. CANNON‡

## 1 Introduction

The growth of computational group theory over the past twenty-five years has stimulated demand for software which would permit the exploitation of these techniques by the wider group theory community. The development over the past decade of machine independent software implementing particular group theory algorithms represented the first step towards meeting this demand. However, it soon became clear that the solution of many problems required the application of several of these programs. Because of incompatibilities between the various programs, this was often a cumbersome business.

Motivated by these and other considerations, in 1975 the author commenced the development of a high level programming language (Cayley) designed for computation in groups and related structures. The features of the language include the following.

(i) The objects definable in the language include the basic entities of modern algebra: set, sequence, algebraic structure (e.g. group, ring, field, module), and mapping.
(ii) The language provides a reasonable notation for the expression of the types of algorithms used in modern algebra.
(iii) The language processor has an elementary knowledge of group theory built into it, so that much of the detail in the description of an algorithm will be taken care of automatically.
(iv) The language is supported by an extensive library containing high quality implementations of most standard group theory algorithms.

This paper is intended to serve as a general introduction to the language. The exposition is built around four Cayley programs. Section 2 of the paper

‡ This research was supported by the Australian Research Grants Scheme.

COMPUTATIONAL GROUP THEORY
ISBN 0-12-066270-1

gives a brief introduction to the major constructs of the language. On a first reading of the paper, this section should be read quickly and then consulted as appropriate when reading the subsequent sections. Section 3 outlines the set and sequence machinery and illustrates its use in a program which attempts to show that a $p$-group is non-regular. Section 4 summarizes some of the library facilities for computing group structure and then uses them in a Cayley program which implements an efficient test for simplicity. Section 5 contains a discussion of some of the mapping machinery and illustrates it with a program for computing transfer. Section 6 describes some of the facilities available for performing combinatorial type calculations in finitely presented groups. These facilities are used in a program which computes the Wirtinger presentation for a knot.

It should be emphasized that only a fraction of the facilities provided by the language are mentioned in this paper. Essentially, the paper confines itself to a discussion of those features of Cayley required for the examples. A full description of the language may be found in "A Language for Group Theory" by the author [2], which will be referred to as ALFGT.

## 2 An Overview of the Language

### 2.1 Identifiers

The Cayley vocabulary consists of special symbols, standard identifiers and ordinary identifiers. *Special symbols* serve to represent operators and various kinds of delimiters. *Standard identifiers* are the names of certain constants, attributes and standard functions. Note that standard identifiers are permitted to have embedded spaces. For example, the phrase *lower central series* is a standard identifier.

*Ordinary identifiers* are names denoting variables and procedures. An identifier must begin with a letter and this letter may be followed by any combination of letters and digits. Although an identifier may be of arbitrary length, only the first six characters will be recognized.

Strings used as special symbols or as standard identifiers are called *reserved words* and they have a fixed meaning. A Cayley programmer may not use them in contexts other than those explicitly permitted in the Cayley syntax. In particular, reserved words may not be used as identifiers. The reader should note that Cayley does not distinguish between upper and lower case letters. In order to elucidate the structure of programs, it is usual to reserve upper case for special symbols and lower case for identifiers.

### 2.2 Types

Each item of data in a program may be categorized as either a *constant* or a *variable*, the distinction being that the value of a variable may change during the execution of a program. Each variable (and constant) has a *type* associated with it. The type of a variable serves to define the possible values and the operations that can be performed on it. The generic types permitted in Cayley include:

*integer*
*Boolean value*
*algebraic structure*
*algebraic element*
*word* (i.e. a string in the generators and generator inverses of some group)
*set*
*sequence*
*structured sequence*
*mapping*
*coset table*
*file*
*character string*

The term *generic type* is used because certain of the above types are actually families of types. For example, if $x$ is an element of a group $G$, and $y$ is an element of a distinct group $H$, then $x$ and $y$ must be regarded as belonging to different types. Since Cayley has been designed as an interactive system, explicit type declarations are avoided except in the case of algebraic structures. The type of a variable is normally determined at execution time whenever it is assigned a value.

The types of algebraic structure permitted in Cayley include:

(i) The ring $Z$ of integers, and the ring $Z_m$ of integer residues;
(ii) The finite field GF($q$), $q$ a prime power;
(iii) The ring $M_n(R)$ of $n \times n$ matrices over the ring $R$, where $R$ is $Z$, $Z_m$ or GF($q$);
(iv) The finitely generated free $R$-module of rank $n$, where $R$ is $Z$ or $Z_m$;
(v) The vector space $V(n, K)$ of dimension $n$ over the field $K$, where $K$ is GF($q$);
(vi) A finitely presented group;
(vii) A group of permutations acting on a finite set;
(viii) A group of matrices over $Z$, $Z_m$ or GF($q$).

The elements of a particular algebraic structure form an algebraic element type. For example, the elements of the symmetric group of degree 6 constitute such a type.

The elements of sets must belong to one of the following types:

*integer*
*ring element*
*field element*
*vector*
*group element*
*word*

All the elements of a particular set must belong to the same type. The reader may wonder at the difference between *group element* and *word,* since *word* is obviously an element of some finitely presented group. This distinction is necessary, for example, in the case of a permutation or matrix group $G$ (which has been specified by means of a set of generating elements), where it is necessary to distinguish between a word in the generators of $G$ and the corresponding permutation or matrix of $G$. The concept of a set in Cayley is rather restrictive in the interests of efficiency: the permitted set types are such that the lists of their elements can be stored very compactly, and the set operations can be performed very rapidly.

In contrast to the situation with sets, the sequence concept is much more liberal, in that terms of a sequence may belong to almost any Cayley type. In particular, the terms of a sequence may themselves be sequences so that very complex data structures can be constructed using this concept. Certain homogeneity conditions are imposed on the terms of a sequence. For example, in the case of a sequence of group elements, all the elements must belong to the same group.

Certain objects, such as the lattice of subgroups of a finite group, do not conveniently fit into either the set or sequence types. Rather than going to the expense of introducing individual types for such objects, it has been found convenient to lump them together as members of an omnibus type known as the *structured sequence* type. The basic assumption here is that the programmer will be content with little more than the facility to extract information from such an object. Objects belonging to the structured sequence type may be created only through use of one of a small number of standard functions. Few operations are defined for this type: essentially a structured sequence may be assigned (see section 2.7), printed and indexed. The indexing operation extracts the $i$th term of the sequence and converts it to the standard representation for its type.

The programmer may define a homomorphism of a group into some other structure by specifying the images of a set of generators for the group. Mappings of algebraic structures or sets can be specified either by giving a rule for the image of an arbitrary element, or, in the case of finite sets, by listing the image of each element of the domain. Further, the Cayley library contains routines which construct many of the standard group theoretic homomorphisms.

## 2.3 Expressions

Objects are defined in a programming language by means of the *expression* construct. An expression is built up out of constants, variables and operators according to rules that are very similar to those used in algebra. An expression can be regarded as a rule for calculating a value where the conventional rules of left-to-right evaluation and *operator precedence* are observed. The rules of operator precedence define the order in which operations are performed in the absence of parentheses. For example, the fact that the operator ↑ (exponentiation) has higher precedence than the operator * (multiplication) means that the value of the expression

$$2 * 3 \uparrow 2$$

is 18 rather than 36 (i.e. $2 * (3 \uparrow 2)$). The precedence of operators is described in ALFGT, [2].

## 2.4 The type Boolean

A *Boolean variable* is a variable which takes as its value one of the two logical truth values which are denoted in Cayley by the symbols *true* and *false*. The three logical operators (*AND*, *OR* and *NOT*) which give a Boolean value when applied to Boolean operands are summarized in Table 1. In the absence of parentheses, the operator *NOT* is applied first, then *AND* and finally *OR*. In addition to the Boolean operators, there are a number of relational operators which are summarized in Table 2. The operators *EQ* and *NE* are applicable whenever $x$ and $y$ are both Boolean values, integers, algebraic structures, algebraic elements, sets, or sequences. The operators *LT*, *LE*, *GT* and *GE* are applicable whenever $x$ and $y$ are both integers, or both words.

*Table 1* Boolean Operators

| Operator | Example | Meaning |
|---|---|---|
| *NOT* | *NOT* $x$ | logical negation of the Boolean variable $x$ |
| *AND* | $x1$ *AND* $x2$ | logical conjunction of the Boolean variables $x1$ and $x2$ |
| *OR* | $x1$ *OR* $x2$ | logical disjunction of the Boolean variables $x1$ and $x2$ |

*Table 2* Relational operators

| Operator | Example | Meaning |
|---|---|---|
| *EQ* | $x$ *EQ* $y$ | true if object $x$ is equal to object $y$ |
| *NE* | $x$ *NE* $y$ | true if object $x$ is not equal to object $y$ |
| *LT* | $x$ *LT* $y$ | true if object $x$ is less than object $y$ |
| *LE* | $x$ *LE* $y$ | true if object $x$ is less than or equal to object $y$ |
| *GT* | $x$ *GT* $y$ | true if object $x$ is greater than object $y$ |
| *GE* | $x$ *GE* $y$ | true if object $x$ is greater than or equal to object $y$ |
| *IN* | $x$ *IN* $S$ | true if the object $x$ is an element of the set or algebraic structure $S$ |
| *IN* | $R$ *IN* $S$ | true if the set or algebraic structure $R$ is contained in the set or algebraic structure $S$ |

## 2.5 The type integer

The type *integer* is an implementation-dependent subset of the set of positive and negative integers. The fixed length representation of integers used by computers implies that only a finite subset of the integers can be represented. Variables of type integer are mainly used as counters and indices. The integer operators are defined in Table 3.

*Table 3* Integer Operators

| Operator | Example | Meaning |
|---|---|---|
| + | $i+j$ | sum of integers $i$ and $j$ |
| + | $+j$ | integer $j$ |
| − | $i-j$ | difference of integers $i$ and $j$ |
| − | $-j$ | negation of integer $j$ |
| * | $i * j$ | product of integers $i$ and $j$ |
| / | $i / j$ | integer quotient upon dividing integer $i$ by integer $j$ |
| *MOD* | $i$ *MOD* $j$ | integer remainder upon dividing integer $i$ by integer $j$ |
| ↑ | $i \uparrow n$ | $n$th power of integer $i$ ($n$ a non-negative integer) |

## 2.6 Statements

The "expression construct" introduced in section 2.3 allows the programmer to define all the different types of object permitted in the language. A *statement*, on the other hand, specifies that certain actions are to be performed. For example, a statement may specify that a certain block of statements is to be repeatedly executed until some condition is met.

A Cayley program consists of a sequence of statements. Each statement *must* be terminated by a semi-colon, i.e. the semi-colon is used as a terminator rather than as a separator.

## 2.7 The assignment statement

The basic form of the assignment statement is

*variable = expression;*

This statement specifies that a certain action is to be performed: it does not express an algebraic equality. The expression on the right hand side is evaluated and the result becomes the current value of *variable*. For example, the statement

*t = (1,2,4) (5,6,8) (3,9,7);*

specifies that the permutation on the right hand side is to be given the name *t*.

If the expression on the right hand side of the assignment involves words in the generators of a permutation or matrix group $G$, then it may be ambiguous as to whether such word subexpressions are to be evaluated as permutations or matrices, or retained as words. The convention adopted is that, in the case of the assignment statement introduced above, word subexpressions will be evaluated whenever this is possible. If the user wishes to prevent such words from being evaluated, then he should use the word assignment statement which has the form

*variable := expression;*

To illustrate the effects of these assignment statements, suppose that $a$ and $b$ are generators of a permutation group $G$. The assignment

*x = a*b;*

assigns a permutation to *x* (the product of *a* and *b*) while the assignment

*x:= a*b;*

assigns the word *a*b* to *x*.

A few operators and standard functions can potentially return more than one value. For example, the standard function *blocks homomorphism* $(G,I)$

constructs the action of a permutation group $G$ on a system of imprimitivity $I$. The function returns the homomorphism, the image of $G$ with respect to this homomorphism, and the kernel of the action. Such functions will be termed *multiple-valued functions.* The assignment of values returned by such functions is accomplished through use of the *multiple assignment statement*, which has the general form

```
variable_1, ..., variable_n = expression;
```

where *variable*$_1$, ..., *variable*$_n$ are the variables to be assigned and *expression* represents the invocation of a multiple-valued function. Thus,

```
f,h,k = blocks homomorphism (g,i);
```

assigns the homomorphism to *f*, the image to *h*, and the kernel to *k*.

## 2.8 Structure definition statements

In order to keep this account to a reasonable length, it will be necessary to restrict discussion to just a few of the ways in which algebraic structures can be defined in Cayley.

The finite field GF$(p^n)$, $p$ a prime, is usually defined as an extension of GF$(p)$:

```
k: field (p);
l: field (k,w,x,f(x));
```

where *w* is the name which will be used to refer to a primitive element of the extension, *x* is an indeterminant, and *f(x)* is a polynomial of degree *n* in *x*, which is irreducible over *k*, and which has *w* as a root. If the programmer does not care to supply a primitive polynomial, then the alternative form

```
l: field(q,w);
```

may be used. Here *q* is an expression whose value is $p^n$, while *w* is the name of a primitive element.

Elements of GF$(p)$ are normally represented by the integers $0, 1, \ldots, p-1$, while the non-zero elements of GF$(p^n)$, $n>1$, are usually represented as powers of the primitive element *w*.

The specification of a vector space $V$ involves supplying the dimension and field:

```
v: vector space (n,k);
```

where *n* is an expression defining the dimension, and *k* is an expression defining the field. In Cayley the standard vector space is the space of row vectors. A vector is represented as a list of expressions defining its components, the components being separated by commas, and the list as a whole being enclosed between the symbols *VEC(* and *)*. As an illustration, the effect of the four statements

```
v: vector space (3,field(4,w));
x = VEC(1,0,0);
y = VEC(0,1,0);
z = VEC(0,0,1);
```

is to define the vector space of dimension 3 over GF(4), and then to define vectors *x*, *y* and *z* belonging to this space.

The *element type* of a group is defined to be the element type of the "universal" group which underlies the definition of $G$. In Cayley there are currently three main element types:

(i) $G$ is a quotient group of the free group on the generating set $X$ (the elements of $G$ are words on the set $X$);
(ii) $G$ is a subgroup of the symmetric group $S_\Omega$ (the elements of $G$ are permutations on $\Omega$);
(iii) $G$ is a subgroup of the general linear group $\mathrm{GL}(n,R)$, $R$ a ring or field (the elements of $G$ are $n \times n$ matrices over $R$).

The first stage in defining a group $G$ is the specification of its element type. This involves stating:

(ia) the names of the generators (set $X$), in the case of a finitely presented group;
(iia) the set $\Omega$ on which $G$ acts, in the case of a permutation group;
(iiia) the vector space or module on which the group acts, in the case of a matrix group.

For some kinds of group theoretic computation, the definition of element type alone is sufficient. Thus, for example, it is possible to calculate products in a permutation group $G$ as soon as the element type of $G$ is known. However, for most computations the second stage (definition of the actual group) must be carried out. This involves supplying:

(ib) defining relations in the case of a finitely presented group;
(iib) generating permutations in the case of a permutation group;
(iiib) generating matrices in the case of a group of matrices.

The effect of the second stage is, therefore, to define the appropriate quotient group of the free group on $X$, or the appropriate subgroup of $S_\Omega$, or the appropriate subgroup of $\mathrm{GL}(n,R)$.

The method of definition is best illustrated by examples. The finitely presented group $G$ having presentation

$$\langle a,b \mid a^6,b^6,c^6,abc,a^2ba^2 = b,b^2cb^2 = c,c^2ac^2 = a\rangle$$

is defined by the two statements

```
g: free (a,b,c);
g.relations: a↑6, b↑6, c↑6, a*b*c, a↑2*b*a↑2 = b,
             b↑2*c*b↑2 = c, c↑2*a*c↑2 = a;
```

The same group may be defined as a permutation group generated by the permutations $a = (1,2)(4,5,6)(8,9)$, $b = (2,3)(4,5)(7,8,9)$ and $c = (1,2,3)(5,6)(7,8)$;

```
g: permutation group (9);
g.generators: a = (1,2) (4,5,6) (8,9),
              b = (2,3) (4,5) (7,8,9),
              c = (1,2,3) (5,6) (7,8);
```

The projective special linear group PSL(2,16) is generated by the matrices

$$\mathbf{x} = \begin{pmatrix} 1 & 1 \\ 0 & 1 \end{pmatrix}, \quad \mathbf{y} = \begin{pmatrix} w & 0 \\ 0 & w^{-1} \end{pmatrix}, \quad \mathbf{z} = \begin{pmatrix} 1 & 0 \\ 1 & 1 \end{pmatrix}$$

where $w$ is a primitive element of GF(16). This group may be defined by means of the statements,

```
v: vector space (2,field(16,w));
g: matrix group (v);
g.generators: x = MAT(1,1:0,1),
              y = MAT(w,0:0,w↑-1),
              z = MAT(1,0:1,1);
```

If $e_1, \ldots, e_r$ are expressions defining elements $x_1, \ldots, x_r$ of the group $G$, then the statement

```
h = ⟨e1, ..., er⟩;
```

defines $H$ to be the subgroup of $G$ that is generated by the elements $x_1, \ldots, x_r$. More generally, this statement is valid when some or all of the expressions $e_i$ ($i = 1, \ldots, n$) define either sets of elements of $G$ or subgroups of $G$.

## 2.9 The conditional statement

The *conditional statement* (or *IF-statement*) specifies that a sequence of statements is to be executed only if a certain condition is met. The condition is expressed in the form of a Boolean expression. For example, if $m$ and $n$ are integer variables, the statement

```
IF m GE n
   THEN m = m - n;
END;
```

will subtract $n$ from $m$ whenever $m$ is greater than or equal to $n$.

Formally, the IF-statement has the structure

```
IF expression
   THEN statements1
   ELSE statements2
END;
```

The sequence of statements *statements*$_1$ will be executed when *expression* (a Boolean expression) is *true*, while the sequence *statements*$_2$ will be executed when *expression* is *false*. If no action is to be taken when *expression* is *false*, the clause "*ELSE statements*$_2$" may be omitted.

## 2.10 Iterative statements

An iterative statement (or loop) specifies that a block of statements is to be executed a number of times. Although the statements do not change, the objects (data) on which they act may change.

The *set FOR-statement* performs iteration over the elements of a finite set:

```
FOR EACH x IN S DO
     statements
END;
```

The variable *x* successively takes the elements of the set *S* as its values. For each value of *x*, the sequence of statements *statements* is executed (however, see *LOOP* and *BREAK* below). Analogous FOR-statements are provided for sequences, structured sequences and finite algebraic structures.

The *integer FOR-statement* performs iteration over an integer range:

```
FOR i = n1 TO n2 BY n3 DO
     statements
END;
```

Here $n_1$, $n_2$ and $n_3$ are expressions defining integers. The sequence of statements *statements* will be executed with *i* taking successively the values

$$n_1, n_1+n_3, n_1+2n_3, \ldots, n_1+kn_3$$

where $n_1+kn_3$ is the largest integer of this form which does not exceed $n_3$, when $n_3$ is positive, and analogously when $n_3$ is negative. If $n_3$ is $+1$, the clause "BY $n_3$" may be omitted.

The *WHILE- and REPEAT-statements* allow a sequence of statements to be repeatedly executed while ever a Boolean expression remains true:

```
WHILE expression DO
     statements
END;
```

Thus, *statements* will be executed while ever the Boolean expression

*expression* is true. The REPEAT-statement is similar, except that *statements* will always be executed at least once:

```
REPEAT
    statements
WHILE expression;
```

The symbols *LOOP* and *BREAK* may appear wherever a statement is permitted in the body of a loop. The effect of *LOOP* is to abort the current iteration and to commence the next iteration immediately. The symbol *BREAK* aborts the execution of the loop so that execution is transferred to the first statement that follows the repetitive statement.

### 2.11 The PRINT-statement

The purpose of the PRINT-statement is to enable the programmer to display the result of some computation. It has the general form

```
PRINT e1, . . . , en;
```

where $e_1, \ldots, e_n$ are expressions defining the objects that are to be printed. It is possible to qualify the printing of certain complex objects, such as a lattice of subgroups, so that only selected information appears. Details of this facility may be found in Chapter 13 of ALFGT, [2].

### 2.12 Procedures

A procedure is a sequence of statements which is associated with an identifier. The appearance of this identifier within a sequence of statements will cause the statements associated with the procedure to be executed at this point. The definition of a procedure has the general form

```
PROCEDURE identifier (u1, . . . , ur; v1, . . . , vs);
    statements
END;
```

where *identifier* is the name of the procedure, $u_1, \ldots, u_r$ are variables representing the input parameters for the procedure, and $v_1, \ldots, v_s$ are variables representing the output parameters for the procedure.

The invocation of a procedure has the form

```
identifier (e1, . . . , er; w1, . . . , ws);
```

where *identifier* is the name of the procedure, $e_1, \ldots, e_r$ are expressions defining the actual values for the input parameters, and $w_1, \ldots, w_s$ are the external variables which are to be assigned the values returned by the procedure.

Excepting those variables appearing in the heading of the procedure definition, all variables used in a procedure are *local* to the procedure in the sense that a value assigned to variable (other than an output parameter) is only accessible from within the procedure. Similarly, the value of a variable defined outside a procedure is accessible from within the procedure, only if the variable appears among the actual input parameters, i.e. among the expressions $e_1, \ldots, e_r$.

The symbol *RETURN* may appear anywhere that a statement is permitted in the body of a procedure and it causes immediate exit from the procedure.

Suppose that $a$ and $b$ are elements of some group $G$. The procedure *engel* given below calculates the commutator

$$c = (\ldots((a,b),b),\ldots b)$$

where the number of $b$s that occur in the commutator is $r$:

```
PROCEDURE engel (a,b,r;c);
    c = a;
    FOR i = 1 TO r DO
        c = (c,b);
    END;
END;
```

If $a$ and $b$ are the permutations (1,3,2) (4,5) and (2,4,5), respectively, then the statement

```
engel ((1,3,2) (4,5),(2,4,5),4;z);
```

will compute the commutator $((((a,b),b),b),b)$ and assign it to the variable *z*.

## 2.13 The standard functions

Much of the power of Cayley as a tool for research in group theory derives from the fact that it is supported by an extensive library of group theory routines. The library routines appear in the language as *standard functions*. For example, if $G$ is a finite group, and $x \in G$, the standard function *centralizer* when applied to $G$ and $x$, as in the expression

*centralizer* $(G,x)$

will find the centralizer of $x$ in $G$, and return it as the value of the function. The programmer may use the expression *centralizer* $(G,x)$ wherever a group is allowed.

The contents of the library of standard functions are outlined below. It should be emphasized that this is only a partial list of standard functions: the reader should consult Chapter 12 of ALFGT, [2], for a more complete list.

The functions fall into three classes:

Group a: Combinatorial type algorithms for finitely presented groups;
Groups b–c: Group structure algorithms which are essentially independent of the manner in which the group is specified;
Groups d–g: Group structure algorithms which are dependent upon how the group is specified.

The authors of those library routines written at institutions other than Sydney, are given in parentheses.

(a) *Finitely presented groups (combinatorial algorithms)*

The functions in this set belong to the area of combinatorial group theory. With the exception of (vi), none of them require a multiplication algorithm to be defined for the group.

(i) Todd-Coxeter methods (Alford and Havas).
(ii) Low index subgroups (Cannon and Gallagher).
(iii) Nilpotent quotient algorithm (Havas and Newman [**6**]).
(iv) Integer matrix diagonalization (Havas and Sterling [**7**]).
(v) Abelianized Reidemeister-Schreier algorithm (Richardson).
(vi) Construction of defining relations from a faithful representation.

(b) *Normal structure*

All these algorithms require a multiplication algorithm for the group. Essentially the same algorithm is used for all the different types of group:

(i) Normal closure.
(ii) Core.
(iii) Derived series.
(iv) Lower central series.
(v) Upper central series.
(vi) Frattini subgroup (of a $p$-group).
(vii) Fitting subgroup.
(viii) Lattice of normal subgroups.
(ix) Tests for simplicity, special (for a $p$-group), extra-special (for a $p$-group) etc.

(c) *General structure*

The algorithms in this set require a multiplication algorithm for the group. The same algorithm is used for all the different types of group:

(i) Lattice of subgroups (Felsch and Sandlöbes [these Proceedings]).

(ii) Automorphism group (Robertz [9]).
(iii) Character table.

(d) *Permutation groups*

Most of these algorithms assume that a base and strong generating set are available (Sims [10, 11]):

(i) Tests for transitivity, regularity and primitivity.
(ii) Tests for the symmetric or alternating group.
(iii) Orbit of a point, set, or sequence.
(iv) Stabilizer of a point, set, or sequence.
(v) Restriction of an intransitive group to an orbit or union of orbits (homomorphism, image, kernel).
(vi) Action of an imprimitive group on a system of imprimitivity (homomorphism, image, kernel).
(vii) Conjugacy classes.
(viii) Centralizer.
(ix) Normalizer.
(x) Sylow $p$-subgroup.
(xi) Intersection of subgroups.
(xii) Transversal and coset table.

(e) *Matrix groups*

Most of these algorithms assume that a base and strong generating set are available (Butler [1]):

(i) Orbit of an affine point or an affine line.
(ii) Stabilizer of a point (line) or a sequence of points (lines).
(iii) Action of a group on a point (line) orbit or a union of point (line) orbits (homomorphism, image, kernel).
(iv) Conjugacy classes.
(v) Centralizer.
(vi) Sylow $p$-subgroup.
(vii) Intersection of subgroups.
(viii) Transversal and coset table.

(f) *Finitely presented p-groups (structure algorithms)*

This set of algorithms assume that a power-commutator presentation is available (Newman [8]):

(i) Conjugacy classes.
(ii) Centralizer.

(iii) Normalizer.
(iv) Intersection of subgroups.
(v) Transversal and coset table.

(g) *Finitely presented groups* (*structure algorithms*)

All these algorithms assume that the group is finite and that its Cayley graph is available:

(i) Conjugacy classes.
(ii) Centralizer.
(iii) Normalizer.
(iv) Intersection of subgroups.
(v) Transversal and coset table.

## 3 Sets, Sequences and Groups

### 3.1 Sets

A set may be either a collection of integers, or a collection of elements, all of which belong to a single algebraic structure. A non-empty set is represented as a list of expressions defining its elements, the elements being separated by commas, and the list as a whole being enclosed within square brackets, i.e.

$$[e_1, \ldots, e_n]$$

where $e_1, \ldots, e_n$ are expressions defining its elements. The null set is represented by the symbol *null*.

*Table 4* Set Operators

| Operator | Example | Meaning |
|---|---|---|
| *JOIN* | $R$ *JOIN* $S$ | union of the sets $R$ and $S$, where $R$ and $S$ are sets of the same type. |
| *MEET* | $R$ *MEET* $S$ | intersection of the sets $R$ and $S$, where $R$ and $S$ are sets of the same type. |
| $-$ | $R-S$ | difference of the sets $R$ and $S$, where $R$ and $S$ are sets of the same type: $R-S = \{x \mid x \in R \text{ and } x \notin S\}$. |
| *card* | *card* $(R)$ | cardinality of the set $R$. |
| *setrep* | *setrep* $(R)$ | an arbitrary element from the set $R$. |

Some basic set operations are listed in Table 4. A complete list may be found in Chapter 6 of ALFGT, [2].

The relational operators *EQ*, *NE* and *IN* (Table 2) may also be used with set operands.

### 3.2 Sequences

Sequences may be constructed from objects of any type. The terms of a sequence must be all of the same type except in the case where the terms are themselves sequences. If $Q = \{q_1, \ldots, q_r\}$ is a sequence of sequences, then the terms of the sequence $q_i$ may be of a different type to the sequence $q_j$, $i \neq j$.

A non-empty sequence is represented as a list of expressions defining its terms, the terms being separated by commas, and the list as a whole being enclosed between the symbols *SEQ(* and *)*, i.e.

*SEQ(e1, . . . , en)*

where $e_1, \ldots, e_n$ are expressions defining its terms. The empty sequence is represented by the symbol *empty*.

*Table 5* Sequence Operators

| Operator | Example | Meaning |
|---|---|---|
| *append* | *append* $(Q,x)$ | append object $x$ to the end of sequence $Q$: $\{q_1, \ldots, q_n, x\}$. |
| *concatenate* | *concatenate* $(Q,T)$ | concatenate the sequence $T$ to the end of sequence $Q$: $\{q_1, \ldots, q_n, t_1, \ldots, t_m\}$. |
| *conseq* | *conseq* $(x,n)$ | form a constant sequence of length $n$, where each term is $x$: $\{x, \ldots, x\}$. |
| *length* | *length* $(Q)$ | the length of sequence $Q$. |
| *position* | *position* $(Q,x)$ | the position of the first occurrence of $x$ in sequence $Q$, or zero if there is no term of $Q$ equal to $x$. |
| *prune* | *prune* $(Q)$ | delete the last term of sequence $Q$: $\{q_1, \ldots, q_{n-1}\}$. |
| *remove* | *remove* $(Q,i)$ | delete the $i$th term of sequence $Q$: $\{q_1, \ldots, q_{i-1}, q_{i+1}, \ldots, q_n\}$. |
| [ ] | $Q[i] = x$ | replace the $i$th term of sequence $Q$ by $x$. |
| [ ] | $x = Q[i]$ | extract the value of the $i$th term of sequence $Q$. |

Table 5 contains a summary of some of the sequence operators. A full account of sequence operators may be found in Chapter 6 of ALFGT, [2].

In this table, $Q$ denotes the sequence $q_1, \ldots, q_n$ and $T$ denotes the sequence $t_1, \ldots, t_m$. In addition to the operators described in Table 5, transfer functions are provided to convert between sets and sequences, and between algebraic elements and sequences.

### 3.3 Groups

The variable *g* is declared to be of type *permutation group* through use of the statement

*g: permutation group (n);*

where *n* is an expression defining the degree of $G$. The group $G$ is assumed to act on the set $\{1, 2, \ldots, n\}$. A permutation of $G$ may be represented either as a mapping or as a product of disjoint cycles. A permutation represented as a product of disjoint cycles is written in the usual way (i.e. cycles enclosed within round brackets, the letters separated by commas), except that cycles of length 1 must be omitted. Having declared $G$ using the above statement, the user may define a set of generators for $G$ using a *generators*-statement as illustrated in the examples of section 2.8.

The variable *g* is declared to be of type *matrix group* through use of the statement.

*g: matrix group (m);*

where *m* is an expression defining the module over which $G$ is defined. An $n \times n$ matrix is represented as a list of $n$ rows, the rows being separated by colons, and the list as a whole being enclosed between the symbols *MAT(* and *)*. The $n$ expressions defining the elements of a row are separated by commas. After declaring $G$, a generating set may be supplied by means of a *generators*-statement.

The variable *g* is declared to be of type *free group* on the set $S = \{x_1, \ldots, x_n\}$ through use of the statement

*g: free* $(x_1, \ldots, x_n)$*;*

The words of $G$ are written in the usual manner as strings of elements of $S$ and their inverses. Multiplication is denoted by $*$ and exponentiation by $\uparrow$. For example, the word $ab^{-1}a^3b$ is represented by the string $a*b\uparrow-1*a\uparrow3*b$. Alternatively, the variable *g* may be declared to be of type free group of rank $n$, through use of the statement

*g: fgrank (n);*

where *n* is an expression defining the rank. In this case, the generators of $G$ may be referred to by the names $G.1, \ldots, G.n$.

*Table 6* Element Operators

| Operator | Example | Meaning |
|---|---|---|
| $*$ | $x * y$ | product of the element $x$ with the element $y$. |
| $*$ | $v * x$ | product of the vector $v$ and the matrix $x$. The vector $v$ must belong to some module on which $x$ acts. |
| $\uparrow$ | $x \uparrow n$ | $n$th power of the element $x$, where $n$ is an integer. |
| $\uparrow$ | $x \uparrow y$ | conjugate of the element $x$ by the element $y$ $(y^{-1}xy)$. |
| $\uparrow$ | $i \uparrow x$ | image of the point $i$ under the action of the permutation $x$, where $x$ is a permutation of the set $\Omega$, and $i \in \Omega$. |
| $\uparrow$ | $S \uparrow x$ | image of the set $S$ under the action of the permutation $x$, where $x$ is a permutation of the set $\Omega$, and $S$ is a subset of $\Omega$. |
| $\uparrow$ | $Q \uparrow x$ | image of the sequence $Q$ under the action of the permutation $x$, where $x$ is a permutation of the set $\Omega$, and $Q$ is a sequence of points from $\Omega$. |
| *( , )* | $(x,y)$ | commutator of the elements $x$ and $y$ $(x^{-1}y^{-1}xy)$. |
| *order* | *order* $(x)$ | order of the element $x$. |

Irrespective as to how $G$ is given, the identity element is represented by the expression

*identity OF g*

The elementary operations that may be performed on group elements are summarized in Table 6. Apart from the operators displayed in Table 6, the relational operators *EQ* and *NE* (Table 2) may be used with group element operands.

## 3.4 Regularity of *p*-groups

A $p$-group $G$ is said to be *regular* if for every pair of elements $a,b \in G$, and for every positive integer $n$, it is true that

$$(ab)^{p^n} = a^{p^n} b^{p^n} c_1{}^{p^n} \ldots c_r{}^{p^n},$$

where $c_1, \ldots, c_r$ are elements of the derived subgroup of $\langle a,b \rangle$.

*Theorem 1.* (Hall [**4**]): A finite $p$-group $G$ is regular if and only if for every pair of elements $a,b \in G$, there exists an element $c$ of the derived subgroup of $\langle a,b \rangle$, such that

$$(ab)^p = a^p b^p c^p.$$

It is difficult to devise an efficient algorithm which decides whether or not a particular finite $p$-group is regular. The procedure *nonregular* given below, attempts to prove that a $p$-group is not regular by examining pairs of random elements in the hope of finding a pair which violates the conditions of Theorem 1. If no such pair has been found after a reasonable number of attempts, the user can then apply a more expensive algorithm which is able to decide the issue definitively. The random test for non-regularity was originally devised and implemented in Cayley by Leedham-Green. The procedure makes use of the standard function *ranelt (g)* which returns a random element of the group *g* as its value.

The reader should note that text enclosed within double quotes (") is treated as documentation and is ignored by Cayley.

```
PROCEDURE nonregular (g,p,limit; flag);
  "Attempt to deduce that the p-group g is non-regular by
  examining limit pairs of random elements of g. If the procedure is
  able to deduce that g is not regular, then flag is set true, otherwise
  it is set false"

  flag = false;

  FOR i = 1 TO limit DO
      a = ranelt (g);
      b = ranelt (g);
      c = a↑p*b↑p*((a*b)↑p)↑-1;
      IF c EQ identity OF g
         THEN LOOP;
      END;

      "The sequence comms will be used to store all the distinct
      commutators of the form (...((a,b),x1),...,xr), where xi is a
      or b (i = 1,...,r). The parallel sequence weights stores the
      corresponding weights of these commutators"

      comms = SEQ((a,b));
      weights = SEQ(2);
      j = 0;
```

```
WHILE j LT length (comms) DO
    j = j+1;
    x = comms[j];
    FOR EACH y IN [a,b] DO
        z = (x,y);
        IF z NE identity OF g
          THEN
              comms = append (comms,z);
              weights = append (weights,weights[j] + 1);
        END;
    END;
END; "construction of commutators"

"If the nilpotency class of ⟨a,b⟩ is less than p, then ⟨a,b⟩ is
regular, so that the current pair of elements can be ignored.
Otherwise, the subgroup h generated by the p-th powers of
the elements of the derived subgroup of ⟨a,b⟩ is formed and
checked to see if it contains c"

IF p LE weights[length (weights)]
  THEN
      h = ⟨identity OF g⟩;

      FOR k = 1 TO length (comms) DO
          x = comms[k]↑p;
          IF NOT (x IN h)
            THEN
                h = ⟨h,x⟩;
          END;
      END; "p-th powers of commutators"

  IF NOT (c IN h)
      THEN
          flag = true;
          RETURN;
  END;

END;

END; "random elements"

END; "procedure nonregular"
```

---

*Application:* Let $G$ be the wreath product $C_5 \operatorname{wr} C_5$. The procedure *nonregular* took 2 seconds to establish that $G$ is not a regular group. (This timing was obtained on a CYBER 170/730 computer.)

## 4 Group Structure

### 4.1 Structure functions

The ability of Cayley to elucidate the structure of a large group derives from its extensive library of standard functions. A large amount of effort has been invested in the development and implementation of efficient group structure algorithms. Thus, functions are provided for computing conjugacy classes, various types of series, normal subgroups, the lattice of subgroups (as described by Felsch and Sandlöbes in these Proceedings), and many other things. The purpose of this section is to demonstrate how the basic structure functions can be used to rapidly implement new structure algorithms. A small selection of the available structure functions is displayed in Table 7. The reader should consult Chapter 12 of ALFGT, [2], for a complete list of such functions.

### 4.2 A test for simplicity

This section contains a Cayley procedure which implements a new algorithm for testing whether or not a group is simple. The algorithm is an application of Theorems 2 and 3 given below. Throughout this section, $G$ will denote a finite group. The maximal normal $p$-subgroup of $G$ is denoted by $O_p(G)$, while the centre of $G$ is denoted by $Z(G)$.

*Theorem 2.* The group $G$ contains a non-trivial abelian normal subgroup if and only if $O_p(G)$ is non-trivial, for some prime $p$ dividing $|G|$.

*Theorem 3.* Let $G$ be a group that has no non-trivial abelian normal subgroups, and let $N$ denote the normal closure in $G$ of $Z(T)$, where $T$ is a Sylow 2-subgroup of $G$. Then the socle of $G$ is equal to $N^{(\infty)}$, the soluble residual of $N$.

The simplicity algorithm may be summarized as follows:

(i) Determine the order of $G$ using the standard function *forder*.
(ii) Attempt to show that $G$ is not simple using number theoretic conditions on the group order.
(iii) Using the function *derived subgroup*, determine whether or not $G$ is perfect. If it is not, then $G$ is not simple.

*Table 7* Some Group Structure Functions

| Function | Value |
|---|---|
| *abelian invariants* $(G)$ | an integer sequence containing the invariants of the abelian group $G$. |
| *centralizer* $(G,x)$ | the centralizer of the element $x$ in the group $G$. |
| *centralizer* $(G,H)$ | the centralizer of the subgroup $H$ of the group $G$. |
| *centre* $(G)$ | the centre of the group $G$. |
| *conjugate* $(G,x,y)$ | given elements $x$ and $y$ belonging to the group $G$, this function returns the Boolean value *true*, if $x$ and $y$ are conjugate, and the value *false* otherwise. |
| *core* $(G,H)$ | the maximal normal subgroup of $G$ that is contained in the subgroup $H$. |
| *derived series* $(G)$ | the derived series of the group $G$. The series is returned as a sequence of groups. |
| *derived subgroup* $(G)$ | the derived subgroup of $G$. |
| *forder* $(G)$ | a sequence $Q$ containing the order of the group in factored form. If $\lvert G\rvert = p_1^{e_1}\ldots p_n^{e_n}(p_1 < \ldots < p_n, e_i \neq 0)$ then $Q$ will be the integer sequence $\{p_1, e_1, p_2, e_2, \ldots, p_n, e_n\}$. |
| *normal closure* $(G,H)$ | the smallest normal subgroup of the group $G$ containing the subgroup $H$ of $G$. |
| *normalizer* $(G,H)$ | the normalizer of the subgroup $H$ in the group $G$. |
| *sylow* $(G,p)$ | the Sylow $p$-subgroup of the group $G$, where $p$ is a prime. |

(iv) Using the functions *sylow* and *core*, determine whether or not $G$ possesses an odd order normal subgroup.
(v) Using the functions *sylow* and *core*, determining whether or not $O_2(G)$ is trivial.
(vi) Using the functions *centre* and *normal closure*, apply the test of Theorem 3.

Although steps (ii) and (iii) are logically unnecessary, they may save a great deal of time when $G$ is not simple. This is particularly true of the number theoretic tests, since such tests are extremely cheap compared to tests requiring the calculation of structure. In the interests of brevity, only a few number theoretic tests have been included in the program reproduced below.

This particular simplicity algorithm is restricted mainly by the limitations of the Sylow subgroup algorithm. The implementation available in the current version of Cayley can handle permutation groups of degree up to 500. If the order of the Sylow $p$-subgroup $P$ of $G$ is $p^n$, the cost of computing $P$ increases with $n$. If $p$ and $q$, $p < q$, are primes dividing $|G|$, it is probable the $|G|$ will involve a smaller power of $p$ than of $q$. For this reason, the odd primes dividing $|G|$ are considered in decreasing order of size.

```
PROCEDURE einfach (g;flag);
  "If the group g is simple, flag is set true; otherwise it is set false"

  flag = false;

  "The first step is to compute the order n of g in factored form"

  n = forder (g);

  "Is g cyclic of prime order?"

  IF (length (n) EQ 2) AND (n[2] EQ 1)
     THEN
          flag = true;
          RETURN;
  END;

  "Now apply various numerical criteria to the order of g in an
  attempt to prove that g is not simple"

  "If n is of the form p↑a or p↑a*q↑b, p and q primes, then g is not
  simple"

  k = length (n)/2;
  IF k LE 2
     THEN RETURN;
  END;

  "If n is not divisible by 8 or 12, then g is not simple"

  IF NOT ((n[1] EQ 2 AND n[2] GE 3) OR
          ((n[1] EQ 2 AND n[2] GE 2) AND (n[3] EQ 3)))
     THEN RETURN;
  END;
```

```
"Since g has passed the elementary numerical tests, it is necessary
to look for proper normal subgroups. First check if the derived
subgroup of g is a proper subgroup. While this is logically
unnecessary, it may save time."

IF g NE derived subgroup (g)
   THEN RETURN;
END;

"Check whether g contains an abelian normal subgroup: if it
does, then O(p,G), for some odd p, must be non-trivial"

FOR i = k TO 1 BY -1 DO
      IF core (g,sylow(g,n[2*i-1])) NE <identity OF g>
         THEN RETURN;
      END;
END;

"Use Theorem 3 to determine if the socle of G is a proper
subgroup"

z2 = centre (sylow(g,2));
n = normal closure (g,z2);
IF derived subgroup (n) NE g
   THEN RETURN;
END;


"The socle is all of G, so check whether G is a direct product"

FOR EACH x in z2 DO
   IF order (x) EQ 2 THEN
      IF normal closure (g,<n>) NE g
         THEN RETURN;
      END;
   END;
END; "elements of z2"

"g is simple"

flag = true;

END; "procedure einfach"
```

*Application:* The Higman-Sims group HS of order 44 352 000 has a permutation representation of degree 100. Using this representation, the procedure *einfach* took 170 seconds to establish that HS is simple. (This timing was obtained on a CYBER 170/730 computer.)

## 5 Mappings

### 5.1 The definition of mappings

A mapping may be defined either through use of a mapping definition statement (the DEFINE-statement) or by the invocation of certain standard functions.

The DEFINE-statement allows the programmer to specify mappings in one of four ways:

(a) *Generator image:* A homomorphism from a group $G$ into a group $H$ is specified by listing the images of a set of generators for $G$;
(b) *General image:* A mapping from a set or structure $A$ into a set or structure $B$ is specified by an expression which defines the image of a generic element of $A$;
(c) *Image list:* A mapping $f$ from a finite set or structure $A$ into a finite set or structure $B$ is specified by listing the pairs $(x,f(x))$, for all $x \in A$;
(d) *Context free:* A mapping from the product set $A_1 \times \ldots \times A_n$ into a set or structure $B$ is specified by an expression which defines the image of a generic element of the domain. Unlike (b), the domain and codomain are not included in the original definition of the mapping.

This discussion will confine itself to the following two variants of the DEFINE-statement:

(i) *DEFINE f: g TO h BY IMAGES e;*
(ii) *DEFINE f: a TO b BY f(x) = e;*

Variant (i) defines the mapping $f$ to be a mapping from the group *g* into the group *h* by listing images for the generators of *g*. If *g* is defined on *n* generators then *e* denotes an expression which must define a sequence $q$ of $n$ elements of *h* under the convention that $g.i \rightarrow q[i]$ $(i = 1, \ldots, n)$. Note that *g* and *h* must be variables. Variant (ii) defines the mapping $f$ to be a mapping from the set or structure *a* into the set or structure *b* by specifying the image *e* of a generic element *x* of *a*. Here *e* is an expression, usually involving *x*, which defines the element of *b* that is to be the image of *x*. Again, *a* and *b* must be variables.

Some useful mappings created by standard functions are summarized in Table 8. In each case the function returns multiple values.

*Table 8* Standard Functions Creating Mappings

| Function | Values |
|---|---|
| *blocks homomorphism* $(G,S)$ | given a transitive permutation group and a system of imprimitivity $S$ for $G$, this function constructs the homomorphism of $G$ onto the group induced by the action of $G$ on $S$. It returns the homomorphism, the induced group, and the kernel of the action. |
| *constituent homomorphism* $(G,S)$ | given a permutation group $G$ and a union of orbits $S$ of $G$, this function constructs the restriction of $G$ to $S$. It returns the homomorphism, the restriction of $G$, and the kernel. |
| *cosact homomorphism* $(G,H)$ | given a group $G$ and subgroup $H$ of $G$, this function constructs the permutation representation of $G$ defined by its action on the cosets of $H$. It returns the homomorphism, the image, and the kernel. |
| *matact homomorphism* $(G,S,l)$ | given a matrix group $G$ (acting on a vector space $V$) this function constructs the permutation group induced by the action of $G$ on the union of $G$-orbits of the objects in $S$. If the Boolean variable *l* is *true*, the elements of $S$ are interpreted as lines; if *false*, as vectors. The function returns the homomorphism, the induced permutation group, and the kernel of the action. |
| *pquotient* $(G,p,n)$ | given a finitely presented group $G$, a prime $p$, and a positive integer $n$, this function constructs the class $n$ nilpotent $p$-quotient $Q$ of $G$. It returns $Q$ and the homomorphism from $G$ onto $Q$. |
| *transversal* $(G,H)$ | given a group $G$ and a subgroup $H$ of $G$, having finite index in $G$, this function constructs a right transversal $T = \{x_1, \ldots, x_n\}$ for $H$ in $G$. Apart from returning $T$, the function also returns the mapping $\phi: G \rightarrow T$ defined by $\phi(x) = x_i$, where $x \in Hx_i$. |
| $G/N$ | given a group $G$ and normal subgroup $N$ of $G$, this function returns the quotient group $G/N$, and the natural homomorphism of $G$ onto $G/N$. |

Table 9 Mapping Operators

| Function | Example | Meaning |
|---|---|---|
| ( ) | f(x) | the image of element $x$ under the mapping $f$, where $x$ belongs to the domain of $f$. |
| @( ) | f@(y) | the preimage of element $y$ under the mapping $f$, where $y$ belongs to the codomain of $f$. |
| *faithful* | *faithful (f)* | true if the kernel of the homomorphism $f$ from the group $G$ to the group $H$ is trivial. The use of this function should be avoided when $G$ is infinite. |
| *image* | *image (f)* | the image of the homomorphism $f$ from the group $G$ into the group $H$. |
| *kernel* | *kernel (f)* | the kernel of the homomorphism $f$ from the group $G$ into the group $H$. |
| *morphism* | *morphism (f)* | true if the mapping $f$ from the group $G$ into the group $H$ is a homomorphism. |

## 5.2 Mapping operators

The operators and functions which apply to mappings are listed in Table 9. It should be noted that the inverse image operator is only available for certain kinds of mappings.

## 5.3 Calculation of transfer

The procedure given below constructs the transfer homomorphism of a finite group $G$ into the section $S/S'$, where $S$ is a Sylow $p$-subgroup of $G$, and $S'$ is the derived subgroup of $S$.

```
PROCEDURE transhom (g,p;q,f);

  "Construct the transfer f of the group g into the section q = s/s',
  where s is the Sylow p-subgroup of g"

  s = sylow (g,p);
  s1 = derived subgroup (s);
  q,f = s/s1;
  t,phi = transversal (g,s);
```

```
"Determine the image of each generator of g under the transfer"

gims = empty;
FOR EACH y IN generators (g) DO
    z = identity OF g;
    FOR EACH x IN t DO
        z = z*x*y*phi(x*y)↑-1;
    END;
    gims = append (gims,f(z));
END;

"Define the transfer f"

DEFINE f:  g TO q BY IMAGES gims;

END; "procedure transhom"
```

# 6 Finitely Presented Groups

## 6.1 Combinatorial operations

Cayley permits the user to work with finitely presented groups at two levels. Firstly, a range of operators and functions is provided for the combinatorial manipulation of words (e.g. Tietze transformations). Secondly, implementations of important algorithms for computing information about subgroups and quotient groups of finitely presented groups are available in the standard function library. These include the low index subgroups algorithm, the Todd-Coxeter algorithm, the nilpotent quotient algorithm (the latter two being implementations developed by Havas and Newman), and others. Only the elementary word operations will be discussed here: the interested reader will find details of the other functions in Chapter 15 of ALFGT, [**2**].

Either the *free*-statement or *fgrank*-statement may be used to define the free group which is to be the setting for the calculations (see section 3.3). In Table 10, some of the elementary word functions are summarized.

The operators *EQ* and *NE* may be used with words. Further, a lexicographic ordering is defined on the elements of the free group of rank $n$, so that the operators *LE*, *LT*, *GE* and *GT* are applicable to pairs of words. Finally, transfer functions between words and sequences are provided.

## 6.2 The Wirtinger presentation of a knot

A *polygonal knot* is one which is the union of a finite number of closed

*Table 10* Some Combinatorial Functions

| Function | Value |
|---|---|
| *eliminate* $(U,x,v)$ | $U$ is a set of words, $v$ is a word, and $x$ is a generator name, all belonging to the group $G$. The value of *eliminate*, is the set of words obtained by taking each element of $U$ and replacing each occurrence of $x$ in that word by $v$ and each occurrence of $x^{-1}$ by $v^{-1}$. |
| *evaluate* $(G,U)$ | $U$ is a set of words belonging to the $n$-generator group $H$, and $G$ is also an $n$-generator group for which a multiplication algorithm can be defined. The value of *evaluate*, is the set of elements of $G$ obtained by taking each member of $U$ in turn and making the substitutions $H.i \rightarrow G.i$ $(i = 1, \ldots, n)$, where the $G.i$ are interpreted as elements of $G$ rather than as generator symbols. |
| *length* $(u)$ | the length of the word $u$. |
| *match* $(u,v,f)$ | $u$ and $v$ are words belonging to the group $G$, while $f$ is an integer. If $1 \leqslant f \leqslant$ length $(u)$ then the value of *match* is the least integer $l$ such that<br>(a) $l \geqslant f$; and<br>(b) $v$ appears as a subword of $u$, starting at the $l$th letter of $u$.<br>If no such $l$ exists or if $f$ lies outside the range [1, length$(u)$] *match* returns the value zero. |
| *rotate* $(u,n)$ | the word obtained by cyclically permuting the word $u$ by $n$ places. If $n$ is positive, the rotation is from left to right, while if $n$ is negative the rotation is from right to left. In the case where $n$ is zero, the function returns $u$. |
| *satisfy* $(Q,U)$ | $U$ is a set of words belonging to the $n$-generator group $H$, and $Q$ is a sequence of $n$ elements $\{e_1, \ldots, e_n\}$ belonging to the group $G$ for which a multiplication algorithm exists. The function *satisfy* takes the Boolean value *true* if each word of $U$ evaluates to the identity under the substitutions $H.i \rightarrow e_i$ $(i = 1, \ldots, n)$. If, on the other hand, some word of $U$ is not the identity, *satisfy* takes the value *false*. |
| *substitute* $(u,f,n,v)$ | $u$ and $v$ are words belonging to a group $G$, while $f$ and $n$ are integers. The function *substitute* replaces $n$ letters of $u$, starting with the $f$th, by the word $v$. |
| *subword* $(u,f,n)$ | the subword of the word $u$ comprising the $n$ consecutive letters commencing at the $f$th letter of $u$. If $f$ lies outside the range [1, length$(u)$] or if $n$ is zero, *subword* returns the empty word. |

straight line segments called *edges*, whose endpoints are called *vertices*. A polygonal knot $K$ is in *regular position* if:

(i) the only multiple points of $K$ are double points, and there are only a finite number of them; and
(ii) no double point is the image of any vertex of $K$.

It can be shown [**3**] that every polygonal knot $K$ is equivalent under arbitrary small rotation in $R^3$ to a polygonal knot in regular position. If $K$ is a knot in 3-dimensional space, $R^3$ and $p_0$ is any point in $R^3 - K$, then the fundamental group $\pi(R^3 - K, p_0)$ is called the *group* of $K$. Since different basepoints yield isomorphic groups, it is usual to speak of the group $\pi(R^3 - K)$ of the knot.

Suppose that the knot $K$ is in regular position and that it contains $n$ double points. Each double point is the image of two points of the knot: an *overcrossing point* and an *undercrossing point*. The $n$ undercrossing points divide $K$ into $n$ arcs $A_1, \ldots, A_n$. Assume that $K$ has been given an orientation. Let $x_j$ denote the element of $\pi(R^3 - K)$ represented by a loop which encloses $A_j$ just once in the direction of a left-hand screw. The set $X = \{x_1, \ldots, x_n\}$ is a set of generators for $\pi(R^3 - K)$. With each crossing of $K$ associate a word $r_k$ according to the following rules:

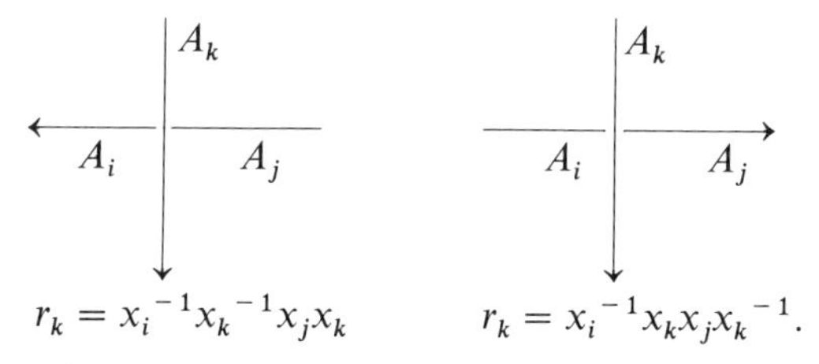


Then the presentation

$$\langle x_1, \ldots, x_m \mid r_1, \ldots, r_n \rangle$$

is the *Wirtinger presentation* for $\pi(R^3 - K)$. Note that any one of the relators can be derived from the remaining $n-1$ relators. The above account follows [**3**], to which the reader is referred for further details.

The program, given in this section, constructs the Wirtinger presentation for an alternating knot. It then attempts to eliminate as many redundant generators as possible by means of Tietze transformations. If the knot has $n$ crossings, it is described by an integer sequence of length $2n$, where the sequence is constructed from the knot as follows:

(i) Give the knot an orientation.
(ii) Number the crossings in any order.
(iii) Starting at any non-crossing point, traverse the knot in the direction determined by the orientation.

(iv) When a crossing (with number $i$, say) is encountered, add the term $\pm i$ to the sequence according to the following rules:

(a) If this is the first visit to crossing $i$, add $+i$ if it is an overpass, and $-i$ if it is an underpass.

(b) If this is the second visit to crossing $i$, add $+i$ if the oriented underpass makes a left hand screw with respect to the oriented overpass, and $-i$ otherwise.

(v) When the starting point is reached, the description will be complete.

This numerical description of a knot is due to Havas [5].

```
PROCEDURE knotgroup (knot;g);

    "Given a knot represented by the integer sequence knot, construct
    the Wirtinger presentation for the group g of the knot"

    "Let n denote the number of crossings. A description of the knot is
    built up in the integer sequences over, enter, and leave together with
    the Boolean sequence screw. The contents of these sequences are
    as follows:

      over[i]:  the number of the generator corresponding to the
                overpass at crossing i;

      enter[i]: the number of the generator corresponding to the
                approaching arc of the underpass at crossing i;

      leave[i]: the number of the generator corresponding to the
                departing arc of the underpass at crossing i;

      screw[i]: true if the oriented underpass makes a left hand screw
                relative to the oriented overpass at crossing i"

    "Initialize the sequences over, enter, leave and screw"

      l = length (knot);
      n = l/2;
      over = conseq (0,n);
      enter = conseq (0,n);
      leave = conseq (0,n);
      screw = conseq (true,n);

    "Define the contents of over, enter, leave and screw by scanning
    the knot"
```

```
FOR i = 1 TO l DO

    "Define cross to be the number of the current crossing and
    plus to be true if the number of the current crossing is
    positive"

    k = knot[i];
    cross = abs (k);

    IF k GE 0
       THEN plus = true;
       ELSE plus = false;
    END;

    "If the crossing is approached on the underpass, then the
    numbers of the entering and leaving arcs can be noted. The
    variable first is used to flag whether or not this is the first visit
    to the crossing"

    first = over[cross] EQ 0;
    IF (first AND NOT plus) OR (NOT first AND
    enter[cross] EQ 0)
       THEN
            IF i EQ 1
               THEN last = l;
               ELSE last = i-1;
            END;
            IF i EQ l
               THEN next = 1;
               ELSE next = i+1;
            END;
            enter[cross] = abs(knot[last]);
            leave[cross]=abs(knot[next]);
    END;

    "On the first visit to a crossing, over[cross] is assigned the
    crossing number to indicate that it has been visited. On the
    second visit screw[cross] is set true if the underpass makes a
    left-hand screw with respect to the overpass"

    IF first
       THEN over[cross] = cross;
       ELSE screw[cross] = plus;
    END;
```

```
    END; "i-th element of knot"

    "Define the group g"
    g: fgrank (n);
    rels = null;
    FOR i = 1 TO n DO
        p = enter[i];
        q = leave[i];
        r = over[i];
        IF screw[i]
          THEN
              rel:= g.p↑ - 1*g.r↑ - 1*g.q*g.r;
          ELSE
              rel:= g.p↑ - 1*g.r*g.q*g.r↑ - 1;
        END;
        rels = rels JOIN [rel];
    END;

    "Remove redundant generators"

    simplify (g,rels;rels);

    "Any one of the relators is redundant so the longest is removed"
    maximum = 0;
    FOR EACH w IN rels DO
        IF length(w) GT maximum
          THEN
              longest = w;
              maximum = length(w);
        END;
    END;

    rels = rels - [longest];

    "Define rels to be the presentation for g"

    g.relations: rels;

END; "procedure knotgroup"


PROCEDURE search (w,x;solve,inv,p);
```

```
"Search sets the Boolean variable solve true if, either,
  (a) the generator x occurs exactly once in the word w and not
      at all in w↑ − 1; or
  (b) the generator inverse x↑ − 1 occurs exactly once in the
      word w and not at all in w↑ − 1.
If condition (a) holds, then the Boolean variable inv is set false
and the integer variable p gives the position of x in w, while if (b)
holds, then inv is set true and p gives the position of x↑ − 1 in w."

solve = false;
inv = false;
p = match (w,x,1);
IF p NE 0
  THEN
      IF match (w,x,p + 1) EQ 0
          THEN
             IF match (w,x↑ − 1,1) EQ 0
                THEN
                     solve = true;
                     inv = false;
                     RETURN;
             END;
      END;

  ELSE
      p = match (w,x↑ − 1,1);
      IF p NE 0
         THEN
             IF match (w,x↑ − 1,p + 1) EQ 0
                THEN
                     solve = true;
                     inv = true;
                     RETURN;
             END;
      END;
  END;

END; "procedure search"

PROCEDURE simplify (g,rels;rels1);

"Eliminate redundant generators from the set of relators rels and
return the simplified presentation as rels 1"
```

```
rels1 = rels;
IF card (rels1) LT 2
   THEN RETURN;
END;

"Perform eliminations until no relator contains either a single
occurrence of some generator, or a single occurrence of the
inverse of some generator"
elgen = null;
REPEAT

      minlen = 100000;

      "Take each relator in turn and look for a single occurrence of a
      generator or a generator inverse"

      FOR EACH w IN rels1 DO

            IF length (w) LT minlen
              THEN

                    FOR EACH x IN generators (g) DO

                          search (w,x;solve,inv,p);

                          "If w can be solved for x, check that the word
                          expressing x in terms of the other generators
                          is shorter than any previously found
                          solution"

                          IF solve
                             THEN
                                   w1 = w;
                                   x1 = x;
                                   inv1 = inv;
                                   p1 = p;
                                   minlen = length(w);
                                   BREAK;
                          END;
                    END; "generators"
            END;
      END; "relators"

      "Now eliminate generator x1"
```

```
    IF minlen LT 100000
      THEN
          u = subword (w1,1,p1 - 1);
          v = subword (w1,p1 + 1,length(w1) - p1);
          IF inv1
            THEN
                xword = v*u;
            ELSE
                xword = u↑ - 1*v↑ - 1;
          END;
          rels1 = eliminate (rels1,x1,xword) - [identity OF g];
          elim = true;
          elgen = elgen JOIN [x1];
      ELSE
          elim = false;
    END;
  WHILE (elim AND (card(rels1) GT 1));

  "Define the group g1 on the non-redundant generators of g"
  g1: fgrank (ngenerators(g) - card(elgen));
  genim = empty;
  j = 0;
  FOR EACH x IN generators(g) DO
    IF x IN elgen
      THEN
          genimg = append (genimg, identity OF g1);
      ELSE
          j = j + 1;
          genimg = append (genimg, g1.j);
    END;
   END;

   DEFINE f: g TO g1 BY IMAGES genimg;
   newrels = null;
   FOR EACH w in rels1 DO
         newrels = newrels JOIN [f(w)];
   END;

   rels1 = newrels;

END; "procedure simplify"
```

*Application:* Assume the three-lead four bight Turk's head knot is numbered and oriented as in Fig. 1. This knot is represented numerically by the sequence

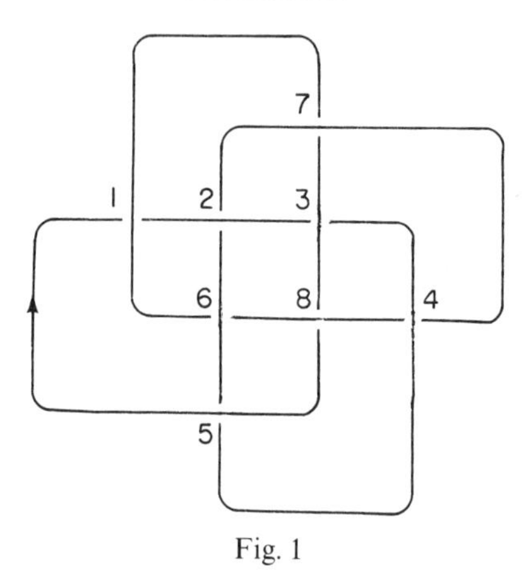


Fig. 1

$\{-1,2,-3,4,-5,6,-2,7,4,8,-6,1,7,-3,-8,5\}$. Starting with this sequence, the procedure *knotgroup* first constructed the Wirtinger presentation:

$$\langle x_1,x_2,x_3,x_4,x_5,x_6,x_7,x_8 \mid x_5^{-1}x_1^{-1}x_2x_1, x_6^{-1}x_2x_7x_2^{-1}, x_2^{-1}x_3x_4x_3^{-1}, x_7^{-1}x_4^{-1}x_8x_4, x_4^{-1}x_5^{-1}x_6x_5, x_8^{-1}x_6x_1x_6^{-1}, x_1^{-1}x_7^{-1}x_3x_7, x_3^{-1}x_8x_5x_8^{-1}\rangle$$

The application of procedure *simplify*, followed by the removal of the longest relator, resulted in the following three-generator, two-relator presentation:

$$\langle x_5,x_6,x_8 \mid x_6^{-1}x_8x_6x_5^{-1}x_6^{-1}x_8^{-1}x_6x_8x_5x_8^{-1}x_5^{-1}x_6x_5x_8x_5^{-1}x_8^{-1}, x_6^{-1}x_8^{-1}x_6x_5^{-1}x_6^{-1}x_5x_8^{-1}x_5^{-1}x_6x_5x_8x_5x_8^{-1}x_5^{-1}x_6^{-1}x_5x_8x_5^{-1}x_6x_5\rangle$$

The calculation of the Wirtinger presentation took 2 seconds, while the simplification took a further 10 seconds. (These timings where obtained using a CYBER 170/730 computer).

## References


1. Butler, G. (1976). The Schreier algorithm for matrix groups. *In* "SYMSAC '76. Proc. of the 1976 ACM Symposium on Symbolic and Algebraic Computation" (Ed. R. D. Jenks), 167–170. Assoc. Comput. Mach., New York.
2. Cannon, J. J. (1982). (Preprint). A Language for Group Theory. University of Sydney.
3. Crowell, R. H. and Fox, R. H. (1963). "Introduction to Knot Theory". Ginn, New York.
4. Hall Jr., M. (1959). "The Theory of Groups". Macmillan, New York.
5. Havas, G. (1974). Computational Approaches to Combinatorial Group Theory. Ph.D. Thesis, University of Sydney.
6. Havas, G. and Newman, M. F. (1980). Applications of computers to questions

like those of Burnside. *In* "Burnside Groups" (Ed. J. L. Mennicke). Lecture Notes in Mathematics, vol. **806**, 211–230. Springer-Verlag, Berlin.

7. Havas, G. and Sterling, L. S. (1979). Integer matrices and abelian groups. *In* "Symbolic and Algebraic Computation" (Ed. E. W. Ng.). Lecture Notes in Computer Science, vol. **72**, 431–451. Springer-Verlag, Berlin.
8. Newman, M. F. (1976). Calculating presentations for certain kinds of quotient groups. *In* "SYMSAC '76, Proc. of the 1976 ACM Symposium on Symbolic and Algebraic Computation". (Ed. R. D. Jenks), 2–8. Assoc. Comput. Mach., New York.
9. Robertz, H. (1976). Eine Methode zur Berechnung der Automorphismengruppe einer endlichen Gruppe, Diplomarbeit, RWTH Aachen.
10. Sims, C. C. (1971a). Determining the conjugacy classes of a permutation group. *In* "Computers in Algebra and Number Theory" (Eds. G. Birkhoff and M. Hall Jr.), *SIAM-AMS Proc.* **4**, 191–195. Amer. Math. Soc.
11. Sims, C. C. (1971b). Computation with permutation groups. *In* "Proc. of the Second Symposium on Symbolic and Algebraic Manipulation" (Ed. S. R. Petrick), 23–28. Assoc. Comput. Mach., New York.

# More on Moonshine

S. P. NORTON

## 1 Introduction

This paper is a continuation of [1]. The topics discussed here will be further developments with replication formulae of the type given in section 8 of [1], and Rademacher constants, which were mentioned only parenthetically in section 11 of [1]. The notation will be essentially the same, but certain concepts will be re-defined for the sake of convenience. Proofs will often be omitted. It is, however, convenient to record here that an existence proof for the Monster simple group of section 1 of [1] has been obtained by Griess [2]. Thus, although an explanation of the "Moonshine" phenomenon would help to explain the existence of the Monster, it would not be required to prove it.

## 2 The Bivarial Transformation

Let

$$f(z) = q^{-1} + H_0 + H_1 q + H_2 q^2 + \dots, \text{ with } q = e^{2\pi i z},$$

where initially the coefficients are arbitrary and we ignore problems of convergence by considering our power series as purely formal. We define $H_{m,n}$ by the formula

$$F(y, z) = \log(f(y) - f(z)) = \log(r^{-1} - q^{-1}) - \sum_{m,n=1}^{\infty} H_{m,n} q^m r^n,$$

$$\text{where } r = e^{2\pi i y}$$

Then $\frac{1}{n} q^{-n} + \sum H_{m,n} q^m$ is the coefficient of $r^n$ in

$$-\log(1 - r/q) + \log(r^{-1} - q^{-1}) - \log(f(y) - f(z)) = -\log r - \log(f(y) - f(z)),$$

COMPUTATIONAL GROUP THEORY
ISBN 0-12-066270-1

so that it is a polynomial in $f(z)$. This polynomial, which we call $P_n(z)$, is of the kind we saw in section 8 of [1], so it is not surprising that the replication formulae can easily be expressed in terms of the $H_{m,n}$. It can be seen that $H_{1,n} = H_n$ (hence the notation) and that $H_{m,n} = H_{n,m}$, and that the expansion of this in the $H_i$ has leading term $H_{m+n-1}$. It is also easy to see that $(m,n)H_{m,n}$ has integer coefficients. We call the transformation from the $H_i$ to the $H_{m,n}$, and from $f$ to $F$, the bivarial transform as it goes from functions of one variable to functions of two.

First we consider the problem of convergence. Within the domain of convergence of $f$, a singularity can only occur in the definition of $F$ where $f(y) = f(z)$, and if $y = z$ and $f$ has no multiple point there, it is cancelled out by the singularity of $\log(r^{-1} - q^{-1})$. In particular, if $f$ is a modular function, the double power series will be convergent if the imaginary parts of $y, z$ are greater than those of any multiple point of $f$. Also, if the automorphism group (called in section 5 of [1] the fixing group) of $f$ has genus 0 it is sufficient that $\mathrm{Im}(y)$ and $\mathrm{Im}(z)$ be greater than the imaginary part of any fixed point of any element of the group. (In practice, this gives the "correct" bounds for the growth rate of the $H_i$.)

## 3 Replicable Functions

We assume for the present that all coefficients are rational, though later we relax this assumption. Let us call a function $f$ replicable if $H_{a,b} = H_{c,d}$ whenever $ab = cd$ and $(a,b) = (c,d)$. The reason for this name is that this is exactly the condition required so that the operation of abstract replication, mentioned in section 9 of [1], can be performed.

To see this, we write $H_m^{(s)}$ for the coefficient of $q^m$ in the $s$th replication power of $f(z)$, which we call $f^{(s)}(z)$. The formula of section 8 of [1] can be written

$$H_{m,n} = \sum_{s|(m,n)} \frac{1}{s} H^{(s)}_{mn/s^2}$$

for which it obviously follows that our definition of replicability is necessary. From the Moebius transform of this formula, it is easy to see that it is also sufficient.

We conjecture that if a function has integral coefficients and zero constant term, it is replicable if and only if either it is the canonical Hauptmodul for a group satisfying the conditions of section 10 of [1] (i.e. of genus 0, containing $\Gamma_0(N)$ for some $N$ and containing $z \to z+k$ only when $k$ is an integer) or it satisfies $H_i = 0$ for $i > 1$. Some computational evidence for this is given in section 4 below. This conjecture will be extended to the case when there are certain irrationalities in the coefficients, for which replicability still has a natural definition.

We now generalize an earlier formula to define a new concept:

$$H_{m,n}^{(s)} = \sum_{t|(m,n)} \frac{1}{t} H_{mn/t^2}^{(st)}$$

Clearly this reduces to $H_m^{(s)}$ or $H_{m,n}$ when $n$ or $s$ is 1. It turns out that, for any $k$, the following conditions, where $s$ ranges over divisors of $k$, are equivalent:

(1) $f^{(s)}$ is replicable.
(2) The bivarial transform of $f^{(s)}$ is the generating function of $H_{m,n}^{(s)}$.
(3) In addition to condition (1), the $t$th replication power of $f^{(s)}$ is $f^{(st)}$.

If these conditions hold for all $s$, then we call $f$ completely replicable. This will be the case when the power series of $f$ is the Thompson series (defined in section 2 of [**1**]) of an element of the Monster, and also in many other cases, an example being $(j(2z)-1728)^{1/2}$, where $j$ is the standard modular function. However there are also replicable functions that are not completely replicable, such as $-j(z+\frac{1}{2})$. Its replicability follows immediately from that of $j$, but one may verify that its replication square is non-replicable.

There are many other formulae of interest. For example, from the definition of $H_{m,n}^{(s)}$, for any prime $p$ we have that

$$H_{m,np}^{(s)} - pH_{m/p,n}^{(ps)} = H_{mp,n}^{(s)} - pH_{m,n/p}^{(ps)}$$

where terms with non-integral suffices are defined to be zero. Note that this can be used to define $H_{m,n}^{(s)}$ inductively. The following equation can be derived from this by taking logarithms and equating coefficients. If $f$ and $f^{(p)}$ are replicable, then

$$(f^{(p)}(y)-f^{(p)}(pz)) \cdot \prod_{i=0}^{p-1} \left( f(y) - f\left(\frac{z+i}{p}\right)\right)$$

is symmetric in $y$ and $z$. Another formula is

$$F(y,z) = \log(f(y)-f(z)) = -2\pi i y - \sum_{a,c=0}^{\infty} \sum_{b=0}^{c-1} F^{(a)}\left(\frac{az+b}{c}\right) \cdot \frac{r^{ac}}{ac}$$

where, as in [**1**], $F(z)$ differs from $f(z)$ only in having a zero constant term in its power series. This formula holds for all replicable $f$.

If $f$ corresponds to a conjugacy class of the Monster whose order is not divisible by $p$, then $f$ is its own $p$th replication power. More generally, it appears that, since the coefficients are assumed to be rational, this is also true if $f$ is the canonical Hauptmodul for a group satisfying the conditions of the replicability conjecture with level (least value of $N$) not divisible by $p$. Under these circumstances, one of the above formulae reduces to

$$H_{mp,n} - pH_{m,n/p} = H_{m,np} - pH_{m/p,n},$$

which we call the self-replication formula (for the prime $p$).

Most of the above can be generalized to the case of cyclotomic coefficients. To do this, we define the Galois automorphism $*n$. For the case $n = p$ (a prime) the operation $*p$ will replace roots of unity not involving $p$ by their $p$th powers, and fix $p^i$th roots of unity. We use this to define $*n$ multiplicatively in terms of the factorization of $n$. With this notation, the condition for replicability is that $H_{a,b} * b = H_{c,d} * d$ whenever $ab = cd$ and $(a,b) = (c,d)$. As $H_{m,n}$ is symmetric in $m$ and $n$, this definition is plausible only if the action of $*n$ is involutory. We shall therefore assume that the Galois groups of all the coefficients have exponent 1 or 2 over the rationals. It follows immediately that $H_{m,n}$ is invariant under the operation $*mn$. If $H_{m,n}$, wherever it occurs, is replaced by $H_{m,n} * n$ (and similarly for terms with superscripts), then all equations written in terms of these symbols go through. As for the ones written in terms of the $f$s, the expression symmetric in $y$ and $z$ in the rational case will now be taken by the interchange of $y$ and $z$ to its image under $*p$. In the next equation, the coefficient of $r^n$ should be replaced by its image under $*n$. And self-replicability of $f$ means that $f^{(p)}$ is equal to $f * p$ rather than $f$.

We may summarize this by saying that

(1) Any modular function whose fixing group is "good" is replicable.
(2) If the power series of the function is the Thompson series of an element of the Monster, then it will be completely replicable, and its replication powers will correspond to the powers of the relevant conjugacy class in the Monster.
(3) If the level of the function is prime to $p$, it will satisfy the self-replication formula for the prime $p$.

Finally, we note that the logarithm of the cross-ratio $(f(x), f(y); f(z), f(t))$ is readily expressed in terms of the bivarial transform of $f$. The significance of this is that the action group of the automorphism group of the cross-ratio on one of the four variables, is the full converting group as defined in section 5 of [**1**]. This will usually be transitive on cusps, though it is not known whether this has any importance.

## 4 Computational Results

From now on we assume all coefficients to be integers. The self-replication formula implies that for any $y$,

$$(f(y) - f(pz)) \prod_{i=0}^{p-1} \left( f(y) - f\left(\frac{z+i}{p}\right)\right)$$

is a polynomial in $f(z)$, from which it follows that the same is true of any symmetric function of $f(pz)$ and the $f((z+i)/p)$. It now follows that $f(z)$ and $f(pz)$ are linked symmetrically by an algebraic equation of degree $p+1$. As this equation has only a finite number of parameters (in fact $\frac{1}{2}p(p+1)+1$), we can deduce all the coefficients from a finite number of them. Also, if we know (modulo $p^i$) the first $p^2n$ coefficients of a function which is self-replicable for the prime $p$, we can deduce the first $n$ coefficients modulo $p^{i+1}$. So if we can use the above method to deduce, from the first $n$ coefficients (modulo any power of $p$), those of the first $p^2n$ (to the same modulus) we can, by $p$-adic approximation, obtain the actual values of the first $n$ (and hence the entire series), if we know them modulo $p$. Furthermore, the same method will enable us to deduce all coefficients from a finite number if we know the $p$th replication power of our function. Finally, if a function is replicable, all the $H_i$ can be deduced from those with $i = 1, 2, 3, 4, 5, 7, 8, 9, 11, 17, 19, 23$. These results follow from the properties of the bivarial coefficients mentioned in section 2 above, together with further details concerning the extent to which the $H_{m,n}$ fail to be integral polynomials in the $H_i$. These results have formed the theoretical basis for the computations mentioned below.

The first of these related to functions satisfying the self-duplication and self-triplication formulae. All 81 possibilities for the values of $H_1$, $H_2$, $H_3$ and $H_5$ modulo 3 were fed in. From these, the values of various $H_i$ for $i$ up to 45 were obtained modulo 3. Then, using the process of 3-adic approximation described above, these coefficients were obtained modulo 19 683. In 29 out of these 81 cases, no contradiction was encountered during the process, and all of these were found to correspond to known functions. Next, for each of these 29 and, iteratively, for all functions produced in the process, replication square roots were extracted. For each extraction, as before, 81 possibilities were tested, and for those that worked, the values of the coefficients modulo 19 683 were obtained. Here, however, some of the series appear to be spurious, in that they do not correspond to replicable functions. All the same, some previously unknown replicable functions were obtained, such as $t_{58|2+}$ and $t_{82|2+}$ (in the notation of section 5 of [**1**]). Also, replicable functions modulo 2 have been completely classified. Here there are 4096 initial possibilities for the 12 basic coefficients mentioned above, but only 72 of these survived the replicability test. For each of these, a unique function of odd level was found, which would therefore also satisfy the self-duplication formula. In the Appendix we show the complete list of 72 functions. All the computations were performed by D. J. Seal.

## 5 A Graded Algebra?

In section 12 of [**1**], reference is made to the possibility of a graded algebra structure, where the dimension of the space corresponding to the $n$th grade is the $n$th coefficient of the $j$-function. We revise this speculation, and now suggest that it would be more natural to expect a doubly-graded algebra structure where the dimensions are the coefficients of the bivarial transform. The reasons for this belief are as follows:

(1) The coefficient $H_0$ of the $j$-function (and the other Thompson series) appears to have no canonical definition. But this constant disappears when we take the difference of two $j$-functions, as in the bivarial transform.
(2) The coefficients of the bivarial transform involve the very terms that appear so naturally in the replication formulae.
(3) Most convincingly, $H_{m,n}$ is the coefficient of $q^m$ in $P_n(z)$, while as $P_n(z)$ is a polynomial of degree $n$, the leading term in the expansion of $P_t(z).P_n(z)$ in terms of the $P_i(z)$ is $P_{t+n}(z)$. It is, therefore, not unnatural to expect a correspondence between, on the one hand, the product of the coefficients of $q^s$ and $q^m$ in $P_t(z)$ and $P_n(z)$ and, on the other hand, the coefficient of $q^{s+m}$ in the leading term of the product of $P_t(z)$ and $P_n(z)$. Therefore, the grading of the algebra—whatever it might be—would be expected to happen in the "right" manner. This is not the case with the $j$-function itself, where for example the product of $H_1$ and $H_2$ happens naturally not in $H_3$ but in $H_6$.

## 6 Rademacher Constants

We now turn to the topic of Rademacher constants. In section 11 of [**1**] it is stated parenthetically that there is a unique space, invariant under the positive elements of $PGL_2(\mathbb{Q})$, that is of co-dimension 1 in the space of all modular functions invariant under groups commensurable with the modular group itself. Every modular function will be expressible uniquely as a sum of a member of this subspace and a constant. The Rademacher constant of a modular function (whose power series may be assumed to have zero constant term) will then be the negative of the constant in the above decomposition of the function. It is the object here to show how this statement can be proved.

In [**3**], Rademacher obtains his famous formula for the coefficients of the $j$-function. He observes that this formula provides a meaningful answer for the constant term (namely 24), which is, however, wrong for the $j$-function as classically defined (for which the constant term is 744). However, as he says, addition of a constant to the $j$-function does not affect its behaviour at cusps, from which his formula was deduced.

The Rademacher method can be applied to any modular function, and it will be fairly clear that it satisfies the additivity and invariance properties. All that remains to be proved is that there can be no other map satisfying these properties.

Let us consider such a map, and assume that its value is known on the $j$-function. The sum of $j(nz)$ and its images under the modular group is a polynomial of degree $n$ in $j(z)$, so by induction on $n$ we can express the $n$th power of the $j$-function as a sum of $j$-functions evaluated at linear transforms of $z$ (plus a constant). We will, therefore, be able to evaluate the map on any function invariant under the full modular group. If a function is invariant under a group of finite index therein, the sum of its images will be invariant under the full group, and therefore the map will have a known value on this sum. As each image is related to all the others by a linear transformation on $z$, the map will have the same value on each, so that its value on the original function will also be known. We have, therefore, shown that there is at most one parameter involved.

Finally, we have the following formulae, where capital letters are, following [**1**], used to denote functions with zero constant terms:

$$J(z) = T_{2+}(z) + T_{2+}(z/2) + T_{2+}((z+1)/2)$$

$$T_{2+}(z) = T_{2-}(z) - T_{2-}(z/2) - T_{2-}((z+1)/2)$$

$$T_{2+}(z) = T_{2-}(z) + T_{2-}(-1/2z) - 24.$$

If the Rademacher constants of $J, T_{2+}$ and $T_{2-}$ are called $a, b, c$ respectively we obtain the equations $a = 3b$, $b = -c$, and $b = 2c + 24$, from which we can deduce $a = 24$. Uniqueness follows immediately.

It should be noted that, although in a sense this defines a canonical value for the constant terms of modular functions, it can be seen from Table 4 of [**1**] that the corresponding class function on the Monster fails to be a character and is, in fact, not even integer valued. The statement, above, that the coefficients of the $j$-function were unlikely to correspond to a graded algebra therefore still stands.

## Appendix

In the following table each row represents a replicable function of odd level. We start with the parities of the first 100 coefficients, with each character representing 4 binary digits (0–9 as in their binary expansion, $A$–$F$ as in the binary expansion of 10–15 respectively). This is followed by the name. We have, where possible, stuck to the naming system employed in [**1**], but there are several functions that fall outside this classification, for which it has been attempted to make the names suggestive. Finally, except for the $j$-function (for which the values are too large) we have given the exact values of $H_1, H_2, H_3$ and $H_5$.

*Table 1*

| | | | | | |
|---|---|---|---|---|---|
| 000000000000000000000000 | 0− | 0 | 0 | 0 | 0 |
| 000002000000000000020000020 | 3¦3 | 0 | 248 | 0 | 4124 |
| 002002002002000000000020020 | 9¦3− | 0 | −4 | 0 | 2 |
| 002080002082082080080000 | 63¦3+ | 0 | 0 | 0 | 2 |
| 020200020002020002000200 | 1+ | 196884 | ETC. | | |
| 022202202022202202200220222 | 13+ | 12 | 28 | 66 | 258 |
| 080000080002000000208000020 | 9¦3+ | 0 | 14 | 0 | 65 |
| 080008208000000000800000800 | 27¦3+ | 0 | 2 | 0 | 5 |
| 082000002000002000000000020 | 15¦3 | 0 | −2 | 0 | −1 |
| 082080002080002080082002020 | (81¦3+)' | 0 | 2 | 0 | −1 |
| 092490002402400080012082 4 | 93¦3+ | 0 | 0 | 0 | 1 |
| 0AA2A8282820AAA280880AAA0 | 15+ | 8 | 22 | 42 | 155 |
| 1E522D4E93803E0B6082C9C79 | 39+39 | 2 | 2 | 4 | 7 |
| 200202200200000202200002 0 | 3− | 54 | −76 | −243 | −1384 |
| 202200222220222002022020 2 | 5+ | 134 | 760 | 3345 | 39350 |
| 20A2A8882A2AA2A8820000228 | 33+ | 2 | 4 | 5 | 14 |
| 220020022000022000202220 0 | 7− | 2 | 8 | −5 | −10 |
| 222020200002200222002200 0 | 5¦5 | −6 | 20 | 15 | 0 |
| 288A88200800A0A028AA082A0 | 19+ | 6 | 10 | 21 | 61 |
| 2A20028A8AA2A0A2280A0A220 | 21+ | 6 | 6 | 15 | 41 |
| 2A802822AA08828202A880022 | 41+ | 2 | 2 | 3 | 7 |
| 3FD0E330B8EBC3949709768A2 | 119+ | 0 | 0 | 1 | 1 |
| 4000120124804004120920124 | 45¦3+15 | 0 | 1 | 0 | 2 |
| 4020924100020820800104804 | 117¦3+ | 0 | 1 | 0 | 0 |
| 4104824124824924804824820 | (81¦3+)" | 0 | −1 | 0 | 2 |
| 4124020904800924024004800 | 45¦15+ | 0 | −1 | 0 | 0 |
| 4124820900024900820800804 | 21¦3+ | 0 | 3 | 0 | 8 |
| 4804104920920004820924120 | 27¦3− | 0 | −1 | 0 | −1 |
| 4900004120824020820000824 | 39¦3+ | 0 | 1 | 0 | 3 |
| 4920020104024820800820824 | 57¦3+ | 0 | 1 | 0 | 1 |
| 4920800900804820020020804 | 9− | 0 | 5 | 0 | −7 |
| 4B2AE0A108282A0003800000C | 25+ | 4 | 5 | 10 | 25 |
| 56DB3487EEBF0EB478730A3DA | 49+ | 2 | 1 | 2 | 4 |
| 61C6E289292AC3A8A580A048C | 35+35 | 2 | 3 | 5 | 10 |
| 694A6E0121AAC28A29A203CA4 | 55+ | 2 | 1 | 1 | 3 |
| 6BC0808B01AA68220020AA804 | 45+ | 2 | 1 | 3 | 5 |
| 741746597251 56C87084B00DB | 87+ | 0 | 1 | 1 | 2 |
| 7C3BA0EB0A2F5CCEAA24621D7 | 23+ | 4 | 7 | 13 | 33 |
| 8000000000000000000000000 | 0+ | 1 | 0 | 0 | 0 |
| 8028020020020008008028028 | (9¦3)+ | 9 | −4 | 0 | 2 |
| 80A00AA20222A020202282220 | 21+21 | 5 | 8 | 16 | 44 |
| 82222A8882208A28822208080 | 35+ | 1 | 4 | 6 | 10 |
| 8882008202088200020280000 | 5− | 9 | 10 | −30 | −25 |
| 88AA22022A8AA2A8222820080 | 51+ | 1 | 2 | 2 | 5 |
| 8A08820A08008208808208800 | (27¦3+)* | −3 | 2 | 0 | 5 |
| 8A20208020208220000000220 | (15¦3)+ | 5 | −2 | 0 | −1 |
| 8AA8088802008A0A0000820A2 | 11+ | 17 | 46 | 116 | 533 |
| 9424094A50040121084204A50 | 25− | −1 | 0 | 0 | 0 |
| 9C8ECD3228EEA74B01A881A93 | (63+9)*63 | −1 | 2 | −2 | 5 |
| A0088A088A82A8820A8082802 | 15¦5+ | 3 | 2 | −3 | 0 |
| A08802A02220A200000082220 | 13− | −1 | 2 | −1 | 2 |
| A0A082800202A002002080022 | 17+ | 7 | 14 | 29 | 92 |
| A20202A002008202022200020 | 9+ | 27 | 86 | 243 | 1370 |
| A80A222AA2028A22A222A2280 | 29+ | 3 | 4 | 7 | 17 |
| A88A0AA00A0080020A28880A0 | 3+ | 783 | 8672 | 65367 | 1741655 |
| AA08A882A08802A88020A2200 | 7+ | 51 | 204 | 681 | 5135 |
| AAA020802028A222002082220 | 15+5 | −1 | 4 | −3 | 11 |
| BD264832EA8B0182226A1A207 | 47+ | 1 | 2 | 3 | 5 |

*Table 1 (cont.)*

| | | | | | |
|---|---|---|---|---|---|
| BE28B34C322A8A2B0A7A20AB3 | 95+ | 1 | 0 | 1 | 1 |
| C7E8432D8B2A6CA229A0A8C36 | 15+15 | 9 | 19 | 42 | 146 |
| CD4C8907854A87E82480E021A | 59+ | 1 | 1 | 2 | 3 |
| CF66C10509C26DA20FA22A416 | 105+ | 1 | 1 | 0 | 1 |
| D24932016CA4C20C321B20325 | (45¦3+15)+ | −1 | 1 | 0 | 2 |
| D84D345B2096104C860B64368 | (27¦3)+ | 3 | −1 | 0 | −1 |
| DA67FABBBCA802AE0699C0620 | 31+ | 3 | 3 | 6 | 13 |
| E58A8B01A088682808A88A81C | 27+ | 3 | 5 | 9 | 20 |
| E728ED292AA84400412AA903C | 33+11 | −1 | 1 | −1 | 2 |
| E780E1EBA00A64A0ABAC0A2BE | 39+ | 3 | 1 | 3 | 6 |
| EF26C30188CA6D808B0008896 | 21+3 | −1 | −1 | 1 | −1 |
| F22D663310B608C02AC22438E | 35¦5+ | 1 | −1 | 1 | 0 |
| F309B305B629683903CA22407 | 71+ | 1 | 1 | 1 | 2 |
| F82140C932ACF6EC8202C83CD | 69+ | 1 | 1 | 1 | 3 |

## References


1. Conway, J. H. and Norton, S. P. (1979). Monstrous Moonshine, *Bull. London Math. Soc.* **11**, 308–339.
2. Griess, R. L. (1982). The Friendly Giant, *Inv. Math.* **69**, 1–102.
3. Rademacher, H. (1938). The Fourier coefficients of the molecular invariant $J(T)$, *Amer. J. Math.* **60**, 501–512.

# CAS; Design and Use of a System for the Handling of Characters of Finite Groups

J. NEUBÜSER, H. PAHLINGS AND W. PLESKEN

## 1 Introduction

The task of constructing the character table of a finite group $G$ may occur in rather different situations. The most direct access is possible if elements of $G$ can be multiplied and compared; then if $G$ is small enough, its conjugacy classes may be determined and the class-multiplication-coefficients can be calculated. The irreducible (in this paper this will always mean $\mathbb{C}$-irreducible) characters of $G$ can then be found as the common eigencolumns of the matrices of class-multiplication coefficients. McKay [**35**] reports that he determined the character table of $J1$ (of order 175 560) in this way. Dixon [**15**] describes an algorithm by which numerical difficulties of this approach are avoided, by first calculating these eigencolumns in some finite field $\mathbb{Z}_p$ and then translating character values back into $\mathbb{C}$. Dixon's method has been implemented several times [**6**], [**14**]; it is available through the CAYLEY system [**7**], and we shall briefly come back to its being available to the CAS system in section 3.

More often, although $G$ is concretely given and its elements can be handled, $G$ is too big for direct calculation of class-multiplication coefficients from the elements (e.g. $G = O_{11}(2)$ of order $2^{25}.3^6.5^2.7.11.17.31$), or even only a (possibly hypothetical) description of $G$ by some properties such as the existence of certain subgroups is known. This was the case, for instance, when the character table of the Monster was determined without any concrete description of its elements, even before the existence of the Monster had been proved [**25**].

The construction of character tables in the last two situations usually starts from the (sometimes only partial) knowledge of only a few characters and makes use of the many strong interrelations between characters. Such constructions involve a great amount of numerical computation with character values, often investigating various alternatives in a trial-and-error fashion. In several cases these computations have been assisted by computer routines

COMPUTATIONAL GROUP THEORY
ISBN 0-12-066270-1

that could perform all or some of the necessary operations; however, rather little information is available on the design and scope of such systems of computer routines [**17**], [**22**], [**23**], [**14**], [**26**], [**4**].

In this paper we are, therefore, going to describe in some detail the system CAS (**C**haracter **A**lgorithm **S**ystem) that has been developed in Aachen since 1979 and the way it can be used. The bulk of the implementation was done by Aasman [**1**], Janißen [**32**] and Lammers [**33**] until March 1982; additional features are presently implemented by E. Cleuvers and M. Heinen. The design of CAS has been influenced in particular by the programs of D. Livingstone's team at Birmingham and those of T. Gabrysch at Bielefeld. We have also used the experience gained in some pilot projects in Aachen which were implemented by Esper [**17**] and Deeken [**14**]. To all of them and to many others who gave us valuable advice, in particular J. H. Conway, J. S. Frame, A. Kerber, J. McKay and W. Wynn, we would like to express our thanks.

As compared with its predecessors known to us, CAS incorporates some innovations; for instance, in the construction of new irreducible characters from known ones, and with respect to the ability to handle irrational character values. The latter in particular allows more information to be obtained from the table, e.g. the *p*-blocks.

CAS is mainly written in FORTRAN EXTENDED IV, its present size is about 60 000 Fortran cards. CAS is built on Cannon's STACKHANDLER [**8**] as storage allocation system. As this (and a few special features) have to be tuned for the host machine, at present there are four versions running on a (CDC-)CYBER 175, on an IBM 370, on an ICL 2980, and on a DEC 2060.

The CYBER-implementation was developed at the Rechenzentrum der RWTH Aachen, the adaption to the IBM at the Rechenzentrum der Kernforschungsanlage Jülich. We thank both for most helpful co-operation. We also thank the Deutsche Forschungsgemeinschaft for generous financial support of the project.

CAS is available on magnetic tape from Lehrstuhl D. A user manual [**3**] and a programmer's handbook [**2**] can be obtained with the tape (we regret that we must ask for refund of expenses).

In section 2 of this paper we sketch the representation of character values in the computer, the design of some main data structures and the operating language of CAS. Section 3 describes the ideas of some main CAS operations. In section 4 we give examples of their use by a number of typical applications in the construction of character tables, while in section 5 we describe how information on decomposition numbers, Molien series and subgroups can be obtained from a character table, using CAS. Finally, in section 6 we report about the project of collecting a comprehensive library of character tables in connection with CAS. As an example of possible output of CAS, a microfiche supplement contains the character tables (with various additional information) of the 26 sporadic simple groups. It also contains some computer output illustrating various ways of using CAS that are described in sections 4 and 5.

Facts on representation theory used in this paper which are not explicitly referenced, can be found e.g. in [**29**]. The notation for simple groups follows [**27**]. When referring to "the character table" of a group $G$, we always mean a table whose columns correspond to the conjugacy classes of $G$ and whose rows correspond to the $\mathbb{C}$-irreducible characters of $G$.

## 2 Arithmetic, Data Structures, and Language of CAS

In this section we only want to point to some main features of the design of the data structures and the operating language for CAS. The user manual [**3**] gives sufficient information to use CAS through its language; while details about the design of its arithmetic can be found in Chapter 2, of its data structures in Chapter 3, of its language in Chapter 4, about syntax recognition and interpretation in Chapter 5 of the "Handbuch" [**2**].

The first decision that had to be made in the design of CAS, concerned the way in which character values were to be represented in the computer. There are two well-known difficulties: character values can be fairly large integers—the highest degrees of irreducible characters of the Monster are 27 digit decimal numbers—and they can be irrational. The first one is solved in practically all systems for the handling of characters by the inclusion of some multiple precision package for integers. The second, however, is usually circumvented by restricting calculations to rational valued characters [**17**], [**22**], [**23**], [**26**]. Therefore, with such systems, often only "rationalized character tables" are produced. A row of such a table is the sum of all algebraic conjugates of an irreducible character while a column corresponds to a conjugacy class of cyclic subgroups. The determination of the "irrationalities" is then left to hand calculation (cf. Example 4.6).

In view of the applications of algebraic number theory in representation theory, we did not want CAS to be restricted in this way nor did it seem advisable to use only FORTRAN floating point representation for irrational numbers. Rather, we decided to represent irrational character values in exact algebraic form in the following way.

Let $\chi$ be a character of the group $G$ and $g \in G$ of order $|g| = n$. Let $\zeta$ be a primitive $n$th root of unity, e.g. $\zeta = \exp(2\pi i/n)$ then $\chi(g)$ can be uniquely expressed as

$$\chi(g) = \sum_{k=0}^{\phi(n)-1} a_k \zeta^k$$

with $a_k \in \mathbb{Z}$ and $\phi(n)$ the Eulerian function of $n$. Hence $\chi(g)$ can be uniquely represented by a sequence of $\phi(n)$ integers and since $\mathbb{Z}[\zeta] \cong \mathbb{Z}[x]/\Phi_n(x)$, where $\Phi_n(x)$ is the $n$th cyclotomic polynomial, addition and multiplication of such irrational character values can be implemented as addition and multiplication of integral polynomials, followed, if necessary, by reduction modulo $\Phi_n(x)$.

Since in an expression

$$\sum_k a_k \zeta^k$$

quite often $a_k = 0$ for many $k$, such an expression is stored in CAS as a list of pairs $(k, a_k)$ with $a_k \neq 0$. For both rational character values and integral coefficients in the above-mentioned polynomials, an arithmetic with practically unlimited precision (in the present CYBER-implementation allowing integers $x$ with $|x| < 2^{48\,000}$) has been implemented. In the same CYBER-implementation, $n$ is allowed the maximum value 4095.

Conversion of this representation of irrational numbers to floating point representation has been implemented and is used to speed up computations, when no information is lost this way.

While the scope for character values thus obtained has proved most valuable for several applications, some of which we are going to describe later—and in fact sufficient for all tasks that we have encountered so far—one has to bear in mind that large integers and irrational numbers are the exception rather than the rule among the character values in the character tables of most groups that one has usually to deal with, e.g. non-abelian simple groups, their covers and extensions. Therefore, in order to save space, a character of a group $G$ is basically represented by the "character rump" which is a sequence of $k$ "small" fields (in the CYBER-implementation consisting of one machine word each) corresponding to the $k$ conjugacy classes of $G$ in some fixed order. Each field first of all contains a mark indicating the "type" of the character value on the corresponding class, namely if it is a "short" integer $x$ (in the CYBER-implementation $|x| < 2^{48}$), a "long" integer $x$ (in the CYBER-implementation $2^{48} \leqslant |x| < 2^{48\,000}$), an irrational number, or if it is the "undefined" symbol $*$, which is used e.g. if a character value is not yet known. In the first case this field will also contain the actual character value, while in the next two cases it will hold a pointer to a place in the "character tail" which is a list in which all the "unhandy" values of this character are gathered. The character rump is preceded by a "character head" which contains general information about the character, e.g. whether all character values are short integers or about the length and structure of the character tail.

In CAS, this same data structure "character head", "character rump", "character tail" is used for various kinds of functions which are defined either on the set of conjugacy classes or on the set of irreducible characters of a group $G$ and which have their values in some $\mathbb{Z}[\zeta] \cup \{*\}$, where $\zeta = \exp(2\pi i/n)$ is some primitive root of unity and $*$ is the "undefined" symbol. Functions defined on the classes can e.g. be irreducible characters, reducible characters, generalized characters, central characters, centralizer orders or class lengths; functions on the irreducible characters can e.g. be the columns of the character table or the Schur-Frobenius indicators.

The "undefined" value $*$ allows calculation with this data structure even if the number of classes, say, is not yet known, by introducing a certain number of fictitious classes on which all class functions get the "undefined" value for the time being. In order to facilitate typical computations with incomplete rows and columns of a character table the "undefined" symbol is included in the arithmetic. Usually its sum or product with any $x \in \mathbb{Z}[\zeta]$ is defined as $*$, except for $0 . * = * . 0 = 0$. However in certain cases, e.g. if one is trying to find missing values of a row or column of the character table by use of orthogonality relations, it can be left out in the formation of a sum.

Because of the various kinds of representations of numbers, division is the technically most complicated part of the arithmetic in CAS; it suffices to mention here that for $x, y \in \mathbb{Z}[\zeta]$ with $x/y \notin \mathbb{Z}[\zeta]$, the quotient is put "undefined".

Functions defined on the same set, for instance irreducible characters, can be gathered in "tables". Each table again has a table head, which inter alia specifies its maximal size, its present contents, the kind of function it is designed for and information necessary to work with these functions. For instance, if the table holds characters, a particular place in the table head holds a pointer to a list of cyclotomic polynomials corresponding to the conjugacy classes which are needed for calculations with the respective character values.

Since CAS is designed for working with groups with up to 1000 conjugacy classes (and possibly more), it is mandatory that such tables are kept on backing store. On a computer without virtual memory, therefore, data structure and programs for tables involve a good deal of file organization, (which in the CYBER implementation is not restricted to standard FORTRAN and hence has to be adapted when CAS is transferred to a different type of computer). This file organization is also reflected in the table head. CAS can handle several (in the present implementation up to 10) tables simultaneously.

There is one more particular kind of data structure in CAS, namely "mappings" from one set of classes or irreducible characters into some other (or the same) set of classes, or irreducible characters respectively. These are used to store power maps and fusions.

CAS is equipped with a problem oriented language of its own which takes into account that CAS

(i) is primarily designed for interactive use,
(ii) should be easily extendable by incorporation of new algorithms,
(iii) should be easy to use through this language.

A number of language elements are distinguished which partly reflect the data structures that have been discussed, *inter alia*: constants (integer, long integer, irrational, and undetermined); names; character tables; maps; lists; texts; identifiers and functions (in the sense of a programming language, i.e. functions like the scalar product of two characters to be used in arithmetic expressions).

Three kinds of statements are distinguished: instructions, assignments and procedure calls. Because of its design for interactive use, the majority of the statements are instructions which invoke the rather global execution of a certain algorithm on specified data structures, like peeling all known irreducible characters from a table of reducible ones. We shall describe the more important ones in Section 3. Assignments, e.g. assignments of arithmetic expressions to an identifier, on the other hand, allow the user to define local operations and to build up procedures. For these, also, simple expressions of identifiers can be formed using logical operators. Procedures are then built from three kinds of procedure blocks: sequential blocks, the statements of which are executed in sequential order; selection blocks whose statements are executed if a condition is satisfied; and repetition blocks which are repeatedly executed depending on a condition.

These facilities allow short programs to be written in the CAS language. However, as CAS is to be used mainly interactively, the language is interpreted rather than compiled.

## 3 Some Main CAS Operations

In this chapter we give an overview of the operations available in CAS. Information on the syntax of the corresponding instructions, sufficient to use them, is found in [**2**]. With some arbitrariness in certain cases we can sort the CAS operations by their tasks.

### 3.1 System control

There are presently 14 instructions for this purpose, used to enter and leave the system, to organize input and output, to handle files and to get intermediate information about the state of the computation.

### 3.2 Table organization

The about 20 instructions of this kind are used to set up a new table or to recover it from file, to add and delete specified parts of a table, to fuse or to split classes, to permute rows or columns according to specification or to certain rules, to transpose a table, to print a table according to specification and to record in a tablehead which operations have already been performed with this table.

### 3.3 Generation of complete character tables

(i) *Generation from generic formulae*

For certain families of groups, the character tables of all groups in the family have been described by generic formulae depending on parameters specifying the group within the family. CAS presently contains implementations of such generic formulae for the following families (the source, from which the formulae were taken, is given in parentheses):

| | |
|---|---|
| $C(n)$ | the cyclic group of order $n$ (the dihedral groups of order $2n$ will be added soon); |
| $PSL(2,q)$ | for all $q = p^r$, $p$ prime, [**36**] ($PSL(2,q) \cong PSU(2,q^2)$, see e.g. p. 194 of [**27**]; |
| $SL(2,q)$ | for all $q = p^r$, $p$ prime, [**16**] ($SL(2,q) \cong SU(2,q^2)$, see e.g. p. 194 of [**27**]; |
| $PGL(2,q)$ | for all $q = p^r$, $p$ prime, [**46**]; |
| $GL(2,q)$ | for all $q = p^r$, $p$ prime, [**46**]; |
| $PSL(3,q)$ | for all $q = p^r$, $p$ prime, [**43**]; |
| $PSU(3,q^2)$ | for all $q = p^r$, $p$ prime, [**43**]; |
| $SL(3,q)$ | for $q = p^r$, $p$ prime, and $q-1 \not\equiv 0 \pmod 9$ ($SL(3,q) \cong PSL(3,q)$ iff $q-1 \not\equiv 0 \pmod 3$), see e.g. p. 178 of [**27**], generic formulae for $q-1 \equiv 0 \pmod 3$ in [**43**], however these yield incorrect results for some $q$ with $q-1 \equiv 0 \pmod 9$); |
| $SU(3,q^2)$ | for $q = p^r$, $p$ prime, and $q+1 \not\equiv 0 \pmod 9$ ($SU(3,q^2) \cong PSU(3,q^2)$ iff $q+1 \not\equiv 0 \pmod 3$), see e.g. p. 242 of [**27**], generic formulae for $q+1 \equiv 0 \pmod 3$ in [**43**], however these yield incorrect results for some $q$ with $q+1 \equiv 0 \pmod 9$). |

In the CAS language, the construction of the table of one of these groups is invoked by an assignment statement, the table can then be handled further in CAS under the name given to it by the assignment.

(ii) *Generation by the Dixon algorithm*

For a "small" group $G$ (of order up to several 1000s, say, and with not too many conjugacy classes, up to about 150, say) Dixon's algorithm, which has already been mentioned in the introduction, can be used to construct the character table of $G$. For CAS this is available only via a link to the CAYLEY system [7]. To CAYLEY, $G$ can either be given by "concrete" generators, e.g. permutations or matrices over $\mathbb{Z}$ or a finite field, or $G$ can be defined by a finite presentation. From these the classes of $G$ will be determined in CAYLEY, the Dixon algorithm is implemented as a "special function" of CAYLEY [**14**], see also Chapter 9 of [**2**]. The complete character table is handed to CAS on a "local file".

(iii) *Generation from other character tables*

CAS contains instructions to form the character table of the direct product $G_1 \times G_2$ from the character tables of $G_1$ and $G_2$ (instruction: DIRECT) and to form the character table of a factorgroup $G/N$ from the character table of $G$, when $N$ is defined as generated by certain classes of $G$. (Instruction: MODULO)

### 3.4 Generation of characters

In this section we summarize methods that are used in the determination of new characters from known ones. For some of the standard operations no further explanation is needed, for some less well-known ones we sketch the mathematical background. For all operations, options are available to apply them to prescribed lists of characters.

(i) Using the exact algebraic representation of irrational character values, Galois conjugates of a given character can be found and the complete set of them can be added up to form a rational character. (Instruction: GALOIS)

(ii) From an outer automorphism $\alpha$ and a character $\chi$ of a group $G$ the character $\chi^\alpha$ defined by $\chi^\alpha(g) := \chi(\alpha(g))$ can be found. (Instruction: OUTO)

(iii) For a given $n \in \mathbb{N}$, the generalized character $\chi^{(n)}$ defined by $\chi^{(n)}(g) := \chi(g^n)$ (power character) can be formed, provided the $n$th power map is known. (Instruction: POWER)

(iv) Products of characters, i.e. characters of tensor products of representations can be formed. (Instruction: TENSOR)

(v) The usefulness of the—perhaps not so well known—operations to be discussed next, is indicated by the following theorem (Burnside-Brauer, see e.g. p. 49 of [**29**]). Let $\chi$ be a faithful character of the finite group $G$, having $m$ distinct values $\chi(g)$, then every irreducible character of $G$ is a constituent of at least one of the tensor powers $\chi^i$ of $\chi$ with $0 \leqslant i < m$.

No general methods are known to determine these irreducible constituents of tensor powers, however, with $\chi$ of degree $n$, say, Schur's formulae for the splitting of the $i$th tensor power of the natural representation of $\mathrm{GL}(n, \mathbb{C})$ into irreducible representations (see e.g. p. 81 of [**37**]) provide a partial splitting of $\chi^i$ into the so-called symmetrizations of $\chi$. If $\chi$ is the character of an orthogonal representation, Weyl's formulae [**47**], for the orthogonal group $O(n, \mathbb{R})$ provide an even finer splitting of the $\chi^i$, in particular for bigger values of $i$.

The symmetrization $\chi^{[\lambda]}$ ($= L_\lambda(\chi)$ in the notation of [**21**] as well as [**2**] and [**3**]) of the character $\chi$ of $G$ corresponding to the partition $\lambda = 1^{l_1} \dots m^{l_m}$ of $i \in \mathbb{N}$ is defined by the following formula. Let $\phi^\lambda$ be the character of the symmetric group $S_i$ corresponding to $\lambda$, and for $\pi \in S_i$, let $a_k(\pi)$ be the number of $k$-cycles of $\pi$, then for $g \in G$

$$(*) \qquad \chi^{[\lambda]}(g) := \frac{1}{i!} \sum_{\pi \in S_i} \phi_\lambda(\pi) \prod_{k=1}^{i} \chi(g^k)^{a_k(\pi)}.$$

It is clear from this, that the $k$th power maps for $2 \leqslant k \leqslant i$ are needed to compute any $\chi^{[\lambda]}$ with $\lambda$ a partition of $i$.

$\chi$ is the character of an orthogonal representation, iff the Schur-Frobenius indicator $v_2(\chi) = 1$. (cf. SCHUR instruction, section 3.6(v)). In this case the components for the orthogonal group are derived from the symmetrizations in [**38**], p. 119ff. Frame [**21**] (see also [**3**] and [**2**]), has given a recursive scheme for the computation of both symmetrizations and "orthogonal" components for $i \leqslant 6$, which greatly reduces the computational effort. This scheme is implemented in CAS as well as the direct evaluation of (*) for a given $i \leqslant 6$. (Instructions: FRAME, SYM, PLUS, MINUS)

A similar scheme for symplectic characters (which can be detected by $v_2(\chi) = -1$) will soon be implemented.

We shall demonstrate the power of these operations in Example 4.1.

(vi) The next three operations are well known again. If the fusion of the conjugacy classes of a subgroup $U < G$ into the conjugacy classes of $G$ is given, characters of $G$ can be formed by induction from characters of $U$. (Instruction: INDUCE)

Likewise, if the fusion of the classes of $G$ into the classes of a group $H > G$ is known, characters of $H$ can be restricted to $G$. (Instruction: RESTRICT)

In Example 4.2 we shall demonstrate the use of these operations and in Example 4.3 we shall show how CAS can be used to find such fusions.

Finally, if the fusion of the classes of $G$ into the classes of a factor group $G/N$ is known, characters of $G/N$ can be extended to characters of $G$ with kernel containing $N$. (Instruction: EXTEND)

(vii) Of course, the scalar product (and in particular the norm) of characters can be calculated in CAS. One main use of it is to find the multiplicity $n_i(\chi)$ with which an irreducible character $\chi_i$ is contained in a character $\chi$. The characters $n_i(\chi)\chi_i$ can then be subtracted from the characters $\chi$ (instruction

REDUCE with several options) in the hope of being left with new irreducibles, which are detected by their norm.

(viii) If the values of a character $\psi$ are not known for some classes, one cannot use the scalar product to decompose $\psi$ into irreducible characters. In such a case it is, however, often possible to find such a decomposition by making an "Ansatz"

$$\psi = \sum_i x_i \chi_i$$

with only a few irreducible characters $\chi_i$, which are e.g. selected because they are the only ones having a degree smaller than the degree of $\psi$. This Ansatz yields a linear equation for each class on which the values of $\psi$ and the $\chi_i$ are known and the system of equations so obtained may have a unique solution. This method has so far often been applied by a mixture of hand calculation and use of CAS; it is presently being implemented in a more automatic form.

(ix) After all known irreducible characters of $G$ have been peeled off from a list $\psi_1, \ldots, \psi_k$ of reducible ones by application of the REDUCE command, it is often possible to find new irreducibles by inspection of the triangle of the scalar products $(\psi_i, \psi_j)$, $1 \leqslant i \leqslant j \leqslant k$, which can be computed and printed for inspection. (Instruction: TRIANGLE)

However, there are also some algorithms for making deductions from it. The easiest case is the following. The difference of two characters, say $\psi_2 - \psi_1$ is irreducible, iff $\psi_2(1) > \psi_1(1)$ and $(\psi_2 - \psi_1, \psi_2 - \psi_1) = 1$, and since the left hand side is equal to $(\psi_2, \psi_2) + (\psi_1, \psi_1) - 2(\psi_2, \psi_1)$, this can be read from the triangle. (Instruction: CHECK)

More generally, Guy's inclusion lemma [**12**] states that a sufficient criterion for the difference of two characters $\psi_2 - \psi_1$ with $0 < (\psi_1, \psi_1) \leqslant (\psi_2, \psi_2)$ to be a proper character, is that $(\psi_1, \psi_1)(\psi_2, \psi_2) - (\psi_1, \psi_2)^2 < (\psi_2, \psi_2)$. Moreover, this condition is sharp in the sense that for each triangle $\begin{pmatrix} a & \\ c & b \end{pmatrix}$ with $0 < a \leqslant b$ and $ab - c^2 \geqslant b$, there exists a group $H$ and characters $\alpha, \beta$ of $H$ such that $(\alpha, \alpha) = a$, $(\beta, \beta) = b$, $(\alpha, \beta) = c$ but $\alpha$ is not a constituent of $\beta$. Guy's condition for $\psi_1$ to be a constituent of $\psi_2$ can be further improved by considering the Frobenius-Schur indicators (computed by the instruction SCHUR) of $\psi_1$ and $\psi_2$ as well [**12**]. These methods that aim at the construction of characters of smaller norm, hoping to obtain characters of norm 1 eventually from those so far known, are at present implemented in CAS following the proposals made in J. Conway's lecture at the Durham conference.

(x) As a last resort, one can try to find new irreducible characters as rational linear combinations of many reducible characters involved in a triangle. In CAS this is done (instruction EXTRACT) as follows.

From the triangle of scalar products, one tries to deduce a matrix whose columns correspond to reducible characters $\psi_1, \ldots, \psi_r$, say, occurring in the triangle and whose rows correspond to the set, $\chi_1, \ldots, \chi_s$, say, of all irreducible constituents of $\psi_1, \ldots, \psi_r$, and whose entry $n_{ij}$ is the multiplicity of $\chi_i$ in $\psi_j$. Having found such a matrix, it is then tested whether some $\chi_k$ can be obtained as a linear combination of the $\psi_j$, $1 \leqslant j \leqslant r$, by trying to solve the inhomogeneous system of $s$ linear equations

$$(**) \qquad \sum_{j=1}^{r} n_{ij} x_j = \delta_{ik} \quad \text{with} \quad i = 1, \ldots, s.$$

Since one is trying to construct a character as a rational linear combination of others, it is of course crucial that the matrix $(n_{ij})$ is uniquely forced by the triangle of scalar products. To ensure this, $\psi_1$ is chosen multiplicity-free (characters with this property are searched for by comparing norm and indicator) either by the user or by the program, which will then choose a character having nontrivial scalar products with as many other characters as possible. By the choice of $\psi_1$ to be multiplicity-free, the first column of the matrix $(n_{ij})$ will, without loss of generality, have the first $(\psi_1, \psi_1)$ entries equal to 1 and all remaining entries, whose number is not known at this stage, equal to 0.

Assuming now, by induction, that $\psi_1, \ldots, \psi_l$ have been selected from the reducible characters occurring in the triangle, in such a way that their columns are uniquely determined (up to additional zeros at the end of each column), one then searches for a $\psi_{l+1}$ among a suitable selection of characters in the triangle that again will allow the unique (up to zeros at the end and permutation of entries in rows which are identical with respect to their first $l$ entries) determination of an $(l+1)$st column of the matrix by evaluation of the consequences of the norm of $\psi_{l+1}$ and the scalar products of $\psi_{l+1}$ with $\psi_1, \ldots, \psi_l$. Some of the entries in this column can immediately be made, e.g. the entry $n_{i,l+1}$ must be 0 if there exists a $\psi_j$, $1 \leqslant j \leqslant l$ such that $n_{i,j} \neq 0$ but $(\psi_j, \psi_{l+1}) = 0$. For the remaining entries that cannot be determined uniquely by one condition, all available conditions are evaluated in a backtrack search with the aim to prove that the column is uniquely determined. If this fails, another reducible character from the triangle is tried.

The search process is first stopped when the number of non-zero rows of the matrix (which in the beginning of course was bigger than the number of columns) becomes equal to the number of columns. If none of the equations can be solved (i.e. no irreducible character can be constructed) the columns of the matrix must be linearly dependent. The attempt to find further uniquely determined columns for the matrix then is resumed until again the number of columns (not counting those known to be linearly dependent on predecessors) becomes equal to the number of rows, and so on, until eventually an irreducible character is found or no further unique enlargement of the matrix is possible.

Before trying to solve it, the system (**) of equations is simplified by leaving out the $i$th row and $j$th column in case $n_{ij}$ is the only non-zero entry in the $i$th row, as well as columns known to be dependent on their predecessors (DECREASED MATRIX). The best to be hoped for is that this decreased matrix is nonsingular, in which case all irreducible characters involved in the remaining reducibles have been found. More often, the system of equations is soluble only for certain irreducibles. Example 4.2 will demonstrate both instances.

(xi) Another method for getting more information from a triangle of scalar products of reducible characters $\psi_1, \ldots, \psi_r$ is to form generalized characters as integral linear combinations of some of the $\psi_i$, in such a way that the norm of this linear combination is smaller than the norm of any of the $\psi_i$ involved in it. The philosophy here is to aim at generalized characters of small norm for which then—possibly using the indicator—it can be shown that they are proper characters or the negative of such. The method was described in a lecture of J. H. Conway at Durham as having been used in Cambridge some time ago [**12**]; it is contained in a microcomputer system for handling characters [**4**] that was demonstrated in Durham by Mrs. R. Hassan. This method, not yet implemented in CAS, will be added to it.

(xii) Another useful tool for the determination of characters is the instruction SQUARE, explained in 3.5.

### 3.5 Tests

When a character table has been constructed, in particular if some manipulation of data by hand and typing in of data is involved, there is always the danger that some errors have crept in. Unfortunately there are no sufficient *a posteriori* tests of the correctness of a character table. However, some necessary conditions have been found to be very useful for the detection of errors.

(i) The two orthogonality relations yield the most widely used, although not very strong, tests for the correctness of a character table. As Isaacs points out, p. 21 [**29**], the second is a consequence of the first, but it is often convenient to use both in order to locate and correct errors or to find missing entries.

Let $g_1, \ldots, g_r$ be representatives of the conjugacy classes and $\chi_i, \chi_j$ absolutely irreducible characters, then

$$\sum_{k=1}^{r} \frac{1}{|C_G(g_k)|} \chi_i(g_k)\overline{\chi_j(g_k)}$$

can be printed (instruction: TEST). If e.g. it is a rational number different from $\delta_{ij}$, its denominator may indicate on which classes errors occur. An application of this technique in the determination of the irrationalities in the table of the Babymonster is described in example 4.6.

If $\chi_1, \ldots, \chi_r$ is a list of absolutely and of rationally irreducible characters, containing each absolutely irreducible character exactly once either by itself or as a constituent,

$$\sum_{i=1}^{r} \chi_i(g)\overline{\chi_i(g)} \cdot (\chi_i, \chi_i)^{-1}$$

can be printed (instruction: SQUARE), which is equal to $|C_G(g)|$, if $g$ is a rational element. A difference of the two values may be used to detect errors or irrationalities in the column of $g$.

(ii) Let $$\nu_n(\chi_i) = \frac{1}{|G|} \sum_{g \in G} \chi_i(g^n)$$

be the $n$th Schur-Frobenius indicator, then

$$r_n(g) = \sum_{\chi \in \mathrm{Irr}\, G} \nu_n(\chi)\chi(g) = |\{h \mid h \in G, h^n = g\}|.$$

This number $r_n(g)$ is computed by the instruction SQUARE with option INDICATOR.

(iii) As a refinement of the second orthogonality relation, block orthogonality relations:

$$\sum_i \chi_i(g)\overline{\chi_i(h)} = 0$$

with the $p$-parts of $g$ and $h$ not conjugate and $\chi_i$ running over all irreducible characters in a $p$-block, can also be used (instruction: BTEST).

(iv) A rather strong test for the correctness of a character table is to form tensor products and decompose these into irreducibles using scalar products. For checking power maps and the character values simultaneously, one can analogously decompose symmetrized powers. For checking power maps, one can also compute the Schur-Frobenius indicator (instruction: SCHUR($n$)).

### 3.6 Getting additional information

The operations to be described in this section primarily derive additional information from a completed character table, but examples in Section 4 will show that some of them can also be useful in the construction of character tables. We leave out, here, some facilities of CAS that are intended to be helpful in finding the decomposition matrix, Molien series and permutation characters. These will be the topic of section 5.

(i) The central characters $\omega_i$, defined by

$$\omega_i(\bar{C}) := \frac{|C|\chi_i(g)}{\chi_i(1)},$$

where $\bar{C}$ is a class sum of a class $C$ of length $|C|$ containing the element $g$, can be computed from the irreducible characters $\chi_i$. (Instruction: OMEGA)

(ii) The structure constants of the center $Z(\mathbb{C}G)$ of the group algebra $\mathbb{C}G$ can be computed from the character table. To compute $a_{ijk}$ defined by

$$\bar{C}_i\bar{C}_j = \sum_k a_{ijk}\bar{C}_k,$$

where $C_i, C_j$, and $C_k$ are classes of lengths $|C_i|$, $|C_j|$, and $|C_k|$ containing elements $g_i, g_j$, and $g_k$, respectively, the formula

$$a_{ijk} = \frac{|C_i||C_j|}{|G|} \sum_{\chi \in \operatorname{Irr} G} \frac{\chi(g_i)\chi(g_j)\overline{\chi(g_k)}}{\chi(1)}$$

is used. (Instruction: STRUCTURE)

(iii) More generally, one can compute arbitrary products of elements of the center $Z(\mathbb{Z}G)$. To do this, one transposes the character table of $G$ (instruction: TRANSPOSE) and treats the table thus obtained as if it were a table of irreducible characters. In particular, one may form linear combinations of these "characters" (i.e. of class sums) and form "tensor products" (i.e. products in the group algebra) storing the result on a table $R$ copied from the transposed character table (instruction: COPY). Reducing $R$ with the transposed character table (instruction: REDUCE) one obtains the decomposition of the product (element) as a linear combination of the "irreducible characters" (i.e. class sums). In order to get the result in a more clearly arranged form one may use, instead of REDUCE, the DECOMPOSE command.

This is just one simple example of how CAS can be used in situations when one is working, strictly speaking, not with character tables but with matrices enjoying similar properties. For more general examples along these lines see [**40**].

(iv) The classes contained in the kernel of a character can be printed. (Instruction: KERNEL)

(v) Given the $n$th power map, the $n$th Schur-Frobenius indicator

$$v_n(g) = \frac{1}{|G|} \sum_{g \in G} \chi(g^n)$$

can be computed (instruction: SCHUR). For the computation of Schur indices in general see [**19**], where CAS is used in some cases.

(vi) The fact that character values are given in the exact algebraic form as described in section 2, allows the calculation of the $p$-blocks of irreducible characters, using the defining equivalence relation of blocks. Let $\chi_1$ and $\chi_2$ be two irreducible characters of $G$ and $\omega_1$ and $\omega_2$ the corresponding central characters. $\chi_1$ and $\chi_2$ belong to the same $p$-block iff for all class sums $\bar{C}$ of $p$-regular classes $C$ of $G$, the values $\omega_1(\bar{C})$ and $\omega_2(\bar{C})$ are congruent, modulo a maximal ideal of algebraic integers in $\mathbb{C}$ containing $p$ (see e.g. p. 190 of [**37**]). With the data structure of CAS, this congruence can effectively be decided by factoring modulo $p$ the cyclotomic polynomials $\Phi$ used in the representation of irrational character values (using Berlekamp's algorithm) and calculating modulo $p$-irreducible factors of $\Phi$ instead of modulo $\Phi$.

$p$-blocks, their defect and the height of each character in a block can be printed. (Instruction: PBLOCK)

(vii) Powermaps can in part automatically be determined, checked or supplemented (instruction: SUPPLEMENT). Let $p$ be a prime, $C$ a class of $G$ and $x \in C$, then the procedure tries systematically to narrow down the set of classes $C'$ that might contain $x^p$, using various conditions until, hopefully, only one class $C'$ remains. With $x' \in C'$ such conditions are: $|x^p| = |x'|$; if $|x^p| = |x|$ then $|C_G(x)| = |C_G(x')|$; if $|x^p| < |x|$ then $|C_G(x)| \leqslant |C_G(x')|$; if $|x^p| = |x|$ and $\chi_i(x)$ is irrational for some irreducible character $\chi_i$, we can apply the Galois-automorphism induced by $p$th powering of roots of unity to obtain $\chi_i(x^p)$ and then have the condition $\chi_i(x^p) = \chi_i(x')$; in any case the congruence $\chi_i(x)^p \equiv \chi_i(x') \pmod{p}$ must hold, and, if $\chi_i(x)$ is rational, $\chi_i(x) \equiv \chi_i(x') \pmod{p}$.

## 4 Examples of the Construction of Character Tables

In this section we wish to demonstrate the use of CAS for constructing character tables. We have chosen what we believe to be some typical situations. Some of the tables computed in this section are not new (e.g. $S_8$, O(7,2), $PO^+(6,3)$).

The examples described in this section and the next one are fully reproduced as computer output—with all commands and tables concerned—on microfiche number 2 enclosed, after the reproduction of the character tables of the 26 sporadic simple groups. The notation of the computer output will be used in the text, when it seems appropriate; for instance S8 will denote the character table of the symmetric group $S_8$ and $X.i$ ($i \in \mathbb{N}$) will always be a specific character.

### 4.1 Reducing tensor powers of a character; $S_8$

In the first example, let us assume that we know the classes, the power maps and at least one faithful character $\chi$ of the group $G$. In order to find new (in general reducible) characters, one may form tensor powers $\chi^m$ of $\chi$ or, much better, symmetrizations of $\chi$ (since the power maps are available) or, still better, in case that $\nu_2(\chi) = \pm 1$, orthogonal or symplectic components of $\chi^m$ as described in 3.4 (v). If the degree of $\chi$ is relatively small (compared with the group order), it is quite feasible to choose $m$ as large as 6. For instance, one may construct the character table of the symmetric group $S_n$ using the irreducible character of degree $n-1$ arising from the natural representation of $S_n$.

Let us illustrate this by constructing the character table S8 of $S_8$ from the natural permutation character $X.1+X.3$ together with the alternating character $X.2$. The 2nd, 3rd and 5th power map are assumed to be stored on the table S8 at the beginning. Since $\nu_2(X.3) = 1$, one may compute the 28 "orthogonal components" of $(X.3)^2, \ldots, (X.3)^6$, corresponding to the 28 partitions of $2, 3, \ldots, 6$. This is done by the command "FRAME, N = 6,..." (cf. 3.4 (v)). If one now subtracts known irreducible characters using "REDUCE" (cf. 3.4 (vii)), one immediately gets all the irreducible characters of $S_8$.

On the other hand, if one just forms the tensor powers $(X.3)^i$, $2 \leqslant i \leqslant 6$, and subtracts the known irreducibles $X.1$, $X.2$, $X.3$, one is left with 5 characters of norm 2, 24, 578, 15 884, 540 061 and it is hard to see how to get a single new irreducible character out of these.

### 4.2 Constructing a table from two subgroups; O(7,2)

In this example, we still assume that the classes of the group $G$ are known (with the order of the centralizers and the power maps) but instead of tensoring a given faithful character of $G$, the character tables of known subgroups are used. Of course, one has to know the fusions of the subgroups into $G$. It is clear that the subgroups should not be too small compared with $G$. On the other hand, in a later stage, when many irreducible characters are already found, it might even be useful to induce up from cyclic subgroups.

In our (easy) example $G = \mathrm{O}(7,2) \cong \mathrm{Sp}(6,2) \cong \mathrm{W}(\mathrm{E}_7)/Z$, and the subgroups chosen are $\mathrm{O}^+(6,2) \cong S_8$ and $\mathrm{O}^-(6,2) \cong \mathrm{W}(\mathrm{E}_6)$. One could do similar things for larger orthogonal groups also; in fact the character table of $\mathrm{O}(11,2) \cong \mathrm{Sp}(10,2)$ (with 198 characters) was computed from the tables of $\mathrm{O}^+(10,2)$ and $\mathrm{O}^-(10,2)$ in joint work with J. S. Frame.

For $G$, we start with a "character table" which contains just the class names, centralizer orders, power maps and the trivial character of O(7,2). On the other hand, on the character tables of $\mathrm{O}^{\pm}(6,2)$ the fusion into $G = \mathrm{O}(7,2)$ is stored. Then the characters of $\mathrm{O}^{\pm}(6,2)$ are induced up to $G$ (cf. 3.4 (vi)). One thus obtains a table $R$ with 47 reducible characters of $G$. Using the "REDUCE" command, the trivial character is subtracted. Two new irreducible characters $X.2$, $X.3$ are found in this way, which is not surprising, since the subgroups are stabilizers of doubly transitive actions of $G$.

Of course, there are now several possibilities for how to proceed. One could try to extract more irreducible characters from the list $R$ using the methods described in 3.4 (ix), (x), (xi). Or one could tensor and symmetrize etc. the characters $X.2$, $X.3$, thereby obtaining a few or even many new characters for the table $R$. The advantage of working interactively becomes apparent.

In the worked-out example it is just checked, whether a difference of characters of $R$ yields an irreducible one (command "CHECK" cf. 3.4(ix)), which is particularly simple and fast and in our case even successful, giving $X.4$. Then the list of reducible characters is enlarged by forming tensor products and symmetrizations of $X.2$, $X.3$, $X.4$. Some new irreducible characters are then found as differences of reducible ones and are symmetrized and tensored. When this process stops, i.e. no further irreducibles are found by reducing and checking differences, the matrix of scalar products of the 30 reducible characters of smallest norm is computed and printed (command "TRIANGLE", see 3.4 (ix)) and the backtracking algorithm (command "EXTRACT", see 3.4 (x)) yields new irreducible characters.

### 4.3 Finding and using fusions; $PO^+(6,3)$

It often happens that one is given a character table of a group $H$ and one wants to find the character table of an extension $G$ of $H$, maybe by some outer automorphism. Of course, the first problem here is to find the conjugacy classes of $G$, with the centralizer orders etc. For this it is very helpful, if one has the character table of some group $S$ containing $G$.

Let us look at a simple example:

$$H = PSO^+(6,3),\ G = PO^+(6,3).$$

Here the character table of the simple group $H$ is given, $[G:H] = 2$, and one might also get information on the classes of $G\backslash H$ by using the Frobenius-Schur count of square roots as mentioned in 3.5 (ii). But computations become a lot easier, if one uses the character table of $S = PSO(7,3)$ together with the known fact that $G \leqslant S$. If one knows only that such an inclusion of groups exists and has got the character tables of both groups (here $H$ and $S$), it is usually not hard to find the fusion of the classes. For some classes the fusion is always obvious, e.g. for the class $1A$. The values of the restrictions $\chi_H$ of an irreducible character $\chi$ of $S$ are then known on such classes. On the other hand $\chi_H = \sum a_i\psi_i$ with $a_i \in \mathbb{Z}_{\geqslant 0}$ and $\psi_i \in \mathrm{Irr}(H)$ and one gets equations for the $a_i$. When these are solved, comparing the values of $\chi_H$ on the classes of $H$ with the values of $\chi$ on the classes of $G$ will give further information on the fusion.

In the present example, both $H = \mathrm{PSO}^+(6,3)$ and $S = \mathrm{PSO}(7,3)$ contain exactly one class $5A$ of elements of order 5, cf. the character tables of the two groups, reproduced on the microfiche. Just looking at the values of the characters on the classes $1A$ and $5A$ one finds for the restriction of the irreducible character $X.2^*$ of degree 78 of $S$ (the irreducible characters of $S$ will be distinguished from those of $H$ by *)

$$X.2^*_H = X.2 + X.5 \quad \text{or} \quad X.3 + X.5,$$

where $X.2$, $X.3$, $X.5$ denote irreducible characters of $H$. There is an automorphism of the character table of $H$, i.e. a pair of permutations, one operating on the classes the other on the characters, leaving the character table of $H$ unchanged, which moves $X.2$ to $X.3$. So one can assume without loss of generality

$$X.2^*_H = X.2 + X.5.$$

Now one compares the values of $X.2^*$ on the classes of elements of order $m$ of $S$ with those of $X.2 + X.5$ on the classes of elements of the same order in $H$, to find corresponding classes. For instance, for the involution classes one has

$$X.2^*(2A) = -34,\ X.2^*(2B) = 14,\ X.2^*(2C) = -2 \quad \text{in} \quad S,$$
$$(X.2 + X.5)(2A) = 14,\ (X.2 + X.5)(2B) = -2 \quad \text{in} \quad H.$$

So one gets the following piece of fusion

| $H$: | $2A$ | $2B$ |
|---|---|---|
| $S$: | $2B$ | $2C$ |

For some classes one is left with an ambiguity, due to the fact that $X.2^*$ has the same value on several classes of elements of equal order. But then the restriction of $X.4^*$ to $H$ gives the missing information.

As mentioned in 3.4 (viii), it is intended to include a command that will do some of these computations automatically.

After having stored the fusion of $H$ into $S$ the next step is to induce the trivial character $1_H$ of $H$ up to $S$. One finds by reducing

$$1_H{}^S = X.1^* + X.4^* + X.6^* + X.7^* + X.10^*,$$

where again the irreducible characters of $S$ are denoted by $X.i^*$ (in the computer printout they appear as $X.i$). On the other hand using $H \subset G \subset S$, $[G:H] = 2$ and the transitivity of induction one obtains

$$1_H{}^S = (1_H{}^G)^S = (1_G + \varepsilon_G)^S = 1_G{}^S + \varepsilon_G{}^S,$$

where $\varepsilon_G$ denotes the linear character of $G$ with kernel $H$. So

$$1_G{}^S + \varepsilon_G{}^S = X.1^* + X.4^* + X.6^* + X.7^* + X.10^*$$

with degrees

$$378+378 = 1+105+182+195+273.$$

This gives the unique solution ($G = \mathrm{PO}^+(6,3)$, $S = \mathrm{PSO}(7,3)$)

$$1_G{}^S = X.1^*+X.6^*+X.7^*,$$

$$\varepsilon_G{}^S = X.4^*+X.10^*.$$

These characters are displayed on the microfiche; they will also be used in Example 4.4.

The classes $C$ of $S$, on which $1_H{}^S(C)$ (resp. $1_G{}^S(C)$) is zero, are exactly those which are disjoint to $H$ (resp. $G$). $C$ contains an element of $G\backslash H$ if and only if $1_G{}^S(C) \neq \varepsilon_G{}^S(C)$. One finds thus, that there are 23 classes of $S$ containing elements of $G\backslash H$.

Now we can count the number $m$ of conjugacy classes of $G$:

$$m \geqslant 23+(29-x),$$

where 29 is the number of classes of $H$ and $x$ is the number of pairs of classes of $H$ which fuse to one class in $G$. On the other hand, $m$ is at the same time the number of irreducible characters of $G$ and $x$ is the number of pairs of characters of $H$, which are conjugate in $G$. Since every pair of conjugate characters of $H$ yields one irreducible character of $G$ by induction, whereas the other irreducible characters of $H$ can each be extended to characters of $G$ in two different ways each, one has

$$m = 2(29-2x)+x.$$

The inequality, above, now gives

$$x \leqslant 3.$$

On the other hand, since there are no elements of order 26 in $G$ (there are also none in $S$) the number of square roots of an element of class $13A$ in $G$ is the same as the number of square roots in $H$, namely 1. By the Frobenius-Schur-count this implies, that the characters $X.23$, $X.24$, and $X.25$, $X.26$ of $H$ are conjugate in $G$ so that for the induced characters $\nu_2(X.23^G) = \nu_2(X.24^G) = +1$. Furthermore, the elements of order 20 have centralizer of order 20 in $S$, hence also in $G$. This implies that the two classes $20A$, $20B$ in $H$ fuse to one class in $G$ and consequently the characters $X.18$, $X.19$ of $H$ are conjugate in $G$.

Thus $x = 3$ and it follows that $G\backslash H$ is a union of 23 conjugacy classes of $G$, no two of which fuse in $S$. So the centralizer orders of these classes are easily computed from $1_G{}^S$ and $1_H{}^S$. Of course, also the fusion of $G$ into $S$ is known as well as the fusion of $H$ into $G$. So one can induce the characters of $H$ to $G$ and restrict those of $S$ to $G$, obtaining thus a large supply of characters of relatively small norm. Using the techniques explained before, it is then quite easy to compute the whole character table of $G = \mathrm{PO}^+(6,3)$. It is displayed on the microfiche and will be used in the next example.

### 4.4 Central extensions—projective characters; 3 · PSO(7,3)

For many purposes it is important to know not only the ordinary representations of a finite group $G$ but also the projective ones. So it is desirable to construct, in addition to the character table of $G$, also the character tables of central extensions ("covering groups") $\tilde{G}$ of $G$ with $\tilde{G}/Z \cong G$, $Z < \tilde{G}' \cap Z(\tilde{G})$ being a factor group of the Schur multiplier of $G$. For instance, the character tables of all the covering groups of the sporadic simple groups have been constructed.

In order to avoid unnecessary computations and in particular (if $|Z| \geqslant 3$) an excessive number of irrationalities, it is usually advisable not to work with the character table of $\tilde{G}$ but as long as possible with tables of projective characters (characters of projective representations) of $G$, one for each factor set considered. Here, for simplicity, we mean by a projective character of $G$ just the restriction of an ordinary character of $\tilde{G}$ to the first $r$ conjugacy classes, these containing representatives of the pre-images of the $r$ conjugacy classes of $G$ under the natural map $\tilde{G} \to G$. Observe that any irreducible character of $\tilde{G}$ is fully determined by its corresponding projective character. It is then easy to obtain the ordinary character table of $\tilde{G}$ from the various projective character tables of $G$. One just forms the table of the direct product $G \times Z$ (command: DIRECT, cf. 3.3 (iii)), where on the table $Z$ one erases all characters not corresponding to the factor set considered. Collecting the characters obtained this way from the various tables of projective irreducible characters of $G$, one obtains the complete character table of $\tilde{G}$ if one uses the command COLLAPSE, which erases duplicated columns (corresponding to classes of $\tilde{G}$ which contain elements $g$ conjugate to $gz$ with $1 \neq z \in Z(\tilde{G})$) and corrects the centralizer orders correspondingly. Of course, the element orders and the power maps in $\tilde{G}$ have to be checked and possibly amended.

In order to get started, one has to obtain, somehow, projective characters of $G$ with nontrivial factor set (characters of $\tilde{G}$, not containing $Z$ in their kernels). If $|Z| = p$ is a prime, one can get such characters without extra work, if $G$ has $p$-blocks of defect zero. Any character $\chi$ of $G$ of $p$-defect zero can be regarded as a character of $\tilde{G}$ of defect 1 and it is easy to find the whole block: it consists of the formal products of $\chi$ with the $p$ linear characters of $Z$. So $\chi$ is, at the same time, a projective irreducible character of $G$ with non-trivial factor set. On the other hand, these projective characters are usually not too helpful for obtaining new irreducible ones, since they vanish on all $p$-singular elements of $G$. So it is in general essential to find some other projective character of $G$ with given factor set.

If one knows in advance that $G$ has a projective character $\chi$ of relatively small degree (compared with the degrees of the ordinary irreducible characters of $G$) it is feasible to find the decomposition of the ordinary character $\chi\bar{\chi}$ (or even $\chi^{|Z|}$) into irreducible characters by the method described in 3.4 (viii). Then the absolute values $|\chi(g)|$ are known for all $g \in G$ and it should be possible to obtain $\chi(g)$ also, although some ambiguities may remain, in particular if $|Z| > 2$.

If some projective characters with fixed factor set are known, one may tensor these with ordinary characters—CAS allows tensoring characters from different tables with the same "table head"—and peel off known irreducible projectives exactly as in the ordinary case. Under favourable conditions one may even use symmetrized powers of projective characters. Of course, one has to make sure that the power maps of $G$ used agree with those of $\tilde{G}$ on all relevant classes. Also, the factor set is usually changed but this often can be compensated by Galois-conjugation.

In our example let $G = \mathrm{PSO}(7,3)$ and $Z = C_3$, the cyclic group of order 3. It is well known, that $G$ has a Schur multiplier of order 6, cf. e.g. [**10**]. Let $\tilde{G}$ denote the perfect central extension of $G$ by $C_3$. Thus, $\tilde{G}/Z \cong G$, $Z = Z(\tilde{G}) \cong C_3$. Let $H = \mathrm{PO}^+(6,3) \leqslant G$, the character table of $H$ and the embedding of $H$ in $G$ were determined in Example 4.3. We denote the inverse image of $H$ in $\tilde{G}$ (under the natural epimorphism $\tilde{G} \to G$) by $\tilde{H}$. The Schur multiplier of $\mathrm{PSO}^+(6,3)$ has order 2 (see e.g. [**10**]) and $[\mathrm{PO}^+(6,3):\mathrm{PSO}^+(6,3)] = 2$. Hence (p. 121 of [**27**]), $\tilde{H}$ is a split extension, i.e. $\tilde{H} = H \times Z$. So $[\tilde{H}:\tilde{H}'] = 6$ and $\tilde{H}$ has six linear characters:

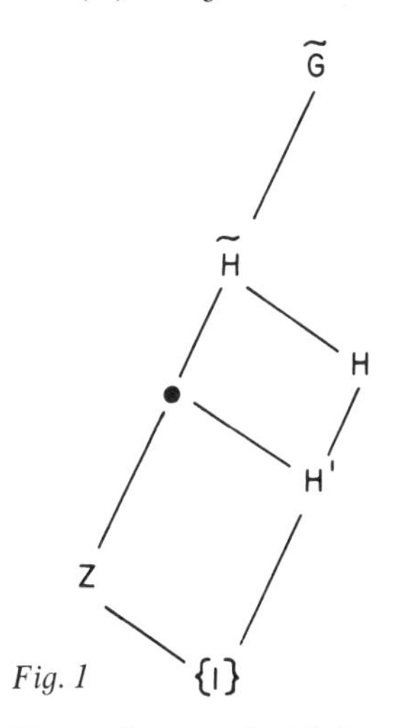

*Fig. 1*

$$1_{\tilde{H}}, \varepsilon_2, \varepsilon_3, \bar{\varepsilon}_3, \varepsilon_6, \bar{\varepsilon}_6,$$

where $\varepsilon_i$ has order $i$. $Z$ is contained in the kernels of $1_{\tilde{H}}$ and $\varepsilon_2$ but not in the kernels of the other linear characters. So one obtains four faithful monomial characters of $\tilde{G}$ of degree $[\tilde{G}:\tilde{H}] = [G:H] = 378$.

If an element $h \in H$ is not conjugate to an element of $\tilde{H} \backslash H$ in $\tilde{G}$, then ${\varepsilon_3}^{\tilde{G}}(h) = {1_{\tilde{H}}}^{\tilde{G}}(h)$ and ${\varepsilon_6}^{\tilde{G}}(h) = {\varepsilon_2}^{\tilde{G}}(h)$. This holds, of course, for all 3'-elements of $H$. Note, that the non-faithful characters ${1_{\tilde{H}}}^{\tilde{G}}$ and ${\varepsilon_2}^{\tilde{G}}$ with norm 3 and 2, respectively are known; they were already determined and used in Example 4.3. The values on the 3'-elements are nearly sufficient to determine $({\varepsilon_3}^{\tilde{G}})_H$ and $({\varepsilon_6}^{\tilde{G}})_H$. There are only two possibilities in each case

$$({\varepsilon_3}^{\tilde{G}})_H = ({1_{\tilde{H}}}^{\tilde{G}})_H = 3X.1 + X.6 + X.8 + X.13 + X.14 + 2X.16$$

or

$$(\varepsilon_3{}^{\tilde{G}})_H = 2X.1+2X.6+X.16+X.17$$

and

$$(\varepsilon_6{}^{\tilde{G}})_H = (\varepsilon_2{}^{\tilde{G}})_H = 2X.2+X.7+X.9+X.11+2X.12+X.15$$

or

$$(\varepsilon_6{}^{\tilde{G}})_H = X.2+X.4+X.9+X.12+X.18$$

where the characters on the right hand side are those of the table $PO^+(6,3)$ computed in Example 4.3.

Now, $\varepsilon_3{}^{\tilde{G}}$ and $\varepsilon_6{}^{\tilde{G}}$ are completely determined by $(\varepsilon_3{}^{\tilde{G}})_H$ and $(\varepsilon_6{}^{\tilde{G}})_H$ respectively.

If $(\varepsilon_3{}^{\tilde{G}})_H = (1_{\tilde{H}}{}^{\tilde{G}})_H$, then $\varepsilon_3{}^{\tilde{G}}$ has the same norm as $1_{\tilde{H}}{}^{\tilde{G}}$, which is 3 (c.f. Example 4.3). So $(\varepsilon_3{}^{\tilde{G}})_H$ should be the sum of three $G$-invariant characters. A glance at the columns corresponding to $2A$ and $2D$ (which fuse to $2B$ of PSO(7,3)) and to $8A$ and $8B$ (which fuse to $8A$) in the table of $H = PO^+(6,3)$ reveals that the only $G$-invariant summands of $(1_{\tilde{H}}{}^{\tilde{G}})_H$ are

$$X.1,\ X.6,\ X.8+X.13+X.16 \quad \text{and} \quad X.14+X.16$$

with degrees 1, 26, 194, 155. But, since $\tilde{G}$ is perfect, the degrees of all faithful irreducible characters of $\tilde{G}$ must be divisible by 3; for otherwise the determinant would yield a non-trivial linear character. Now the only possible decomposition of $(1_{\tilde{H}}{}^{\tilde{G}})_H$ into three $G$-invariant characters with degrees divisible by 3 is into

$$X.1+X.6,\ X.1+X.8+X.13+X.16,\ X.1+X.14+X.16$$

with degrees 27, 195 and 156. But $156 \equiv 2 \bmod 7$ whereas $27 \equiv 195 \equiv -1 \bmod 7$. This gives a contradiction to Brauer's theorem on blocks of defect one; for $\varepsilon_3{}^{\tilde{G}}$ is a character of a projective module for the prime 7, and 7 divides the order of $\tilde{G}$ only to the first power. Also, $\tilde{G}$ must contain six 7-blocks of defect one.

Thus $(\varepsilon_3{}^{\tilde{G}})_H = 2X.1+2X.6+X.16+X.17$. In particular $\varepsilon_3{}^{\tilde{G}}(3A) = 0$. This shows that elements $g$ of $3A$ are conjugate in $\tilde{G}$ to $gz$ with $\langle z \rangle = Z(\tilde{G})$. This implies $\varepsilon_6{}^{\tilde{G}}(3A) = 0$ and hence

$$(\varepsilon_6{}^{\tilde{G}})_H = X.2+X.4+X.9+X.12+X.18.$$

It follows that $\varepsilon_6{}^{\tilde{G}}$ has norm 1 and $\varepsilon_3{}^{\tilde{G}}$ has norm 2. Let $\varepsilon_3{}^{\tilde{G}} = \chi+\chi'$ with characters $\chi,\chi'$ of $\tilde{G}$. Since $\chi(3A) = \chi'(3A) = 0$ one finds

$$\chi_H = X.1+X.6 \quad \text{and} \quad \chi'_H = X.1+X.6+X.16+X.17$$

because $X.1$ must be a summand in both characters.

So $\chi$ is a faithful irreducible character of degree 27 and the values of $\chi$ on all classes of $\tilde{G}$ which intersect $\tilde{H}$ are known. To find the remaining few entries of $\chi$, one may decompose the non-faithful character $\chi\bar{\chi}$ of degree $27^2 = 729$.

Since many values of $\chi$ are already known, it is trivial to see that

$$\chi\bar{\chi} = X.1 + X.6 + X.11,$$

where the characters on the right hand side are those of the character table of $G = \mathrm{PSO}(7,3)$ reproduced on the microfiche in Example 4.3 and $\chi$ is considered as a projective character of $G$ for simplicity. One finds that $\chi$ is zero on all classes of $G$ not intersecting $H$ except for

| | $7A$ | $8B$ | $14A$ | $15A$ | $18A$ | $18B$ |
|---|---|---|---|---|---|---|
| $\lvert\chi\rvert^2$ | 1 | 1 | 1 | 1 | 3 | 3 |

Using congruences (as in 3.6 (vii)) one finds

| | $7A$ | $8B$ | $14A$ | $15A$ | $18A$ | $18B$ |
|---|---|---|---|---|---|---|
| $\chi$ | $-1$ | $\pm 1$ | 1 | $-1$ | $\pm\sqrt{-3}$ | $\mp\sqrt{-3}$ |

The choice for the signs of $\sqrt{-3}$ is unimportant; this amounts to fixing some ordering for the classes. The sign at the class $8B$ can be settled using orthogonality with the irreducible projective character of $G$ obtained from the character of 3-defect zero. It turns out that $\chi(8B) = -1$.

The procedure, as illustrated on the microfiche, is now as follows. From the character table of $G = \mathrm{PSO}(7,3)$ the character $X.56$ of 3-defect zero is copied on the table $PI$ which also receives the projective irreducible characters of degree 27 and 378 just constructed. These characters are then tensored with the ordinary irreducible characters of $G$. The characters of $PI$ are peeled off yielding new projective irreducibles, which again may be tensored with ordinary irreducibles. When this process stops and SQUARE shows that all projective irreducibles with the given factor set are found, the projective characters are sorted according to their degrees (command: SORT) and printed. Then the direct product of $PI$ with a character table of $C_3$ containing only one faithful character of $C_3$ is formed, as well as the direct product of the character table of $G$ with that of $C_3$ containing only the trivial character. The command GALOIS now produces all the remaining irreducible characters of $\tilde{G}$ and using COLLAPSE duplicated columns are deleted. The power maps can be retained (from the direct product-table) for all classes of $\tilde{G}$ which intersect $\tilde{H}$. For the few classes remaining, the power maps can be checked and (for $p = 3$) amended, using the symmetrizations $\chi^{[1^2]}$ and $\chi^{[1^3]}$ of the faithful character $\chi$ of degree 27. Note that these characters are again computed using 3.4 (viii).

Using similar technics and the table just constructed (as well as the known character tables of $\mathrm{O}^+(8,2)$ and the sixfold cover of $\mathrm{U}(6,2^2)$), the character table of the triple cover of the Fischer group F22 has been computed.

### 4.5 Normal subgroups of index 2; F24′

Assume that $G$ has a normal subgroup $H$ of prime index, say of index 2, for simplicity, and that the character table of $G$ is known. How can one find the character table of $H$? This problem arises rather frequently, e.g. think of the symmetric groups or orthogonal groups in even dimensions over GF(2), where it is a lot easier to construct the character table of $G$ than to obtain the character tables of the corresponding simple groups $H = G'$ directly.

If $\chi$ is any irreducible character of $G$, then by a well-known theorem of Clifford, either $\chi_H$ is irreducible and $\chi_H = (\varepsilon\chi)_H$, $\chi \neq \varepsilon\chi$, $\varepsilon$ denoting the linear character of $G$ with kernel $H$, or $\chi_H = \chi' + \chi''$ for two irreducible characters $\chi', \chi''$ of $H$ conjugate in $G$ and $\chi = \varepsilon\chi$, i.e. $\chi$ vanishes outside of $H$. Since $\chi'$ and $\chi''$ are conjugate in $G$ one has

$$\chi'(h) = \chi''(h) = \tfrac{1}{2}\chi(h) \tag{1}$$

for those $h \in H$ which belong to a conjugacy class of $H$ which is at the same time a conjugacy class of $G$.

In order to obtain the character table of $H$, one first has to delete all columns of the table of $G$ which belong to classes not contained in $H$, i.e. those, on which $\varepsilon$ is negative. Then one deletes duplicated rows, retaining just one row for each pair of characters $\chi, \varepsilon\chi$. Finally one has to solve the following two problems:

(i) Find the classes $C_1, \ldots, C_s$ of $G$ which split into two conjugacy classes $C_i'$, $C_i''$ of $H$.
(ii) If $\chi_1, \ldots, \chi_s$ are the irreducible characters of $G$, which split into two conjugate characters $\chi_i' + \chi_i''$ upon restriction to $H$, find the values of $\chi_i'$ on the classes $C_j'$ $(i, j = 1, \ldots, s)$. The values of $\chi_i'$ on the non-splitting classes are given by (1).

In order to solve problem (i), one may use the following tests.

(a) If $\chi_i(h)$, $i \in \{1, \ldots, s\}$ is not twice an algebraic integer, then $h$ must belong to one of the splitting classes $C_1, \ldots, C_s$. This follows from (1) and the fact that character values are algebraic integers.

(b) If the 2-part of $|C_G(h)|$ equals the 2-part of $|\langle h \rangle|$ for $h \in H$, then $h$ belongs to one of the splitting classes $C_1, \ldots, C_s$. On the other hand, all even powers of elements of $G \backslash H$ belong to non-splitting classes. For, $|C_G(h)| = 2^\delta |C_H(h)| (h \in H)$ with $\delta \in \{0, 1\}$, and $h$ belongs to some splitting class if and only if $\delta = 0$.

(c) If $h \in H$ is the product of two involutions of $G \backslash H$, then the $G$-conjugacy class $h^G$ of $h$ is a conjugacy class of $H$ or $h^G$ splits into non-real conjugacy classes of $H$. In particular, if $h$ is also the product of two involutions of $H$ then $h^G$ does not split.

*Proof.* Let $h = a.b$, $a, b$ being involutions in $G\backslash H$. If an element $g \in H$ inverts $h$ by conjugation, then $g.a \in C_G(h)$, because $a$ also inverts $h$. So $C_G(h) \nsubseteq H$ in this case, and consequently, $h^G$ does not split into two conjugacy classes of $H$.

(d) If for an irreducible character $\chi$ of $G$ with $\chi_H$ irreducible, the scalar product $(\chi_i{}^2, \chi)_G$ formed with respect to $G$ is odd for some $i, 1 \leqslant i \leqslant s$, then among the classes of $G$ in $H$, on which $\chi$ does not vanish, there is at least one splitting class $C_i$ $(1 \leqslant i \leqslant s)$. For,

$$(\chi_i{}^2, \chi)_G = \frac{1}{|G|} \sum_{h \in G} \chi_i{}^2(h)\overline{\chi(h)} =$$

$$= 2(\chi_i'^2, \chi_H)_H + \frac{1}{|G|} \sum_{j=1}^{s} \sum_{h \in C_j} (\chi_i{}^2(h) - 4\chi_i'^2(h))\overline{\chi(h)} \tag{2}$$

By far the most powerful test seems to be (c). Test (d) is useful only when the number of candidates for the splitting classes has been reduced. But then the argument often gives even the values of the $\chi_i'$ on the classes $C_j$ $(1 \leqslant j \leqslant s)$.

Let us look at the example $G = \text{F24}$, $H = G'$. The character table of the Fischer group F24 had been computed by D. Hunt. We thank him for sending us a pre-publication copy which is reproduced on the microfiche. There are 11 irreducible characters of $G$ which each split into two irreducible characters when restricted to $H$; these are $X.87, \ldots, X.97$. Six splitting classes are easily detected using test (a) (29*A*, 39*B*, 21*C*, 42*B*, 45*A*, 27*B*). For one further class (24*D*) the 2-part of the centralizer order equals the 2-part of the element order. But even after checking carefully the powermaps, one is still left with 23 candidates for the remaining 4 splitting classes. On the other hand, looking at the class-multiplication coefficients of the involution classes, which are also reproduced on the microfiche, one immediately sees using test (c) that the only candidates for splitting classes are

$$\underline{29A}, (23A), (23B), \underline{39C}, \underline{21C}, \underline{42B}, \underline{45A}, 36A, \underline{27B}, 18G,$$
$$24E, 12J, \underline{24D}, 16A, \tag{3}$$

of which the underlined ones had already been shown to split. 18*G* consists of products of involutions of $G\backslash H$ so if 18*G* splits, the corresponding columns must be complex conjugate. The classes 23*A*, 23*B* can immediately be discarded because they consist of squares of elements of $G\backslash H$. So one is left with 12 candidates for the 11 splitting classes.

$X.76$ is a character, which vanishes on all the classes of the list (3) except for 16*A*, with $X.76(16A) = 1$. Looking at the decomposition of $(X.i)^2$ for $i \in I = \{87, \ldots, 97\}$ (cf. the microfiche) one finds that $((X.i)^2, X.76)_G$ is even for all $i \in I$. From (2) it follows that

$$\frac{1}{|G|} \sum_{h \in 16A} 4(X.i)^2(h) \in 2\mathbb{Z}$$

*Table 1*

| | $21C$ | $42B$ | $36A$ | $27B$ | $18G$ | $24E$ | $12J$ | $24D$ |
|---|---|---|---|---|---|---|---|---|
| $X.82$ | 0 | 0 | 0 | $-1$ | 0 | 0 | 0 | 0 |
| $X.49$ | 0 | 0 | 0 | 0 | 0 | 1 | 0 | 0 |
| $X.61$ | 0 | 0 | 3 | 0 | 0 | 1 | 0 | 0 |
| $X.78$ | 0 | 0 | $-3$ | 0 | 1 | 0 | 0 | 0 |
| $X.17$ | 0 | 0 | 2 | $-1$ | 0 | 0 | 2 | 0 |
| $X.84$ | 0 | 0 | $-1$ | 0 | 0 | 0 | $-1$ | 1 |
| $X.55$ | $-1$ | $-1$ | 0 | 0 | 0 | 1 | 0 | 0 |

where again $(X.i)_H = X.i' + X.i''$ $(i \in I)$. Thus, since $|C_G(h)| = 2^6$, $(X.i')^2(h) \in 2^5\mathbb{Z}$ for $h \in 16A$, and $|X.i'(h)| \in 2^3\mathbb{Z}$, hence $X.i'(h) = 0$ by the orthogonality relations for the column $16A$. So the class $16A$ can be taken off from the list (3).

Having found the splitting classes we now turn to the second problem, to find the character values $\chi_i'(C_j')$. Looking at the character degrees one finds that $X.92$, $X.95$, $X.96$ must split into pairs of exceptional (algebraically conjugate) characters with respect to the primes 13, 29 and 5 respectively. Observe that these characters have defect 1 for the given primes, so Brauer's results may be used, cf. [**18**]. Thus $X.92$ splits only on $39C$, $X.95$ on $29A$, and $X.96$ on $45A$, and the values of the irreducible components of the restrictions on these classes are easily computed using the orthogonality relations for the corresponding rows. One thus finds, for example

$$X.92'(39C') = \frac{1+3\sqrt{13}}{2}, \quad X.92'(39C'') = \frac{1-3\sqrt{13}}{2},$$

where $39C'$ and $39C''$ are the conjugacy classes of $H$ with union $39C$, and all the other values of $X.92'$ are obtained from those of $X.92$ by dividing by two. Also it follows from orthogonality, that the columns corresponding to $39C'$, $39C''$ are known; there is no further splitting, i.e. $(X.i')(39C') = (X.i')(39C'') = \frac{1}{2}(X.i)(39C)$ for $i \neq 92$. The rows $X.95'$, $X.95''$, $X.96'$, $X.96''$ and columns $29A'$, $29A''$, $45A'$, $45A''$ are obtained correspondingly.

In order to obtain the remaining characters, one may look (see Table 1) at the scalar products of $(X.i)^2$ $(i \in I)$ with the characters $X.82$, $X.49$, $X.61$, $X.78$, $X.17$, $X.84$ and $X.55$ (cf. microfiche). The characters are chosen because of their values on those classes which are still in question.

Thus, using equation (2) one gets successively the values of $X.i'$ $(i \in I)$ on the classes $27B, 24E, 36A, 18G, 12J, 24D, 21C$ and $42B$. For instance, for $\chi = X.82$

$$(\chi, (X.i)^2) = \begin{cases} \text{odd} & \text{for } i = 91 \\ \text{even} & \text{for } i \neq 91 \end{cases}$$

So by (2), one gets

$$\frac{1}{|G|} \sum_{h \in 27B} 4(X.i')^2(h) \in 2\mathbb{Z} \quad \text{for} \quad i \neq 91,$$

$$\frac{1}{|G|} \sum_{h \in 27B} (1-4)(X.i')^2(h) \in 1+2\mathbb{Z} \quad \text{for} \quad i = 91.$$

Since $|27A| = \dfrac{|G|}{3^4}$ one has

$$(X.i')^2(h) \in \frac{3^4}{2}\mathbb{Z} \quad \text{for} \quad i \neq 91$$

and with

$$a = (X.91')(27B') \quad \text{and hence} \quad 1-a = (X.91')(27B''),$$
$$1-4a^2+1-4(1-a)^2 \in 2.3^4(1+2\mathbb{Z}),$$

so, $\quad 2a-1 \in 3^2(1+2\mathbb{Z}).$

It follows that $(X.i')(h) = 0$ for $i \neq 91$ and $a \in \{5, -4\}$ since solutions with larger absolute value are impossible because of the orthogonality relations.

The other character values are computed similarly. Scalar products with $X.55$ give the values on the classes $21C$ and $42B$ simultaneously. For instance (2) yields

$$(2X.90'(21C')-1)^2+(2X.90'(42B')-1)^2 \in 42(1+2\mathbb{Z})$$

which again implies

$$X.90'(21C') = \tfrac{1}{2}(1 \pm \sqrt{21})$$
$$X.90'(42B') = \tfrac{1}{2}(1 + \sqrt{21}).$$

Using the orthogonality relations in order to fix the signs in the various rows, one gets the following table (Table 2), which gives the splitting of the characters $X.i$ $(i \in I)$. In this table

a block $\begin{matrix} a & a \\ a & a \end{matrix}$ is abbreviated by $a$, and

$a \pm b$ stands for $\begin{matrix} a+b & a-b \\ a-b & a+b \end{matrix}$.

*Table 2* (F24)′

| | $29A$ | $39C$ | $21C$ | $42B$ | $45A$ |
|---|---|---|---|---|---|
| $X.87$ | 0 | 1 | 0 | 0 | 0 |
| $X.88$ | 0 | 1 | 0 | 0 | 0 |
| $X.89$ | 0 | 1 | 0 | 0 | 0 |
| $X.90$ | 0 | 0 | $\frac{1}{2}(1\pm\sqrt{21})$ | $\frac{1}{2}(1\pm\sqrt{21})$ | 0 |
| $X.91$ | 0 | $-1$ | 0 | 0 | 0 |
| $X.92$ | 0 | $\frac{1}{2}(1\pm 3\sqrt{13})$ | 0 | 0 | 0 |
| $X.93$ | 0 | 0 | $\frac{1}{2}(-1\pm\sqrt{21})$ | $\frac{1}{2}(1\mp\sqrt{21})$ | 0 |
| $X.94$ | 0 | 0 | 1 | $-1$ | 0 |
| $X.95$ | $\frac{1}{2}(1\pm\sqrt{29})$ | $-1$ | 0 | 0 | 0 |
| $X.96$ | 0 | 0 | 0 | 0 | $\frac{1}{2}(1\pm 3\sqrt{5})$ |
| $X.97$ | 0 | 1 | 0 | 0 | 0 |

| | $36A$ | $27B$ | $18G$ | $24E$ | $12J$ | $24D$ |
|---|---|---|---|---|---|---|
| $X.87$ | 1 | 0 | $\pm 3\sqrt{-3}$ | 0 | 1 | 1 |
| $X.88$ | $\begin{matrix}7 & -5\\ -5 & 7\end{matrix}$ | 0 | 0 | 0 | $\begin{matrix}7 & -5\\ -5 & 7\end{matrix}$ | $\begin{matrix}3 & -3\\ -3 & 3\end{matrix}$ |
| $X.89$ | $\begin{matrix}-7 & 5\\ 5 & -7\end{matrix}$ | 0 | 0 | 0 | $\begin{matrix}14 & -10\\ -10 & 14\end{matrix}$ | 0 |
| $X.90$ | $-1$ | 0 | $-1$ | 0 | 2 | 0 |
| $X.91$ | 0 | $\begin{matrix}5 & -4\\ -4 & 5\end{matrix}$ | 0 | 0 | 0 | 0 |
| $X.92$ | 0 | 0 | 0 | 0 | $-3$ | 1 |
| $X.93$ | 0 | 0 | 0 | 0 | 0 | 0 |
| $X.94$ | $\begin{matrix}7 & -5\\ -5 & 7\end{matrix}$ | 0 | 0 | 0 | $\begin{matrix}7 & -5\\ -5 & 7\end{matrix}$ | $\begin{matrix}-3 & 3\\ 3 & -3\end{matrix}$ |
| $X.95$ | 0 | 0 | 0 | 0 | 3 | 1 |
| $X.96$ | 1 | 0 | 0 | 0 | 1 | $-1$ |
| $X.97$ | 0 | 1 | 1 | $\pm 2\sqrt{3}$ | 0 | 0 |

Finally the powermaps have to be checked, as far as the splitting classes are concerned. The only ambiguity results from the fact that the squares of the elements of $24D$ are in $12J$ and it is not obvious from the beginning, whether the squares of elements of $24D'$ ($24D''$) are in $12J'$ ($12J''$) or vice versa. But this can easily be settled by computing the scalar product $(X.88'^{[2]}, X.88')_H$, using for the symmetrization $X.88'^{[2]}$ both alternatives. One arrives at an integer only if the squares of $24D'$ ($24D''$) are assumed to be in $12J'$ ($12J''$, respectively).

### 4.6 Finding irrationalities; BM

It sometimes happens that a character table of a group $G$ is given only in rational form. This means, that the table contains characters of norm $r > 1$ (usually of norm 2) which are sums of Galois-conjugates of irreducible complex characters. Also some of the given "classes" may be unions of conjugacy classes of $G$.

Whereas it is trivial to find the characters of norm $r > 1$, it might not be obvious, which "classes" split into different conjugacy classes. In fact, first proposals sometimes turn out to be false. Of course, if a character of norm $r$ has a value not divisible by $r$ on a class, it is apparent that this class has to split. Also one often sees from the degree whether a character of norm $r$ is the sum of the exceptional characters for some $p$-block of defect 1. In this case it is known that the character values are all $p$-rational, so that the values on the $p'$-elements can be found just by dividing by $r$.

In order to obtain further information on possible "classes" on which a reducible character $\psi$ of norm $r$ splits, it is useful to compute powers of $\psi$ and scalar products $(\psi^i, \chi_j)_G$ with the known absolutely irreducible characters $\chi_j$. If $\chi_j$ vanishes on all those "classes" which are not conjugacy classes, then

$$\sum_{g \in G} (\psi(g)^i)\overline{\chi_j(g)} = \sum_{g \in G} (r\phi(g))^i\overline{\chi_j(g)},$$

where it is assumed that $\psi$ is the sum of the $r$ Galois conjugates of the (unknown) irreducible character $\phi$. So, in this case

$$(\psi^i, \chi_j)_G = r^i(\phi^i, \chi_j)_G \tag{1}$$

of course, for $i = 1$, one gets 0 on both sides of (1) because of the orthogonality relations. But for $i > 1$ one may get information out of (1). Namely, whenever $(\psi^i, \chi_j)_G$ is not divisible by $r^i$, there must be among the "classes" where $\chi_j$ does not vanish, at least one on which $\psi$ splits. Since usually there is just one "class" on which $\psi$ splits and since in a big table there are plenty of $\chi_j$ to test, this method is likely to help identifying the "class". If $\chi_j$ does not vanish on the "classes" on which $\psi$ splits, one has to add the corresponding terms on the right hand side of the equation (1), in which the differences $\psi(g) - r\phi(g)$ enter. If there is just one such "class", this gives a congruence for $\phi(g)$ which is usually sufficient to compute $\phi(g)$ itself.

Let us consider as a specific example the Babymonster BM. The rational character table was computed by D. Hunt. We wish to thank him for sending us a copy prior to publication. There are 9 rational irreducible characters ($\psi_1, \ldots, \psi_9$) of norm 2 (called $X.5, \ldots, X.13$ on the microfiche). Also reproduced on the microfiche are the first four absolutely irreducible characters $X.1,\ldots,X.4$ of the rational character table. Four of these characters $(\psi_1,\psi_2,\psi_4,\psi_5)$ contain odd values on just 4 "classes" in total $(47A, 31A, 23A, 46A)$ and are easily seen to be sums of exceptional characters for the primes 47, 23, 31, 23 (there are two blocks of defect 1 for the prime 23). Thus the splitting of these characters is fully determined. Of course one uses the fact that the character values are algebraic integers in cyclotomic fields. So, for instance, the only possible irrational entries for a class of elements of order 47 are $\frac{1}{2}(a+b\sqrt{(-47)})$ with $a,b\in\mathbb{Z}$ both even or both odd, and it follows readily from the orthogonality relations that only $a,b\in\{1,-1\}$ are feasible. Thus the 8 columns corresponding to $47A$, $31A$, $23A$, $46A$ are easily obtained.

The splitting of $\psi_3$ can be obtained just from the following bit of information:

$$\psi_3(15B) = -2,\ \psi_3(30G) = 0,$$

and $15B$ contains the squares of the elements of $30G$. If $30G$ did not split into two conjugacy classes, then $15B$ could not split either, since squares of conjugate elements are conjugate (and $15B$ could only split into two non-real classes). So it would follow that

$$\phi_3(15B) = -1,\ \phi_3(30G) = 0,$$

$\phi_i$ being an irreducible summand of $\psi_i$. But, on the other hand, one should have

$$\chi(15B) \equiv \chi(30G) \bmod 2,$$

for any character $\chi$ of $G$; a contradiction. So $30G$ splits into two conjugacy classes. Since all the other $\phi_i$ are of defect 0 either for the prime 5 or for the prime 3, we have $\phi_i(30G) = 0\ i \neq 3$. So from orthogonality $|\phi_3(30G)| = \sqrt{15}$, and hence $\phi_3(30G) = \pm\sqrt{(-15)}$, since $\phi_3(30G)$ must be an algebraic integer in $Q(\exp(2\pi i/15))$. It also follows from orthogonality, that $\psi_3$ splits only on $30G$, so that $\phi_3$ is known.

$\psi_8$, $\psi_9$ are sums of exceptional characters for the prime 17 and 7, respectively. So they split each on one "class" of elements of order divisible by 17 and 7, respectively. Since there are no elements of order $4.17$, the (quadratic) irrationalities of $\psi_8$ are in $Q(\exp(2\pi i/17))$, hence $\psi_8$ splits into real characters. There are two characters left, $\psi_6,\psi_7$, which are not sums of exceptional characters for any prime. The Frobenius-Schur count of involutions shows that $\psi_6$ splits into two complex conjugate characters, whereas $\psi_7,\psi_9$ split into real ones. So each character splits on one "class" only.

Now, if $\psi = \phi+\phi'$ splits on exactly one "class" $C$, with $\phi(g) = a+b\sqrt{d}\ (g\in C, a,b,d\in\mathbb{Z})$, then

$$1 = (\phi,\phi)_G = \frac{1}{4}(\psi,\psi)_G + \frac{|C|}{|G|}b^2|d|,$$

hence
$$2\frac{b^2|d|}{|C_G(g)|} = \frac{1}{2}. \tag{2}$$

So $|C_G(g)|$ is $|d|$ times a perfect square (but observe that $|C_G(g)|$ may differ from the value listed in the rationalized character table by the factor 2). So $\psi_8$ can only split on $34B$. Furthermore,

$$\left(\left(\frac{1}{2}\psi\right)^2, \chi_i\right)_G = (\phi^2, \chi_i)_G - \frac{|C|}{|G|} b^2 d\overline{\chi_i(g)}$$
$$= (\phi^2, \chi_i)_G \mp \frac{1}{2}\overline{\chi_i(g)}$$

$$\left(\left(\frac{1}{2}\psi\right)^{[2]}, \chi_i\right)_G = (\phi^{[2]}, \chi_i)_G - \frac{|C|}{|G|}\frac{1}{2} b^2 d\overline{\chi_i(g)}$$
$$= (\phi^{[2]}, \chi_i) \mp \frac{1}{4}\overline{\chi_i(g)}$$

where the minus sign corresponds to the real case $d > 0$. So, for example, if $((\frac{1}{2}\psi)^{[2]}, \chi_i)_G$ is an integer, $\chi_i(g)$ must be divisible by 4. To compute these scalar products in a convenient manner a table is produced, which contains some of the irreducible characters $\chi_i$ (here $X.1, \ldots, X.4$) and $(\frac{1}{2}\psi)^{[2]}$; this table is then tested for orthogonality (cf. microfiche). (Of course, the last class functions are not characters, so a warning is printed, which may be neglected in our case.) The results may be summarized in Table 3, which tabulates the values of

$$((\tfrac{1}{2}\psi_j)^{[2]}, X.i)_G \bmod \mathbb{Z}.$$

*Table 3*

| $j$ \ $i$ | 1 | 2 | 3 | 4 |
|---|---|---|---|---|
| 6 | $\frac{1}{4}$ | $-\frac{1}{4}$ | $\frac{1}{4}$ | 0 |
| 7 | $-\frac{1}{4}$ | $-\frac{1}{4}$ | $\frac{1}{4}$ | 0 |
| 9 | $-\frac{1}{4}$ | 0 | $\frac{1}{4}$ | 0 |

So the first four entries in the column corresponding to the "class" $C_j$, on which $\psi_j$ splits must be

$$\begin{array}{llll} 1, -1, & 1, 0 \bmod 4 & \text{for} & j = 6, \\ 1, & 1, -1, 0 \bmod 4 & \text{for} & j = 7, \text{ and} \\ 1, & 0, -1, 0 \bmod 4 & \text{for} & j = 9. \end{array}$$

The only solutions are

$$\begin{aligned} &\text{for } j = 6\text{: } 15A,\ 30D,\ 32A, \\ &\text{for } j = 7\text{: } 32B, \\ &\text{for } j = 9\text{: } 28E,\ 56A, \end{aligned}$$

where in the last case, only elements of order divisible by 7 were considered. Now the classes $15A$, $30D$, $28E$ can immediately be discarded using (2).

*Table 4* Irrationalities of the characters of the Babymonster

| | $47A$ | $31A$ | $23A$ | $46A$ | $34B$ | $56A$ | $30G$ | $32A$ | $32B$ |
|---|---|---|---|---|---|---|---|---|---|
| $\phi_1$ | $\frac{-1 \pm i\sqrt{47}}{2}$ | 0 | 0 | 0 | 1 | 0 | 0 | $-1$ | 1 |
| $\phi_2$ | $-1$ | 0 | $\frac{-1 \pm i\sqrt{23}}{2}$ | $\frac{-1 \pm i\sqrt{23}}{2}$ | 0 | 0 | 0 | $-1$ | 1 |
| $\phi_3$ | $-1$ | 0 | 0 | 0 | $-1$ | 0 | $\pm i\sqrt{15}$ | 1 | 1 |
| $\phi_4$ | 0 | $\frac{-1 \pm i\sqrt{31}}{2}$ | 0 | 0 | 0 | 0 | 0 | 0 | 0 |
| $\phi_5$ | 0 | 0 | $\frac{-1 \pm i\sqrt{23}}{2}$ | $\frac{1 \mp \sqrt{23}}{2}$ | 0 | 0 | 0 | 1 | 1 |
| $\phi_6$ | 0 | 0 | 0 | 0 | $-1$ | 0 | 0 | $\pm 4i\sqrt{2}$ | 0 |
| $\phi_7$ | $-1$ | $-1$ | 1 | 1 | 1 | 1 | 0 | 0 | $\pm 4\sqrt{2}$ |
| $\phi_8$ | $-1$ | 0 | 0 | 0 | $\pm\sqrt{17}$ | 0 | 0 | 1 | $-1$ |
| $\phi_9$ | 1 | 0 | 0 | 0 | 0 | $\pm 2\sqrt{7}$ | 0 | 0 | 0 |

## 5 Obtaining Information from a Character Table

### 5.1 Trying to find the decomposition matrix

As explained in 3.6 (vi), the CAS-command PBLOCK can be used to compute how the irreducible (Frobenius) characters of a group $G$ (with given character table) are distributed into $p$-blocks for any specified prime number $p$. Let $B$ be a $p$-block of $G$ and $\chi_1, \ldots, \chi_r$ the irreducible Frobenius characters in $B$. Denote the irreducible Brauer characters in $B$ by $\phi_1, \ldots, \phi_s$, and the restriction of a Frobenius character $\chi$ of $G$ to the $p'$-classes by $\hat{\chi}$. The decomposition numbers $d_{ij}$ of $B$ are defined by

$$\hat{\chi}_i = \sum_{j=1}^{s} d_{ij}\phi_j \qquad (1 \leqslant i \leqslant r)$$

and according to Brauer's reciprocity theorem, the projective indecomposable characters $\psi_j$ of $B$ are given by

$$\psi_j = \sum_{i=1}^{r} d_{ij}\chi_i \quad (1 \leqslant j \leqslant s).$$

Accordingly, there are two dual approaches for finding the decomposition matrix

$$\mathbf{D} = (d_{ij}) \in \mathbb{Z}_{\geqslant 0}^{r \times s}$$

of $B$. Either one looks for the irreducible Brauer characters $\phi_j$ in $B$ or for the projective indecomposable characters $\psi_j$ in $B$. In practice, of course, both approaches are used simultaneously. Since there are two detailed accounts of the methods with applications, namely [**30**], and [**31**] Chapter 6.3, we need not go into every detail but only explain how CAS can be used in this context. As for the first approach, the reader is also referred to Parker's article [**39**] in these Proceedings on his powerful method to construct irreducible modular representations of finite groups from reducible ones. Here, however, we are restricted to characters rather than representations.

The CAS-command BRAUER produces a matrix $\tilde{\mathbf{D}}$ which is a first approximation of the decomposition matrix $\mathbf{D}$ of $B$, namely a "pre-decomposition matrix" in the following sense.

*Definition.* A matrix $\tilde{\underset{\sim}{\mathbf{D}}} \in \mathbb{Z}^{r \times s}$ is called a predecomposition matrix of the block $B$, if the columns of $\tilde{\underset{\sim}{\mathbf{D}}}$ generate the same $\mathbb{Z}$-submodule of $\mathbb{Z}^{r \times 1}$ as those of the decomposition matrix $\mathbf{D}$, i.e. $\tilde{\underset{\sim}{\mathbf{D}}} = \mathbf{D}U$ for some $U \in \mathrm{GL}(s, \mathbb{Z})$.

Hence, the number $s$ of irreducible Brauer characters in $B$ is equal to the number of columns of $\tilde{\mathbf{D}}$. Other possibilities to use $\tilde{\mathbf{D}}$ will be discussed below. The computation of $\tilde{\mathbf{D}}$ is based on the fact that a generalized character

$$\chi = \sum_{i=1}^{r} a_i\chi_i \quad \text{of} \quad G, \quad (a_i \in \mathbb{Z})$$

is a generalized projective character, i.e.

$$\chi = \sum_{i=1}^{s} b_i\psi_i \quad \text{for some} \quad b_i \in \mathbb{Z},$$

iff $\chi(g) = 0$ for all $p$-singular elements $g \in G$, cf. p. 133 of [**42**]. For blocks with many characters, it might happen that the solution of the linear equations

$$\sum_{i=1}^{r} a_i\chi_i(g) = 0, \; g \in G \; p\text{-singular},$$

over $\mathbb{Z}$ causes difficulties. In this case, CAS still can compute the number $s$ of columns of $\mathbf{D}$ by solving these equations modulo suitable primes.

If one follows the second approach via the projective indecomposable characters, one can now try to produce projective characters and to prove indecomposability of some of them. First note, if $\psi$ is a projective character of $G$ and $\psi = \psi_B + \psi_{B'} + \ldots$ is a decomposition into characters $\psi_B, \psi_{B'}, \ldots$ where the constituents of $\psi_B$ belong to the $p$-block $B$, those of $\psi_{B'}$ to the $p$-block $B' \neq B$ etc., then also $\psi_B, \psi_{B'}, \ldots$ are projective characters. The CAS-command DECOMPOSE with the option BLOCKS prints the multiplicities of the irreducible characters of $G$ in the (possibly reducible) characters of a second table of $G$-characters in the form of a matrix. The columns of this matrix correspond to the characters of the second table and the rows are indexed by the irreducible $G$-characters which are arranged according to the $p$-blocks of $G$. If this command is applied to a table of projective characters, the $B$-section of the resulting matrix of multiplicities consists of columns which are sums of columns of the (desired) decomposition matrix $\mathbf{D}$.

There are two widely used methods to produce projective characters; the first one consists of inducing (command INDUCE, cf. 3.4 (vi)) projective (indecomposable) characters of subgroups $U$ up to $G$, to obtain projective characters of $G$. As a rule, one chooses a $p'$-subgroup $U$ of $G$ with index as small as possible. The condition $p \nmid |U|$ makes it superfluous to compute the $p$-decomposition matrix of $U$; one simply induces up the irreducible characters of $U$. The small index of $U$ in $G$ is desirable because it keeps the induced characters "close" to being projective indecomposable. For hints on how to find subgroups cf. section 5.3. The second way for producing projective characters is based on an old result by Brauer and Nesbit: if $\psi$ is a projective (indecomposable) character and $\phi$ an (irreducible) Brauer character of $G$, then

$$\psi\phi : G \to \mathbb{C} : g \to \begin{cases} \psi(g)\phi(g) & \text{if } \; g \in G \; p\text{-regular} \\ 0 & \text{if } \; g \in G \; p\text{-singular} \end{cases}$$

is a projective character of $G$, cf. p. 83 of [**18**]. Here again, the indecomposability resp. irreducibility of $\psi$ resp. $\phi$ is only a condition to keep $\psi\phi$ as close as possible to being projective indecomposable. The CAS-command for forming the product is TENSOR (cf. 3.4 (iv)). The Brauer character may be stored (in the present state of CAS) on a table of ordinary characters of $G$ with values "undefined" on $p$-singular classes. (Note $\psi$ and $\phi$ above need not belong to the block $B$.)

To prove indecomposability of a projective character

$$\psi = \sum_{i=1}^{r} a_i \chi_i \text{ in } B,$$

one can sometimes use the pre-decomposition matrix $\tilde{\mathbf{D}}$ of $B$.

*Remark 1.* Let

$$\psi = \sum_{i=1}^{r} a_i\chi_i \text{ and } \tilde{\mathbf{D}} \text{ as just described,}$$

and let $\tilde{\mathbf{D}}_\psi$ be the submatrix of $\tilde{\mathbf{D}}$ consisting of those rows of $\tilde{\mathbf{D}}$, which are indexed by the $\chi_i$ with $a_i = 0$. If $\tilde{\mathbf{D}}_\psi$ has rank $s-1$, then $\tilde{\psi} = a^{-1}\psi$ with $a = gcd(a_1, \ldots, a_r)$, is a projective indecomposable character.

*Proof.* Let

$$\tilde{\psi} = \sum_{i=1}^{r} b_i\chi_i$$

be a projective indecomposable character which is a summand of $\psi$ (i.e. $\psi - \tilde{\psi}$ is a projective character). Then $b_i = 0$ whenever $a_i = 0$. Since both $(b_1, \ldots, b_r)^{tr}$ and $(a_1, \ldots, a_r)^{tr}$ are (integral) linear combinations of the columns of $\tilde{\mathbf{D}}$, rank $(\tilde{\mathbf{D}}_\psi) = s-1$ implies $(b_1, \ldots, b_r) = x(a_1, \ldots, a_r)$ for some $x \in \mathbb{Q}^*$. Since all elementary divisors of $\mathbf{D}$ are equal to 1, one gets $x = (gcd(a_1, \ldots, a_r))^{-1}$, cf. also Corollary (6.3.8) of [**31**].

In the case that the rank of $\tilde{\mathbf{D}}_\psi$ is $s-2$ or $s-3$, one is usually left with several possibilities for the projective indecomposable summands of $\psi$. Sometimes other projective characters help to prove decomposability.

If one pursues the first approach to determine the decomposition matrix $\mathbf{D}$ via the irreducible Brauer characters, one again can use subgroups and the pre-decomposition matrix $\tilde{\mathbf{D}}$. By restricting a (Brauer) character to some (big) subgroups, partial knowledge of the irreducible Brauer characters of these subgroups may help to prove irreducibility, or at least to exclude possibilities for the degrees of the constituents. The CAS-command for restricting characters to subgroups is RESTRICT (cf. 3.4 (vi)). On the other hand, the pre-decomposition matrix $\tilde{\mathbf{D}}$ helps to find equations

$$(*) \quad \sum_{i=1}^{r} a_i\hat{\chi}_i = 0 \qquad (a_i \in \mathbb{Z})$$

between the restrictions of the Frobenius characters $\chi_i$ to $p$-regular classes, because $(*)$ is tantamount to

$$\sum_{i=1}^{r} a_i d_i = 0$$

where $d_1, \ldots, d_r$ are the rows of $\tilde{\mathbf{D}}$. One rewrites the above equation as

$$\sum_{\substack{i=1 \\ a_i > 0}}^{r} a_i\hat{\chi}_i = \sum_{\substack{i=1 \\ a_i < 0}}^{r} -a_i\hat{\chi}_i$$

and uses this to discuss the constituents of the $\hat{\chi}_i$. For instance, if $a_1 > 0$ and $\chi_1(1) = 1$, then $\hat{\chi}_1$ is an irreducible Brauer character which occurs as a constituent of one of the $\hat{\chi}_i$ with $a_i < 0$. One can certainly obtain equations of the type (*) from the character table directly. But usually it is more convenient to obtain them via the pre-decomposition matrix which is easier to survey and contains only integers as entries.

Finally, the combination of the two dual approaches via Brauer characters and via projective characters works in ideal cases as follows. One knows one row and one column of the decomposition matrix **D** and uses this information to produce new projective characters by subtracting the projective indecomposable corresponding to this column from suitable other projective characters. For instance, assume we know that $\hat{\chi}_1$ is an irreducible Brauer character and that

$$\psi_1 = \sum_{i=1}^{r} a_i \chi_i \quad \text{with} \quad a_1 = 1$$

is a projective indecomposable character, then $\psi - a\psi_1$ is a projective character for each projective character $\psi$ with $(\chi_1, \psi) = a$.

## 5.2 Molien series

Invariant theory of finite groups, in particular Molien series, has recently found applications in coding theory [**44**], combinatorics [**45**], and discrete geometry (spherical designs) [**24**]. CAS can compute Molien series and manipulate them for groups with conjugacy class lengths not too big. (This part of CAS presently works with ordinary FORTRAN complex numbers.) The definitions are as follows. Let the finite group $G$ act on the $K$-vector space $V$, $\mathrm{char}(K) = 0$, $K$ a splitting field of $G$; denote the character associated to $V$ by $\psi$. $V$ can be embedded in a polynomial ring $K_V$ (with $\psi(1)$ indeterminates) such that $V$ corresponds to the homogeneous polynomials of degree 1 in $K_V$. The action of $G$ on $V$ induces an (algebra automorphism) action of $G$ on $K_V$ such that the natural grading of $K_V$ is respected: the subspace $K_V{}^d$ of homogeneous polynomials of degree $d \geqslant 0$ of $K_V$ is a KG-module isomorphic to the $d$th symmetrized power of $V$. The corresponding character is $\psi^{[d]}$ in the terminology of 3.4 (v). The Molien series

$$M_{V,\chi_i}(z) = M_{\psi,\chi_i}(z)$$

of $V$ (resp. of $\psi$) with respect to the irreducible character $\chi_i$ of $G$, is the generating function ($\in \mathbb{Z}[[z]]$) of the multiplicities $(\chi_i, \psi^{[d]})$ of $\chi_i$ as constituent of $\psi^{[d]}$:

$$M_{\psi,\chi_i}(z) = \sum_{d=0}^{\infty} (\chi_i, \psi^{[d]}) z^d$$

CAS computes $M_{\psi,\chi_i}(z)$ according to Molien's formula (cf. e.g. [**45**])

$$M_{\psi,\chi_i}(z)=\frac{1}{|G|}\sum_{g\in G}\frac{\bar{\chi}_i(g)}{\det(I-zD(g))}=\frac{(-1)^{\psi(1)}}{|G|}\sum_{g\in G}\frac{\chi_i(g)\det D(g)}{\det(zI-D(g))}$$

where $D$ is the matrix representation associated with $V$ (command: MOLIEN). Of course, the sum over $G$ is replaced by a sum over the conjugacy classes $C_i$ weighted with the lengths $|C_i|$. The denominator is

$$\det(zI-D(g))=\prod_{i=1}^{\psi(1)}(z-\varepsilon_i(D(g)))$$

where $\varepsilon_1(D(g)),\ldots,\varepsilon_{\psi(1)}(D(g))$ denote the eigenvalues of $D(g)$ in some extension field of $K$. Since the computation of the $\varepsilon_i(D(g))$ amounts to the decomposition of the restriction of $\psi$ to $\langle g\rangle$ into irreducible (complex) characters of $\langle g\rangle$, the power maps for the prime divisors $p$ of $|G|$ must be known. $M_{\psi,\chi_i}(z)$ is computed as a quotient of two polynomials $P(z)$ and $Q(z)$, where the denominator $Q(z)$ is a product of cyclotomic polynomials. After this, CAS can print out the first $n$ coefficients

$$(\chi_i,\psi^{[d]})\quad(0\leqslant d\leqslant n)\quad\text{of}\quad M_{\psi,\chi_i}(z),$$

where $n\in\mathbb{N}$ has to be specified by the user (third MOLIEN command in CAS). Of particular interest is $M_\psi(z)=M_{\psi,\chi_1}(z)$, where $\chi_1=1$ is the 1-character of $G$. Namely among the vector spaces

$$K_V{}^{\chi_i}=\bigoplus_{d=0}^{\infty}e_{\chi_i}K_V{}^d$$

where $e_{\chi_i}$ is the central primitive idempotent of $KG$ corresponding to $\chi_i$ $(1\leqslant i\leqslant h)$, $K_V{}^1=K_V{}^{\chi_1}$ is the subring of $G$-invariant polynomials in $K_V$. The other vector spaces $K_V{}^{\chi_i}$ are $K_V{}^1$-modules. It is known (cf. Chapter 3 of [**45**]) that $K_V{}^1$ is a Cohen-Macaulay ring and that the $K_V{}^{\chi_i}$ are Cohen-Macaulay modules, i.e. for each set $\theta_1,\ldots,\theta_{\psi(1)}$ of homogeneous polynomials in $K_V^1$ such that $K_V^1$ is a finitely generated $K[\theta_1,\ldots,\theta_{\psi(1)}]$-module, one has the following: for each $i$, $1\leqslant i\leqslant h$, there are homogeneous polynomials $\eta_{1i},\ldots,\eta_{s(i)i}$ in $K_V{}^{\chi_i}$ such that

$$K_V{}^{\chi_i}=\bigoplus_{j=1}^{s(i)}\eta_{ji}K[\theta_1,\ldots,\theta_{\psi(1)}].$$

(The $\theta_i$ exist and can be chosen in such a way that their degrees divide the group order $|G|$.) This implies that

$$M_{\psi,\chi_i}(z)=\frac{\sum_{j=1}^{s(i)}z^{d_{ji}}}{\prod_{k=1}^{\psi(1)}(1-z^{d_k})}=\sum_{j=1}^{s(i)}z^{d_{ji}}\prod_{k=1}^{\psi(1)}\sum_{n=0}^{\infty}z^{nd_k}$$

where $d_k$ is the degree of $\theta_k$ and $d_{ji}$ the degree of $\eta_{ji}$, (cf. Chapter 2 of [**45**]). CAS provides a help to guess these degrees. After $M_{\psi,\chi_i}(z)$ has been computed as a quotient $P(z)/Q(z)$ as described earlier, the user may specify an $\psi(1)$-tuple $(\tilde{d}_1,\ldots,\tilde{d}_{\psi(1)})$ of candidates of degrees for the $\theta_i$ and CAS reduces the fraction $P(z)/Q(z)$ to higher terms:

$$\frac{\tilde{P}(z)}{\prod_{k=1}^{\psi(1)}(1-z^{\tilde{d}_k})} = \frac{P(z)}{Q(z)},$$

if possible (second MOLIEN command in CAS). The user can try new tuples $(\tilde{d}_1,\ldots,\tilde{d}_{\psi(1)})$, until he is satisfied, e.g. until the coefficients of $P(z)$ are non-negative.

We remark without proof that, also, Stanley's generalized Molien series, p. 492 of [**45**], for reducible KG-modules can be obtained from the Molien series just discussed and from multiplicities of irreducible characters in products of irreducible characters, which can of course also be computed in CAS.

### 5.3 Detecting subgroups

Throughout this section, it is assumed that the character table of a group $G$ together with the powermaps for all prime divisors of the group order $|G|$ are given. Some ways of getting evidence—in some cases even necessary and sufficient criteria—for the existence of subgroups with certain properties will be discussed. There are certainly different types of information one can have about a subgroup $H$ of $G$, e.g.

(a) (the isomorphism type and) the character table of $H$ and the fusion of the conjugacy classes of $H$ into those of $G$; or
(b) the character of $G$ of the permutation representation of $G$ on the cosets of $H$ in $G$.

Certainly (b) contains less information than (a) and there is a range of possibilities between (a) and (b). Accordingly we discuss different strategies $A1$, $A2$ and $B$ to find subgroups.

($A$) Specific isomorphism types of subgroups
($A1$) Using class multiplication constants

CAS supplies a routine for computing the class multiplication constants $a_{ijk}$ $(1 \leqslant i,j,k \leqslant h)$ defined in (3.6). For the subgroup search it is useful to note that $a_{ijk}|C_k|$ is equal to the number of homomorphisms $\alpha$ of the group (cf. [**13**])

$$(l,m,n) := \langle x,y \mid x^l = y^m = (xy)^n = 1\rangle$$

into $G$ with $x\alpha \in C_i$, $y\alpha \in C_j$, and $(xy)\alpha \in C_k$, where $l$, $m$, and $n$ resp. are the orders of the elements in $C_i$, $C_j$, and $C_k$ resp.

If
$$\frac{1}{l}+\frac{1}{m}+\frac{1}{n} > 1,$$

the group $(l,m,n)$ is a finite polyhedral group (cf. [**13**]) and $\alpha$ is a monomorphism under this hypothesis, since $(l,m,n)$ has no proper normal subgroup $N$ such that $xN$, $yN$, and $xyN$ have orders $l$, $m$, and $n$ resp. The following result is well known among experts.

*Proposition 1.* Let $C_i, C_j, C_k$ be fixed conjugacy classes of $G$ with elements of order $l$, $m$, and $n$ resp., and let $u(l,m,n)$ denote the number of subgroups $U$ of $G$ with $U$ isomorphic to $(l,m,n)$ and having nontrivial intersections with $C_i$, $C_j$ and $C_k$. For

$$\frac{1}{l}+\frac{1}{m}+\frac{1}{n} > 1$$

the number $u(l,m,n)$ can be computed according to the list below, and for each subgroup $U$ just described, the fusion can easily be obtained from the power maps of $G$.

(i) $(U \cong V_4)$ $\quad u(2,2,2) = \frac{1}{6} a_{ijk}|C_k|s_{ijk}$

with $$s_{ijk} = \begin{cases} 1 & \text{if } \; i=j=k \\ 3 & \text{if } \; |\{i,j,k\}| = 2 \\ 6 & \text{if } \; |\{i,j,k\}| = 3 \end{cases}$$

(ii) $(U \cong D_{2n}, n > 2)$ $\quad u(2,2,n) = \frac{(2-\delta_{ij})}{n\phi(n)} a_{ijk}|C_k|\text{gal}(C_k)$

with $\delta_{ij}$ = Kronecker delta, $\phi$ = Euler $\phi$-function,
$\text{gal}(C_k)$ = number of columns in the character table of $G$, which are Galois conjugate to the one of $C_k$;

(iii) $(U \cong A_4)$ $\quad u(2,3,3) = \frac{1}{24} a_{ijk}|C_k|\text{gal}(C_k)$

note, $j = k$ in this case;

(iv) $(U \cong S_4)$ $\quad u(2,3,4) = \frac{1}{24} a_{ijk}|C_k|$;

(v) $(U \cong A_5)$ $\quad u(2,3,5) = \frac{1}{120} a_{ijk}|C_k|\text{gal}(C_k)$.

*Proof.* All conjugacy classes of $(l, m, n)$ for

$$\frac{1}{l}+\frac{1}{m}+\frac{1}{n}>1$$

are obtained by those of $x, y$ and $xy$ by taking powers. This proves the statement about the fusion. The proofs of (i)–(v) are all similar; therefore (i) might serve as a representative example. If $C_i \neq C_j \neq C_k \neq C_i$, i.e. $i \neq j \neq k \neq i$, then the number $\mu$ of monomorphisms $\alpha:(2,2,2)\rightarrow G$ such that the image has a non-empty intersection with $C_i$, $C_j$, and $C_k$ is given by

$$\mu = \sum_{\pi} a_{i\pi j\pi k\pi}|C_{k\pi}|$$

where the sum is taken over all 6 permutations $\pi$ of $\{i,j,k\}$. But,

$$a_{i\pi j\pi k\pi}|C_{k\pi}| = \frac{|C_i||C_j||C_k|}{|G|}\sum_{\chi}\frac{\chi(g_i)\chi(g_j)\chi(g_k)}{\chi(1)}$$

where $g_i \in C_i$, $g_j \in C_j$, $g_k \in C_k$ and the sum runs over all irreducible characters $\chi$ of $G$, (cf. p. 45 of [**29**]). Note the rationality of the characters on classes with element orders 2 was used, since otherwise $\chi(g_{k\pi})$ has to be replaced by $\overline{\chi(g_{k\pi})}$ in the above formula. Hence, $\mu = 6a_{ijk}|C_k|$.

Similarly one sees $\mu = 3a_{ijk}|C_k|$ resp. $\mu = a_{ijk}|C_k|$ if $\{i,j,k\}$ consists of 2 resp. 1 element. $\mu$ has to be divided by the order of the automorphism group of $(l, m, n)$, which is 6 in case (i). This yields the desired formula in the case of the Klein four group.

Under favourable conditions one might even be able to say in how many conjugacy classes the groups of the last proposition split and to get some information about the normalizers.

If

$$\frac{1}{l}+\frac{1}{m}+\frac{1}{n}\leqslant 1,$$

the group $(l, m, n)$ is infinite (cf. [**13**]). In this case $m_{ijk} \neq 0$ indicates that some finite epimorphic image of $(l,m,n)$ is a subgroup of $G$ intersecting $C_i$, $C_j$ and $C_k$ as described above. It might be useful to have some insight into what the finite epimorphic images of $(l, m, n)$ are—at least up to a certain order. Note, abelianizing $(l, m, n)$ shows that $(l, m, n)$ (and hence all its epimorphic images) is perfect if and only if $l$, $m$, and $n$ are pairwise relatively prime. The first of these cases occurs for $(l, m, n) = (2, 3, 7)$. Since it is not generally available, we include a complete list (in Table 5) of all finite epimorphic images $U$ of $(2, 3, 7)$ which have a faithful permutation representation of degree $\leqslant 40$. This list was computed using an implementation [**34**] of Sims' method for finding all subgroups up to prescribed low index.

*Table 5*

| isomorphism type | $PSL_2(7)$ | $PSL_2(8)$ | $PSL_2(13)$ | $2^3 \cdot PSL(3,2)$ |
|---|---|---|---|---|
| degree(s) of faithful permutation representation | 7, 8, 21, 24, 28 | 9, 28, 36 | 14, 28 | 14, 28 |

| $A_{15}$ | $A_{21}$ | $3^7 \cdot 2^3 \cdot PSL_3(2)$ | $A_{22}$ | $PSL(2,27)$ | $A_{28}$ | $2^{14} \cdot 3^7 \cdot 2^3 \cdot PSL_3(2)$ | $A_{29}$ |
|---|---|---|---|---|---|---|---|
| 15 | 21 | 21 | 22 | 28 | 28 | 28 | 29 |

| $PSL_2(29)$ | $2^{14} \cdot A_{15}$ | $A_{35}$ | $A_5 \wr PSL_2(7)$ (of order $60^7 \cdot 168$) | $A_{36}$ | $A_{37}$ |
|---|---|---|---|---|---|
| 30 | 30 | 35 | 35 | 36 | 37 |

(Notation: $n \cdot G$ means an extension of a group of order $n$ by $G$)

It should be noted that Conder has determined all $n \in \mathbb{N}$ for which the alternating group $A_n$ is an epimorphic image of $(2, 3, 7)$ (cf. [**11**]).

In this context, also the paper [**9**] is of interest, in which, for each simple group of order less than $10^6$ and not isomorphic to $PSL_2(q)$ for some $q$, a presentation as epimorphic image of $(l, m, n)$ is given with the triple $l, m, n$ as small as possible in the lexicographical ordering. What has been said about subgroups of $G$ which are epimorphic images of $(l, m, n)$, applies *mutatis mutandis* to those which are epimorphic images of $\langle x, y \mid x^l = y^m = (x^s y^t)^n = 1 \rangle$ for some $l, m, n, s, t \in \mathbb{Z}$, since the powermaps of $G$ are known. The approach via epimorphic images of some infinite group has the disadvantage that one has to make sure that the subgroup is not $G$ itself. If any group $U$ is given (which is not necessarily an epimorphic image of one of the groups discussed above), one can sometimes conclude that $U$ cannot be embedded into $G$ by comparing the class multiplication constants.

*Remark 2.* Let $U$ be a finite group with conjugacy classes $\tilde{C}_i$ $(i = 1, \ldots, \tilde{h})$ and class multiplication constants $\tilde{a}_{ijk}$ $(i, j, k = 1, \ldots, \tilde{h})$. If $\phi: U \to G$ is a monomorphism with $\tilde{C}_i \phi \subseteq C_{i'}$ $(i = 1, \ldots, \tilde{h})$, then the class multiplication constants $a_{i'j'k'}$ of $G$ satisfy

$$\sum_{\substack{s' = i' \\ t' = j'}} \tilde{a}_{stk} \leqslant a_{i'j'k'} \quad (1 \leqslant i, j, k \leqslant \tilde{h}).$$

The proof is straightforward. For the application one usually has to check the various possibilities for the fusion $(i \to i')$ of the conjugacy classes. These possibilities are usually restricted by the fact that the powermaps of $U$ and $G$ have to be compatible.

### (A2) *Using the character table directly*

The natural possibility to prove the nonexistence of an embedding is to pretend that an embedding $\phi: U \rightarrow G$ exists and to restrict the irreducible characters of $G$ to those of $U\phi$ and check whether the "restricted characters" decompose properly into irreducibles of $U\phi$ (with non-negative integral multiplicities), i.e. one checks, whether the class functions of $U$ defined via the assumed fusion of the conjugacy classes of $U\phi$ into those of $G$ are characters. CAS supplies a routine for both, restricting the characters of $G$ to those of $U\phi$, and for computing the scalar products of the resulting class functions with the irreducible characters of $U$.

### (B) *Finding candidates for permutation characters*

There are various necessary conditions for a character of $G$ to be the character of a transitive permutation representation. Unfortunately no sufficient criterion is known, and for big groups one gets the impression that all the necessary conditions taken together are far from being sufficient.

There are two problems to discuss in this section. The first is concerned with the fast generation of characters of a given degree or within a certain range of degrees, which satisfy some easily testable properties of transitive permutation characters (= characters of transitive permutation representations). The second problem, then, is to apply other, more sophisticated and more time consuming tests to these characters, to eliminate some of them, or possibly even to prove that they are transitive permutation characters. In the present version of CAS, the first problem is taken care of by generating all characters

$$\pi = \sum_{i=1}^{h} a_i \chi_i$$

($\chi_1, \ldots, \chi_h$ irreducible characters of $G$, $a_i \in \mathbb{Z}_{\geqslant 0}$) which satisfy:

(i) $a_1 = 1$, i.e. the multiplicity of the trivial character in $\pi$ is 1;

(ii) if $\chi_i$ and $\chi_j$ are algebraically conjugate, then $a_i = a_j$;

(iii) the $a_i$ are bounded by parameters which can be defined by the user. If they are not, they are bounded by

$$\min\left(\chi_i(1), \frac{\pi(1)}{\chi_i(1)}\right).$$

(Name of the command: PERMUT).

The program first generates all characters $\pi$ satisfying (i), (ii), and (iii) of specified degree $d_0 \,||\, |G|$ or of degrees $d_0 \,||\, |G|$ within an interval defined by the user. To do this the program needs the rationalized character table of $G$, which can be computed from the original character table of $G$ by the command GALOIS (cf. 3.4 (i)) in CAS. If the user happens to know the Schur indices (over the rationals or some bigger field) of the $\chi_i$, he is well advised to multiply the characters in the rationalized character table by these Schur indices because they divide the corresponding $a_i$s. The program will then produce integral linear combinations of these characters.

CAS also takes care of part of the second problem by applying tests (which can be chosen by the user) to the computed characters $\pi$. However, there usually remains some work to be done to discuss the remaining characters in more detail.

The tests can be divided into three types, namely:

(I) tests based on the fact that the restriction to a subgroup must be (a not necessarily transitive) permutation character;
(II) tests resulting from properties of the (hypothetical) stabilizer;
(III) tests coming from modular representation theory.

*Type (I).* Certainly the restriction of $\pi$ to each (maximal) cyclic subgroup must be a permutation character. Fortunately permutation characters of cyclic groups are easy to recognize.

*Remark 3.* The Burnside ring (=Grothendieck ring of permutation representations) of a cyclic group $U$ is isomorphic to the ring of rational characters of $U$.

*Proof.* The number of subgroups of $U$ is equal to the number of rational irreducible characters of $U$. For $d \,||\, |U|$ let $\psi_d$ denote the $\mathbb{Q}$-irreducible character of $U$ of degree $\phi(d)$ ($\phi$ = Euler $\phi$-function) and $\pi_d$ the transitive permutation character of degree $d$. One has

$$(*) \quad \pi_d = \sum_{r|d} \psi_r \text{ and hence, } (**) \quad \psi_d = \sum_{r|d} \mu\left(\frac{d}{r}\right)\pi_r$$

for each $d \,||\, |U|$ ($\mu$ = Möbius function). The statement follows.

The resulting tests (Test 1 and Test 7) in CAS are

(T1) $a_1(g) = \pi(g) \geqslant 0$ and

(T7) $a_i(g) = \dfrac{1}{i}\sum_{r|i} \mu\left(\dfrac{d}{r}\right)\pi(g^r) \geqslant 0$

for each $g \in G$, $i \mid |g|$, $i > 1$. (If $\pi$ is a permutation character $a_i(g)$ is the number of $i$-cycles of the permutation corresponding to $g$.)

Unfortunately, Test 7 is rather time consuming. For groups containing only elements of prime power order or of order $pq$ with prime numbers $p$ and $q$, it suffices to check (T1) and

(T2) $\pi(g^p) \geqslant \pi(g)$, for all prime divisors $p$ of $|g|$ for all $g \in G$,

since we have the following.

*Remark 4.* Let $q$ and $r$ be (different) prime numbers. A rational character $\pi$ of a cyclic group of order $q^\alpha$ ($\alpha \in \mathbb{N}$) or of order $qr$ is a permutation character if and only if (T1) and (T2) are satisfied.

*Proof.* For cyclic groups of order $q^\alpha$ this follows immediately from the formula for $a_i(g)$ in (T7) and $\mu(q^a) = 0$ for $a \geqslant 2$. Let the group order be $qr$. By Remark 3, $\pi$ can be written as $\pi = a_1\pi_1 + a_q\pi_q + a_r\pi_r + a_{qr}\pi_{qr}$ with $a_1, a_q, a_r$, $a_{qr} \in \mathbb{Z}$. Since $\pi$ is a character and $\psi_{qr}$ occurs only as constituent of $\pi_{qr}$, one gets $a_{qr} \geqslant 0$. Since $\pi(g) = a_1$ for a generator $g$ the group, one has $a_1 \geqslant 0$. Finally, one easily checks for a generator $g$ by inspecting character values that $qa_q = \pi(g) - \pi(g^q)$ and $ra_r = \pi(g) - \pi(g^r)$. Hence $a_q \geqslant 0$ and $a_r \geqslant 0$, showing that $\pi$ is a permutation character. The converse is trivial.

For cyclic groups of order $qr^2$, $q$ a prime number different from $r$, Remark 4 does not hold any more (in the notation of the proof of Remark 3, $f(\pi_{qr^2} + \pi_q + \pi_r) - \pi_{qr}$ is our counter example for $f \in \mathbb{N}$ sufficiently big). Nevertheless, in practice (T2) forms a good substitute for (T7).

The next simplest class of subgroups $U$, one might restrict $\pi$ to, consists of non-cyclic groups which contain only cyclic subgroups. For these groups, one can still easily decide whether a character is a permutation character, though there might be several permutation representations with the same character (unlike in Remark 3). Say $U$ is of order $pq$, $p$ and $q$ prime numbers with $q \mid p-1$, then the transitive permutation characters of $U$ are given by Table 6.

With this table, one can check whether $\pi$ restricts to a permutation representation of $U$ ($|U| = pq$). The dihedral groups of order $2p$, $p$ an odd

*Table 6*

| | $g_1$ | $g_p$ | $g_q$ |
|---|---|---|---|
| $\pi_{pq}$ | $pq$ | 0 | 0 |
| $\pi_q$ | $q$ | $q$ | 0 |
| $\pi_p$ | $p$ | 0 | 1 |
| $\pi_1$ | 1 | 1 | 1 |

($|g_x| = x \in \mathbb{N}$)

prime, belong to the class of groups just discussed and can be looked for by using the class multiplication constants as discussed earlier. CAS does not yet have an automatic routine to go through this test.

*Type (II).* Under this point we have three automatic tests, the first two of which are:

(T3) $|g| \,\Big|\, \dfrac{|G|}{\pi(1)}$ for all $g \in G$ with $\pi(g) > 0$; and

(T4) $\pi(1) \,|\, |C_i| \pi(g_i)$ for $i = 1, \dots, h$, where $g_i$ lies in the $i$th conjugacy class $C_i$.

The property tested in (T3) comes from the fact that an element $g \in G$ with $\pi(g) > 0$ lies in the stabilizer of some element of the (hypothetical) $G$-set belonging to $\pi$ and hence, the order $|g|$ of $g$ divides the order

$$\frac{|G|}{\pi(1)}$$

of this stabilizer.

As for (T4), one uses the elementary fact that

$$|C_i \cap H| = \frac{\pi(g_i)|C_i|}{\pi(1)},$$

where $H$ is the stabilizer in $G$ of some point of the $G$-set belonging to $\pi$. If (T4) is applied in CAS, the orders of the intersections of $H$ with all conjugacy classes $C_i$ of $G$ are printed after $\pi$ has passed all applied tests. This information is often quite valuable for the further discussion of $\pi$ (and $H$). One of the possible tests based on these data is automatized in CAS as Test 5.

(T5) $$p(p-1) \,|\, (s-p+1) \quad \text{and} \quad \frac{s}{p-1} \,\Big|\, \frac{|G|}{\pi(1)}$$

for each prime number $p$ dividing

$$\frac{|G|}{\pi(1)}$$

exactly to the first power, where

$$s = \sum \frac{\pi(g_i)|C_i|}{\pi(1)},$$

the sum taken over the representatives $g_i$ of the conjugacy classes $C_i$ with order $|g_i| = p$.

This test is a simple application of the Sylow theorems to the subgroup $H$, namely $s$ is equal to the number of elements in $H$ of order $p$. Since $p^2 \nmid |H|$, $s/(p-1)$ is equal to the number of $p$-Sylow subgroups of $H$. The two divisibility tests correspond to the facts that this number is congruent to 1 modulo $p$ and divides the order $|H|$ of $H$.

There are a number of additional not implemented tests one can perform with the information from (T4). For example, (T5) can easily be generalized to the case where $H$ has a cyclic $p$-Sylow subgroup. Another simple test is

$$\phi(|g_i|) \left| \frac{\pi(g_i)|C_i|}{\pi(1)} \operatorname{gal}(C_i) \qquad (g_i \in C_i)\right.$$

where $\phi$ is the Euler $\phi$-function and $\operatorname{gal}(C_i) = \phi(|g_i|)\,|\langle g_i\rangle \cap C_i|^{-1}$ (as in Proposition 1). The number

$$\frac{\pi(g_i)|C_i|}{\pi(1)} \quad (=|H \cap C_i|)$$

might even help to guess the isomorphism type of $H$ and the fusion of the classes of $H$ into those of $G$. For further investigation of such a guess, the methods discussed earlier in this chapter can be applied.

*Type* (*III*). One general test (T9) using modular representation theory, is taken from Corollary A, p. 113 of [**41**]. Let $\pi = \sum \pi_B$ the sum over all $p$-blocks $B$ of $G$ (for some prime number $p$), where $\pi_B$ is zero or a sum of irreducible characters in $B$.

(T9) $|\pi_B(g)| \leqslant \pi_B(g^n) \leqslant \pi(g^n)$, for each $g \in G$, $n \in \mathbb{N}$ such that $g^n$ is a $p$-element of $G$.

Whereas this test can be applied for every degree, the other commonly known tests based on $p$-modular representation theory are restricted to degrees divisible by the order $p^n$ of a Sylow $p$-subgroup of $G$. Let $d_0 = \pi(1)$ be such a degree ($p^n | d_0, d_0 || G|$). Then $\pi_B$ (defined above) is the character belonging to a projective indecomposable $\mathbb{Z}_{(p)}G$-lattice, $\mathbb{Z}_{(p)}$ denoting the $p$-adic integers. If the $p$-decomposition numbers of $G$ are known, one can therefore proceed as follows. Instead of starting out with the rationalized character table of $G$, one starts out with the table of characters of the projective indecomposable $\mathbb{Z}_{(p)}G$-lattices. One rationalizes these characters and replaces the character $\psi_1$ belonging to the projective cover of the trivial $\mathbb{Z}_{(p)}G$-lattice by the two characters $\chi_1 = 1$ (= trivial character) and $\psi_1 - 1$ (make sure 1 is the first character of the resulting table). With this new table one then can proceed as usual. This approach has the advantage that the resulting characters have properties which are not actually tested but given *a priori*; it has the disadvantage that the decomposition numbers have to be known and that it is restricted to special degrees.

If one does not know all decomposition numbers or does not want to go through the above procedure for other reasons, one can prescribe a character which is supposed to be contained in $\pi$ (e.g. $\psi_1 - 1$ above). Sometimes one can use a special property of the Brauer tree. Namely, if $p^2 \nmid |G|$ but $p \mid |G|$ then $\psi_1 - 1$ (in the terminology above) is an irreducible character $\lambda_p$. If the user (or in simple cases even CAS) can identify $\lambda_p$, he can apply

(T8) $(\pi, \lambda_p) > 0$ for all prime numbers $p$ with $p \mid \pi(1)$ and $p^2 \nmid |G|$, for which $\lambda_p$ is known.

Using the PERMUT-command in CAS, one will soon realize two things. To keep the time for searching permutation characters $\pi$ (in a reasonable range of degrees) low, one has to put drastic bounds on the multiplicities of the irreducible constituents of $\pi$. Secondly, at least 95% of the characters which do fail in one of the tests, fail because of (T1), which therefore should always be used as first test.

### 5.4 Examples

(i) Decomposition matrix of $SU(4,2^2) \cong PSp(4,3)$ at the prime $p = 3$.

The group $G = SU(4,2^2) \cong PSp(4,3)$ (name of character table on microfiche: SU(4,4)) is well known to be isomorphic to the subgroup of index 2 of the Weyl group $W(E_6)(\cong O^-(6,2))$. Just to convince ourselves of this fact, we compute the Molien series $M_{X.4,X.1}(z)$ of the unique character $X.4$ of degree 6 of $G$. As the first transformation of the Molien series (command MOLIEN) shows, it is equal to

$$\frac{1+z^{36}}{(1-z^2)(1-z^5)(1-z^6)(1-z^8)(1-z^9)(1-z^{12})}.$$

This is indeed the Molien series of the restriction of the natural representation of $W(E_6)$ to $W(E_6)'$ (cf. [**13**], [**45**]). By the command PBLOCK we find two 3-blocks of $G$, one of which is of defect 0. By the command BRAUER we find a pre-decomposition matrix $\tilde{\mathbf{D}}$ for the principal 3-block; there are five irreducible Brauer characters in this block. As a $3'$-subgroup to induce up from we choose a Sylow 2-subgroup $S$ of $G$. By using a presentation of $S$ one could easily generate the character table of $S$ by Dixon's algorithm. We choose a different way.

*Claim.* $S$ has a faithful rational representation of degree 4.

*Proof.* $X.4$ belongs to a rational representation $D_1$ of $G$ with $\det D_1(g) = 1$

for all $g \in G$ by the introductory remarks. Since $1_S + X.2_S = X.4_S$, $S$ has a rational representation $D_2$ of degree 5 with $\det D_2(g) = 1$ for all $g \in S$. Since $S$ is a 2-group, $D_2$ has a 1-dimensional rational constituent (the complex constituents occur in pairs and 5 is odd). Because of $\det D_2(g) = 1$ for all $g \in S$, we get a faithful rational representation of degree 4 of $S$.

The character tables of groups with faithful rational representations of degree 4 are listed in [**5**]. To decide which is the character table of $S$ and what is the fusion, we first produce the permutation character $\psi$ for the permutation representation on the cosets of $S$ in $G$. The degree is 405 which is relatively big compared with the degrees of the irreducible characters of $G$. Hence we better derive bounds for the multiplicities of the $X.i$ in $\psi$. Therefore, we first produce the permutation character $\tilde{\psi}$ of the cosets of $G$ on a suitable subgroup $U$ of $S$ by other means. From the structure constants of the class $2A$ (produced with the two commands TRANSPOSE and STRUCTURE) we find a dihedral group of order 8 intersecting the two classes $2A$ and $2B$. We define a character table $D8$ for $D_8$ with the trivial character as only character. After defining the fusion of $D8$ in SU(4,4), we induce up this trivial character and decompose the induced character into irreducibles (commands INDUCE and REDUCE). The resulting multiplicities are upper bounds for the multiplicities of the $X.i$ in $\psi$. Now we rationalize the character table of $G$ by the command GALOIS and apply PERMUT to the rationalized

*Table 7* Decomposition matrix

| degrees | | 1 | 5 | 10 | 14 | 25 |
|---|---|---|---|---|---|---|
| 1 | $X.1$ | 1 | | | | |
| 5 | $X.2$ | | 1 | | | |
| 5 | $X.3$ | | 1 | | | |
| 6 | $X.4$ | 1 | 1 | | | |
| 10 | $X.5$ | | | 1 | | |
| 10 | $X.6$ | | | 1 | | |
| 15 | $X.7$ | | 1 | 1 | | |
| 15 | $X.8$ | 1 | | | 1 | |
| 20 | $X.9$ | 1 | 1 | | 1 | |
| 24 | $X.10$ | | | 1 | 1 | |
| 30 | $X.11$ | | 1 | | | 1 |
| 30 | $X.12$ | 1 | 1 | 1 | 1 | |
| 30 | $X.13$ | 1 | 1 | 1 | 1 | |
| 40 | $X.14$ | | 1 | 1 | | 1 |
| 40 | $X.15$ | | 1 | 1 | | 1 |
| 45 | $X.16$ | | | 2 | | 1 |
| 45 | $X.17$ | | | 2 | | 1 |
| 60 | $X.18$ | 1 | 2 | 1 | 1 | 1 |
| 64 | $X.19$ | | 1 | 2 | 1 | 1 |

table, to find exactly one permutation character of degree 405. Since we also used Test 5 in the command PERMUT, we get also the lengths of the intersections of $S$ with the conjugacy classes of $G$. In particular, we see that $S$ is a group of exponent 4 with 27 elements of order 2. With this information we know which character table in [**5**] is the one of $S$ and also the fusion is easily determined. We define a new table $P$ with the same table head as SU(4,4), induce up the characters of $S$, and apply the command DECOMPOSE with the option BLOCKS to the table $P$ of induced characters. The columns of the resulting matrix are indexed by the induced characters $Y.1, \ldots, Y.15$. Call the section of $Y.i$ belonging to the principal block $\overline{Y.i}$.

Application of Remark 1, shows that $\overline{Y.2}$, $\overline{Y.3}$, $\overline{Y.4}$, $\overline{Y.9}$ are projective indecomposable. Moreover, either $\overline{Y.1}$ is projective indecomposable or has $\overline{Y.2}$ as projective indecomposable summand. In the latter case, the same argument as above shows that $\overline{Y.1} - \overline{Y.2}$ is projective indecomposable. The degree of $\overline{Y.1} - \overline{Y.2}$ is $2.3^4 = 162$; by using PERMUT we see that $\overline{Y.1} - \overline{Y.2}$ might be a permutation character. One can, indeed, find a subgroup $U$ of $G$ of order $2^5.5$ yielding this permutation character. ($U$ is an extension of an elementary abelian group of order $2^4$ by a dihedral group $D_{10}$ of order 10.)

Hence, the projective indecomposable characters in the principal 3-block of $G = \mathrm{SU}(4,2^2)$ are $\overline{Y.1} - \overline{Y.2}$, $\overline{Y.4}$, $\overline{Y.9}$, $\overline{Y.2}$, $\overline{Y.3}$ and the decomposition matrix is as shown in Table 7.

(ii) Molien series for reflection group [3,3,5].

The real four dimensional reflection group [3,3,5] (notation of [**13**]), with the diagram in Fig. 2, has 34 irreducible characters. The name of the table on the microfiche is $H4$. The character of the reflection representation is $X.3$. We write a CAS-internal procedure to find all Molien series $M_{X.3,X.i}(z)$ for $i = 1, \ldots, 34$.

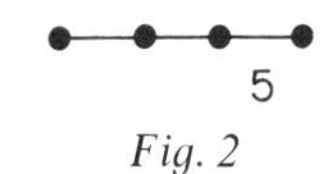


*Fig. 2*

*Procedure*

```
EXEC:
I = I + 1; DISPLAY, I;
MOLIEN, T = H4, CHARAC = 3, C = I, SERIES = SS;
MOLIEN, SERIES = SS(2,12,20,30);
DELETE, T = SS;
```

Procedure call

```
? i = 0; ende = 34;
? repeat (exec) i = ende;
```

In this procedure, first $M_{X.3,X\cdot i}(z)$ is computed by the first MOLIEN command of CAS, then this result is transformed by the second MOLIEN command such that the denominator becomes

$$(1-z^2)(1-z^{12})(1-z^{20})(1-z^{30}).$$

All these Molien series together show how the symmetrization $X.3^{[i]}$ decomposes into irreducible characters for each $i \in \mathbb{Z}_{\geq 0}$.

## 6 A Library of Character Tables in Connection with CAS

As already pointed out by Hunt [**26**], the utility of a system like CAS will be greatly increased by the easy availability through it of as many of the known character tables as possible. Therefore, already while CAS was being implemented, in addition to the implementation of some generic formulae (cf. section 3.3 (i)), character tables gathered from the literature or generated using CAS were put on file. We also asked colleagues and, in addition to hints to the literature, obtained a number of unpublished tables, some of course several times. We thank all who sent us tables, on paper or tape, in particular: J. S. Frame (orthogonal groups over GF(2)); D. C. Hunt (*i.a.* rationalized tables of F24 and BM); A. Kerber (symmetric groups and wreath products); B. Olsson, M. P. Thorne (Monster); T. Gabrysch who sent us a tape with over 100 tables, including e.g. extension and 2-cover of Fischer's F22; and for the by far biggest contribution, J. H. Conway, S. P. Norton and collaborators, who sent a tape with all tables so far prepared for the Cambridge Group Atlas, which contained more than 250 single tables in the form of "compound tables". Such a compound table combines, for the purpose of printing, in a highly condensed format the table of a simple group with those of all its covers and its extensions by cyclic outer automorphism groups.

At present, we have altogether got about 450–500 tables available to us, among them:

(i) The tables of all 26 sporadic simple groups together with all their covers and extensions by outer automorphisms;
(ii) the tables of some, often several, specimens from all the series of finite simple groups except $E_6(q)$, $E_7(q)$, and $E_8(q)$, again in many cases together with covers and extensions by outer automorphisms and, in a few cases, extensions of elementary abelian groups by the ones mentioned;
(iii) a—naturally somewhat accidental—selection of tables of soluble groups, e.g. of some subgroups of the above mentioned groups and of all finite subgroups of $GL(4,\mathbb{Z})$.

We are still—and will be for some time—in the process of bringing all these tables into the internal format of CAS, to supplement some of them, e.g. by determining irrationalities, power maps and $p$-blocks and to apply the checks mentioned in section 3.5. The list of the tables of the 26 sporadic simple groups on the microfiche supplement gives an idea of the format that is eventually intended for the whole collection.

In the library, each table gets a CAS-identifier, which we try to choose close to some standard name under which the respective group is known. As often different names are in use for the same group as well as there are isomorphisms between differently defined groups, a directory to the library will list such different notations for the same table as well as some "bibliographic" information, e.g. its size, the source from which we got it and—to the best of our knowledge—its origin.

We would be grateful for any further tables sent to us for inclusion into the library. A list of the tables at a particular time available to use—and in particular of those already in the library—will be available on request.

## References

1. Aasman, M. (1981). CAS, ein Programmsystem zum Erzeugen und Rechnen mit Charakteren—Gruppentheoretische Invarianten und Sprachbeschreibung, Diplomarbeit, RWTH Aachen.
2. Aasman, M., Janissen, W., Lammers, H., Neubüser, J., Pahlings, H. and Plesken, W. (1981a). Das CAS-System. Handbuch (vorläufige Fassung). Lehrstuhl D für Mathematik, RWTH Aachen.
3. Aasman, M., Janissen, W., Lammers, H., Neubüser, J., Pahlings, H. and Plesken, W. (1981b). The CAS system. User manual (provisional version). Lehrstuhl D für Mathematik, RWTH Aachen.
4. Atkinson, M. D., Hassan, R. A. and Thorne, M. P. (1983). Group Theory on a Micro-Computer, (these Proceedings).
5. Brown, H., Bülow, R., Neubüser, J., Wondratschek, H. and Zassenhaus, H. (1978). "Crystallographic groups of four-dimensional space". Wiley, New York.
6. Cannon, J. J. (1976). The Cayley library of built-in functions. Technical Report No. 13, Computer-Aided Mathematics Project, University of Sydney.
7. Cannon, J. J. (1982). A Language for Group Theory (Preprint). University of Sydney.
8. Cannon, J. J., Gallagher, R. and McAllister, K. (1972, revised 1974). Stackhandler: A language extension for low level set processing. Programming and implementation manual. Technical Report No. 5, Computer-Aided Mathematics Project, University of Sydney.
9. Cannon, J. J., McKay, J. and Young, K.-C. (1979). The non-abelian simple groups $G, |G| < 10^5$—presentations, *Comm. in Alg.* **7**, 1397–1406.
10. Collins, M. J. (1980). Introduction: A survey of the classification project. *In* "Finite simple groups II", 3–40. Academic Press, London, Orlando and New York.
11. Conder, M. D. E. (1980). Generators for alternating and symmetric groups, *J. London Math. Soc.* (2) **22**, 75–86.

12. Conway, J. H. (1983). Character Calisthenics, (these Proceedings).
13. Coxeter, H. S. M. and Moser, W. O. J. (1980). "Generators and Relations for Discrete Groups". 4th ed. Springer-Verlag, Berlin.
14. Deeken, G. (1978). Methoden zum Umgang mit Charakteren und ihre Implementation, Diplomarbeit, RWTH Aachen.
15. Dixon, J. D. (1967). High speed computation of group characters, *Numerische Mathematik* **10**, 446–450.
16. Dornhoff, L. (1971). "Group representation theory, Part A". Marcel Dekker, New York.
17. Esper, N. (1974). Ein interaktives Programmsystem zur Erzeugung der rationalisierten Charakterentafel einer endlichen Gruppe. Staatsexamensarbeit, RWTH Aachen.
18. Feit, W. (1982). "The representation theory of finite groups". North-Holland, Amsterdam.
19. Feit, W. The computations of some Schur indices, (to appear).
20. Frame, J. S. (1949). Congruence relations between the traces of matrix powers, *Canad. J. Math.* **1**, 303–304.
21. Frame, J. S. (1982). Recursive computation of tensor power components, *Bayreuther Mathematische Schriften* **10**, 153–159.
22. Gabrysch, T. (1977). Ein Computerprogramm zur Berechnung von Charaktertafeln und einige Anwendungen, Diplomarbeit, Universität Bielefeld.
23. Gabrysch, T. (1978). Ein Computerprogramm zur Berechnung von Charaktertafeln. Beschreibung des Programms 'CHARAC'. Mimeographed Notes, version 78–2. Fakultät für Mathematik, Universität Bielefeld.
24. Goethals, J. M. and Seidel, J. J. (1979). Spherical designs, *AMS Proc. Symp. Pure Math.* **34**, 255–272.
25. Gorenstein, D. (1979). The classification of finite simple groups: I. Simple groups and local analysis, *Bull. Amer. Math. Soc.* (*New Ser.*) **1**, 43–199.
26. Hunt, D. C. (1980). A computer-based atlas of finite simple groups, *AMS Proc. Symp. Pure Math.* **37**, 507–510.
27. Huppert, B. (1967). "Endliche Gruppen I". Springer-Verlag, Berlin.
28. Hurley, J. F. and Rudvalis, A. (1977). Finite simple groups, *Amer. Math. Monthly* **84**, 693–714.
29. Isaacs, I. M. (1976). "Character theory of finite groups". Academic Press, London, Orlando and New York.
30. James, G. D. (1973). The modular characters of the Mathieu groups, *J. Alg.* **27**, 57–111.
31. James, G. D. and Kerber, A. (1981). "The representation theory of the symmetric groups". Addison-Wesley, Reading, Mass.
32. Janißen, W. (1981). CAS, ein Programmsystem zum Erzeugen und Rechnen mit Charakteren. Erzeugungsalgorithmen und Datenstrukturen, Diplomarbeit, RWTH Aachen.
33. Lammers, H. (1981). CAS, ein Programmsystem zum Erzeugen und Rechnen mit Charakteren. Arithmetik und Verbindung zur modularen Theorie, Diplomarbeit, RWTH Aachen.
34. Lepique, E. (1972). Ein Programm zur Berechnung von Untergruppen von gegebenem Index in endlich präsentierten Gruppen, Diplomarbeit, RWTH Aachen.
35. McKay, J. (1970). The construction of the character table of a finite group from generators and relations. *In* "Computational Problems in Abstract Algebra" (Ed. J. Leech), 89–100. Pergamon Press, Oxford.

36. McKay, J. (1979). The non-abelian simple groups $G$, $|G| < 10^6$—character tables. *Comm. in Alg.* **7**, 1407–1445.
37. Müller, W. (1980). "Darstellungstheorie von endlichen Gruppen". Teubner, Stuttgart.
38. Murnaghan, F. D. (1958). The orthogonal and symplectic groups. *Comm. Dublin Inst. Adv. Studies, Series A*, No. 13.
39. Parker, R. A. (1983). The computer calculation of modular characters, (these Proceedings).
40. Plesken, W. (1982). Counting with groups and rings, *J. Reine Angew. Math.* **334**, 40–68.
41. Scott, L. L. (1973). Modular permutation representations, *Trans. Amer. Math. Soc.* **175**, 101–121.
42. Serre, J.-P. (1977). "Linear Representations of Finite Groups". Springer-Verlag, New York.
43. Simpson, W. A. and Frame, J. S. (1973). The character tables for $SL(3,q)$, $SU(3,q^2)$, $PSL(3,q)$, $PSU(3,q^2)$, *Canad. J. Math.* **25**, 486–494.
44. Sloane, N. J. A. (1977). Error-correcting codes and invariant theory: New applications of a nineteenth-century technique. *Amer. Math. Monthly* **84**, 82–107.
45. Stanley, R. P. (1979). Invariants of finite groups and their applications to combinatorics, *Bull. Amer. Math. Soc. (New Ser.)* **1**, 475–511.
46. Steinberg, R. (1951). The representations of $GL(3,q)$, $GL(4,q)$, $PGL(3,q)$, and $PGL(4,q)$, *Canad. J. Math.* **3**, 225–235.
47. Weyl, H. (1938). "The classical groups", 1st ed. Princeton University Press.

# Character Calisthenics

J. H. CONWAY

## 1 Introduction

This talk was prompted by some little details in Professor Neubüser's description of the CAS system at the Symposium. Around 1970, M. J. T. Guy implemented a small system of this kind, in which one could compute tensor products and symmetrized powers etc., and then attempt to reduce them in various ways. Using this, we computed the character tables of various groups (the sporadic groups of Conway, McLaughlin, Suzuki, in particular), often using information about the conjugacy classes and particular characters supplied by other people (Thompson and Patterson, Livingstone,...).

Although this Cambridge experience is rather dated, it might still be of value to others, since we learnt quite a few tricks along the way. For instance, Professor Neubüser remarked that the CAS system contains the facility to test whether the difference of two characters is an irreducible. We used a strictly stronger test (*Guy's Inclusion Lemma,* Theorem 2($\leqslant$) below) which gives the exact condition under which one can determine, from their inner products, whether the difference of two characters is a proper (rather than a virtual) character. The condition is very simple, though not at all obvious, and it suggests the more general problem of determining just what information there is in the inner product matrix of a family of characters. More recently, Guy has found a way of extending his criterion so as to make maximal use of the indicator information.

Let us set the stage. Given the value of characters $\alpha, \beta, \gamma, \ldots$, we can compute their inner products, using the formula

$$(\alpha,\beta) = \sum \alpha(x) . \beta(x^{-1})/|G| \text{ (summed over all of } G)$$
$$= \sum \alpha(x) . \beta(x^{-1})/|C(x)| \text{ (summed over class representatives).}$$

Of course $\beta(x^{-1})$ is just the complex conjugate of $\beta(x)$. The norm of a character is just its inner product with itself:

$$\text{norm}(\alpha) = (\alpha, \alpha).$$

COMPUTATIONAL GROUP THEORY
ISBN 0-12-066270-1

If the square map is available (and it is well worthwhile taking some effort to make it so), one can also compute the (Frobenius-Schur) *indicator*, $\operatorname{ind}(\alpha)$, using

$$\operatorname{ind}(\alpha) = \sum \alpha(x^2)/|G| \text{ (summed over all of } G)$$
$$= \sum \alpha(x^2)/|C(x)| \text{ (summed over class representatives).}$$

We have $\operatorname{ind}(\alpha)$ = the inner product $(\alpha, \iota)$, where $\iota$, the *indicator character*, is a virtual character whose typical value $\iota(x)$ is the number of group elements $y$ with $y^2 = x$ (square roots of $x$).

The group $G$ will have irreducible characters $\chi_1, \chi_2, \ldots$ of which any character is an integral linear combination

$$\alpha = a_1\chi_1 + a_2\chi_2 + \ldots.$$

The $a_t$ are called the *multiplicities* of $\alpha$. For *proper* characters, the $a_t$ are integers $\geqslant 0$, but it is often convenient to allow also *virtual* characters, for which the $a_t$ may be integers of any signs.

Our problem can now be stated. Suppose that we are given more or less complete information about the conjugacy classes of $G$, including probably the square map and other power maps, and a starting stock of characters of $G$, which are probably not irreducible. Find, using as little further group-theoretical properties of $G$ as we can get away with, as many irreducible characters as we can.

## 2 Finding Characters

Our topic is really the reduction of characters once found, but in practice this process usually takes place simultaneously with the generation of new characters to be reduced, so we had better describe the main ways of finding characters. For the initial stock these will probably include:

(i) Explicit evaluation of traces in a known matrix representation;
(ii) Computation of characters of known permutation representations;
(iii) Induction from a known subgroup of $G$;
(iv) Restriction from a known supergroup of $G$;
(v) Use of Clifford theory from a known quotient of $G$.

One must not forget, either, that nowadays there are many families of groups for which explicit formulae for characters are available in the literature. Even if the entire character table is theoretically known in this way, it can still be cheaper in practice to ignore most of this information. One simply quotes a few characters from the general theory (leaving open the possibility of returning for more if the going gets tough), and reconstructs the rest from these.

However, when a few starting characters have been found, by far the easiest way to get more is to compute tensor products and symmetrized tensor powers of already known ones. The typical value of a *tensor product character* $\alpha\beta$ is just the product of values of $\alpha$ and $\beta$:

$$\alpha\beta(x) = \alpha(x) \,.\, \beta(x)$$

and the *tensor powers* defined as iterated tensor products

$$\alpha^1 = \alpha, \alpha^{n+1} = \alpha^n \,.\, \alpha$$

have well known decompositions into *symmetrized tensor powers*, for example

$$\alpha^2(x) = \alpha^{2+}(x) + \alpha^{2-}(x),$$

where

$$\alpha^{2+}(x) = \frac{1}{2}(\alpha^2(x) + \alpha(x^2))$$

$$\alpha^{2-}(x) = \frac{1}{2}(\alpha^2(x) - \alpha(x^2)).$$

The square, cube, and fifth power maps enable us to compute all symmetrized tensor powers up to the sixth degree (beyond which point they tend to be too large to be useful), and any computer system for working with characters should be designed so that tensor products and symmetrized tensor powers are routinely available. If a character is known to be the character of a representation writable over the real field (the orthogonal case below), there is a slight extra decomposition of the tensor powers (due to Murnaghan) which should also be made available. (An analogue for the symplectic case also exists, but is less useful.)

## 3 The Type Symbol

For an irreducible character $\chi$, the Frobenius-Schur indicator, which is readily computable when the square map is known, gives complete information about the reality properties of $\chi$:

$\text{ind}(\chi) = +1$ if $\chi$ is the character of a representation writable over $\mathbb{R}$;
$\text{ind}(\chi) = -1$ if not, but still $\chi$ takes only real values;
$\text{ind}(\chi) = 0$ if some value of $\chi$ is non-real.

The three cases are usually called *orthogonal*, *symplectic*, and *unitary* since we can respectively regard $\chi$ as obtained by restriction from a character of the orthogonal, symplectic, or unitary group. The abbreviations +, −, o are often used for the indicator values +1, −1, 0 of an irreducible representation.

Even when a character is not irreducible, it is sensible to store information about its reality, since this can give useful restrictions on the multiplicities. A convenient way to do this uses our *type symbol* $(r,s)$, which also tells us whether the character is known to be a proper character (multiplicities $\geqslant 0$) or may be virtual.

The *reality mark* $r$ takes the values:

$r = 0$, which indicates nothing about the reality of $\alpha$;
$r = 1$, which indicates that all the values $\alpha(x)$ are real;
$r = 2$, which indicates that $\alpha$ is an orthogonal character.

A proper character is called *orthogonal* if it is the character of a representation writable over the reals, and a virtual character is orthogonal if it can be expressed as the difference of two proper orthogonal characters.

The *sign mark* $s$ takes the values

$s = +$, which indicates that $\alpha$ is a proper character
$s = -$, which indicates that $\alpha$ may be virtual.

### 3.1 How the type symbol restricts multiplicities

If $r = 0$, no information.

If $r = 1$, the multiplicity of any irreducible whose indicator is 0 must be the same as that of its complex conjugate.

If $r = 2$, then also the multiplicity of any irreducible whose indicator is $-$ must be *even*.

If $s = -$, no information.

If $s = +$, all multiplicities must be non-negative.

### 3.2 How to combine type symbols

If $\alpha$ has been obtained from characters $\beta, \gamma, \delta, \ldots$ by induction, restriction, tensor multiplication, or taking symmetrized tensor powers, then we can take

$$r(\alpha) = \min(r(\beta), r(\gamma), r(\delta), \ldots)$$
$$s(\alpha) = \min(s(\beta), s(\gamma), s(\delta), \ldots).$$

### 3.3 Upgrading type symbols

There are many occasions when examination of a character enables us to provide it with a rather stronger type symbol than that with which it originally arrived. For example, if its values are all real, we can obviously upgrade from $r = 0$ to $r = 1$. If a proper character has its indicator and norm equal, we can upgrade to $r = 2$. If two proper characters $\alpha$ and $\beta$ are known to satisfy $\alpha \leqslant \beta$ (perhaps by Guy's inclusion criterion) and $r(\beta) = 2$, then we can take $r(\alpha) = r(\beta - \alpha) = 2$. If in the course of backtracking it is found that the multiplicities and indicators of the irreducibles occurring in $\alpha$ satisfy the conditions that enable us to increase $r$ or $s$ to some new value, this should always be done, since although the extra information may be "trivial" for $\alpha$ itself, it can often be of considerable value for later characters derived from $\alpha$.

### 3.4 The type symbol and the indicator

We have already remarked that for an irreducible, the indicator gives complete information about reality:

$$\text{ind} = + \text{ implies } r = 2, \text{ ind} = - \text{ implies } r = 1, \text{ else } r = 0.$$

(Of course $s$ may be taken as $+$ for an irreducible.) For other characters the relation is not quite so direct but, nevertheless, the indicator does give useful information about the type. The basic idea is that if $\alpha = \sum a_t \chi_t$, then $\text{ind}(\alpha) = \sum a_t \text{ind}(\chi_t)$, and that $\text{ind}(\chi_t)$ is restricted to the three values $1, -1, 0$. Taken together with the formula $\text{norm}(\alpha) = \sum a_t^2$, and the fact that $a_t^2 \geqslant a_t$ with equality only for $a_t = 0$ or 1, we can often deduce a nugget of information. The above remark about the case when the indicator equals the norm is one example, for then we must have $a_t \text{ind}(\chi_t) = 1$ for each constituent of $\alpha$, and so $a_t = \text{ind}(\chi_t) = \pm 1$. Again, a character $\alpha$ of norm 3 must be the sum of three distinct irreducibles (if proper) and so, if its indicator is less than 3, it cannot have $r = 2$. A norm 2 character whose indicator is 0 and which is proper and has $r = 2$, must be the sum of an irreducible and its complex conjugate. And so on.

## 4 Square-Dancing

We have come close to describing the final stage of the character reduction process as normally performed. This is a backtracking process in which we attempt to determine the unknown multiplicities of the irreducible constituents of given characters from the matrix of inner products of these characters, together with their indicators and type symbols when available. We call this the *square-dancing* process since it makes heavy use of the equation $\sum a_t^2 = \text{norm}(\alpha)$. If a unique solution is obtained, one hopes to be able to solve some of the resulting linear equations so as to recover the irreducibles involved. Even if this is impossible, it can often happen that the information gives us some new characters of small norm which can be input to the character generation process so as to give us a simpler problem.

We shall not describe the square-dancing process in much detail here (see

[2]), but will just concentrate on some easy examples to illustrate the use of the indicator and type-symbols.

If the inner product matrix of the proper characters $\alpha$ and $\beta$ is

$$\begin{bmatrix} 2 & 1 \\ 1 & 2 \end{bmatrix}$$

then the condition $\sum a_t^2 = 2$ implies that $\alpha$ is the sum of two distinct irreducibles, say $\alpha = \chi_1 + \chi_2$. A similar result holds for $\beta$, and the condition $\sum a_t b_t = 1$ implies that $\alpha$ and $\beta$ have just one constituent in common, so that we have, without loss of generality

$$\alpha = \chi_1 + \chi_2, \ \beta = \chi_2 + \chi_3, \ \chi_1, \chi_2, \chi_3 \text{ distinct.}$$

Since only the vector of multiplicities is important, we shall express this by writing

$$(+) \quad \begin{matrix} \alpha \\ \beta \end{matrix} \begin{bmatrix} 2 & 1 \\ 1 & 2 \end{bmatrix} \Rightarrow \alpha = (1, 1, 0), \ \beta = (0, 1, 1).$$

(The $(+)$ at the left indicates that all characters are supposed proper.)

If we are also given the indicators, say $\mathrm{ind}(\alpha) = 1$, $\mathrm{ind}(\beta) = 2$, we could further deduce that $\chi_2$ and $\chi_3$ must both have indicator $+$, and then that $\chi_1$ has indicator $\circ$. We should write this implication as:

$$(+) \quad \begin{matrix} \alpha \\ \beta \end{matrix} \begin{bmatrix} 2 & 1 \\ 1 & 2 \end{bmatrix} \begin{matrix} 1 \\ 2 \end{matrix} \Rightarrow \alpha = (1, 1, 0), \quad \beta = (0, 1, 1), \quad \mathrm{ind} = (\circ, +, +).$$

In this case the indicator information about $\alpha$ and $\beta$ was only used to derive the indicators of the constituents, but often it more significantly restricts the reduction. For instance the $1 \times 1$ matrix $[6]$ for a proper character admits two solutions:

$$(+) \quad \alpha[6] \Rightarrow \alpha = (2, 1, 1) \quad \text{or} \qquad \alpha = (1, 1, 1, 1, 1, 1),$$

but if the indicator is also known to be 6, only one of these survives:

$$(+) \quad \alpha[6]6 \Rightarrow \alpha = (1, 1, 1, 1, 1, 1), \quad \mathrm{ind} = (+, +, +, +, +, +).$$

An indicator of 5 would again give a unique solution:

$$(+) \quad \alpha[6]5 \Rightarrow \alpha = (1, 1, 1, 1, 1, 1), \quad \mathrm{ind} = (+, +, +, +, +, \circ).$$

Backtracking is notoriously an expensive occupation, and it is therefore desirable to postpone it for as long as possible and to take measures to reduce the size of the eventual square dancing problem if it cannot be postponed indefinitely. There are two basic strategies for doing this; one which works freely with virtual characters and the other which takes some pains to keep all characters proper. The first was mentioned by Atkinson at this Conference, and the second is that suggested and used by Guy. The two strategies are in no sense rivals and both should be pursued as a matter of course in any difficult situation.

Experience suggests that a good policy is to try the general Euclidean algorithm approach, using virtual characters, first, but then to keep only the irreducibles it yields, together with a few characters of very small norm. Discarding the rest of the information, one then tries the positive Euclidean algorithm, using only proper characters, and retains the answers to a much larger norm. The reason is that square-dancing problems with virtual characters have an enormous number of solutions, and so virtual characters of quite "moderate" norms, like 4 or 5, cause combinatorial explosions which are hard to contain. Indeed, except in very large problems, it seems a good idea at first to retain only irreducibles obtained from the virtual character calculations, since even norm 2 virtual characters can cause headaches:

$$(+)\quad \begin{matrix}\alpha\\ \beta\end{matrix}\begin{bmatrix}2 & 0\\ 0 & 2\end{bmatrix} \Rightarrow \alpha=(1,1,0,0),\ \beta=(0,0,1,1)\text{ only, but}$$

$$(-)\quad \begin{matrix}\alpha\\ \beta\end{matrix}\begin{bmatrix}2 & 0\\ 0 & 2\end{bmatrix} \Rightarrow \text{this or}\ \ \alpha=(\pm1,\pm1),\ \beta=(\pm1,\mp1).$$

On the other hand, the method often does produce irreducibles, and irreducibles should be treasured no matter what their provenance. (A norm 1 virtual character is $\pm$ an irreducible, and the sign can be determined from the degree.)

## 5 Working with Virtual Characters

The *general Euclidean algorithm* should be available in "push-button" fashion. We suppose that $\alpha, \beta$ are any two of the given characters, with inner product matrix

$$\begin{bmatrix}a & h\\ h & b\end{bmatrix} \quad \text{and} \quad a \leqslant b.$$

If $|h| \geqslant a$, then nothing further is to be done with these two characters, which we call a *reduced pair*. But if $|h| > a$, then for a suitable choice of sign, we have $\text{norm}(\beta \pm \alpha) < \text{norm}(\beta)$, and so can replace $\beta$ by $\beta \pm \alpha$. (Alternatively, one can directly find an integer $r$ which minimizes $\text{norm}(\beta - r\alpha)$, and replace $\beta$ by this in one move.) The process is continued until every pair of characters is reduced. Since the norms are continually being reduced, one can hope that the process will come up with some irreducibles.

There is a connection between this reduction algorithm and the classical theory of quadratic forms. If $\alpha, \beta$ have matrix

$$\begin{bmatrix} a & h \\ h & b \end{bmatrix},$$

then norm$(x\alpha + y\beta)$ is the quadratic form $ax^2 + 2hxy + by^2$, and our definition makes $\alpha, \beta$ a reduced pair just if this form is reduced in the classical sense (see [**1**]).

The appropriate generalization to higher dimensions is in terms of the notion of Minkowski reduced form. If the linearly independent characters $\alpha_1, \alpha_2, \ldots, \alpha_n$ have inner product matrix $[a_{ij}]$, the corresponding $n$-ary quadratic form is Minkowski-reduced just if

$$a_{11} \leqslant a_{22} \leqslant \ldots \leqslant a_{nn}$$

and, for each $i$, $1 \leqslant i \leqslant n$, $a_{ii} = \text{norm}(\alpha_i)$ is the minimal norm of any character of the form

$$\alpha_i + \sum_{j<i} c_j \alpha_j \quad \text{(all } c_j \text{ being integers)}.$$

This condition amounts to a system of linear inequalities on the matrix entries, and it is known that a certain finite subset of these inequalities imply the rest ("the fundamental region has finitely many walls"). Unfortunately, there is an implied combinatorial explosion here too, which makes the Minkowski reduced form impractical for large $n$. Indeed, the number of walls to the fundamental region increases so rapidly with $n$, that they have only been written down in the cases $n \leqslant 7$.

In practice this does not concern us too much, since our procedure will usually be a somewhat tentative one for other more urgent reasons. A good compromise, if one wants more in this direction than our general Euclidean algorithm, is to test all sums of the form (with at least two terms),

$$\pm\alpha_{i_1} \pm \alpha_{i_2} \pm \ldots \pm \alpha_{i_k}$$

for sufficiently small $k$, supposing that

$$\text{norm}(\alpha_{i_1}) \leqslant \text{norm}(\alpha_{i_2}) \leqslant \ldots \leqslant \text{norm}(\alpha_{i_k}),$$

and if any such sum has norm strictly less than that of $\alpha_{i_k}$, to replace $\alpha_{i_k}$ by this sum. This has the merit that for $k \leqslant 4$ it ensures that every $k \times k$ submatrix is Minkowski-reduced. Of course for $k = 2$ it is just our general Euclidean algorithm.

It is worthwhile pointing out that if a set of characters is Minkowski-reduced, but still has large norms, then there is unlikely to be much to be gained by applying more subtle methods, since then the minimal norm of *any* integral linear combination of the $\alpha_i$ will be $a_{11}$. So if one has followed the above recommendation for all $k \leqslant 4$, one has the dubious pleasure of knowing that any character with a strictly smaller norm than any yet found, must involve at least 5 of the given characters (if it is an integral linear combination of them).

We end this section with an example of one of the horrible problems associated with the use of virtual characters, and a theorem which occasionally helps to mitigate such problems.

*Horrid example.* The implication

$$(+) \quad \begin{bmatrix} 3 & & & \\ & 3 & & \\ & & 3 & \\ & & & 3 \end{bmatrix} \Rightarrow \begin{array}{l} \alpha = 111000000000 \\ \beta = 000111000000 \\ \gamma = 000000111000 \\ \delta = 000000000111 \end{array}$$

holds for proper characters, but for virtual characters there are many other solutions, one of which only involves 4 distinct irreducibles!

$$\alpha = 101\bar{1} \qquad \beta = 1\bar{1}01 \qquad \gamma = 11\bar{1}0 \qquad \delta = 0111 \qquad (\bar{1} \text{ means } -1).$$

(These four words define the *tetracode.*)

*Nice theorem.* If a character $\alpha$ has been obtained as an integral linear combination of characters with reality mark 2, and has the same indicator as its norm, then it is a proper character.

*Proof.* Then we have $a_t i_t = a_t^2$ for every $\chi_t$, forcing $a_t = i_t = \pm 1$ if $\chi_t$ is involved. But if $i_t = -1$, $a_t$ would have to be even, since $r(\alpha) = 2$.

## 6 Reduction of Proper Characters. Possibility Problems

At last we approach the real topic of this paper! The sort of problem we want to discuss is the following. Given the inner product matrix $[a_{st}]$ of (say) some proper characters $\alpha_1, \ldots, \alpha_n$, and also perhaps their indicators $i_1, \ldots, i_n$, under what conditions can we deduce *just from this information* that a given rational linear combination

$$c_1\alpha_1 + c_2\alpha_2 + \ldots + c_n\alpha_n$$

of these characters is necessarily a proper character?

An important particular case is the *inclusion problem*: when can we deduce, just from the type of information given above, that necessarily $\alpha_1$ must be *included* in $\alpha_2$ (written $\alpha_1 \leqslant \alpha_2$); in other words, that the difference $\alpha_2 - \alpha_1$ is a proper character? When just these two characters are involved, Guy has solved the inclusion problem completely, and his methods show that it is essentially equivalent to certain *possibility problems*: these ask when is a given matrix (and possibly some extra information) the inner product matrix of some set of characters in some group?

Although, of course, the information we are faced with in character reduction problems is known to be possible in the relevant sense, it seems that it is really the possibility problems, rather than the rather more complicated problems listed above, which are important in understanding our task. The following argument shows why. Imagine that we have reached the square-dancing phase, for a set of proper characters with inner products $[a_{st}]$, and we want to know whether the numbers $m_1, m_2, \ldots, m_n$ can be the multiplicities of some irreducible in $\alpha_1, \alpha_2, \ldots, \alpha_n$. Obviously this can happen if and only if the matrix $[a_{st} - m_s m_t]$ can be the inner product matrix of some proper characters $\beta_1, \ldots, \beta_n$. Notice that the possibility problem being asked here is *not* that corresponding to the given matrix $[a_{st}]$ (for which possibility is known), but for a matrix derived from it by making some assumption about how the characters $\alpha_1, \alpha_2, \ldots, \alpha_n$ might reduce.

We make no further apology for studying possibility problems (even though not all of them have immediate applications!), but proceed straight to the formal definitions. We shall say that a matrix $[a_{st}]$ (perhaps together with a list of indicators) is $(\pm)$-*possible*, or *virtually possible*, if it is the inner product matrix of a set $\alpha_1, \ldots, \alpha_n$ of virtual characters in some group (and these characters have the given indicators). It is $(+)$-*possible*, or *properly possible* if also $\alpha_1, \ldots, \alpha_n$ are proper characters. It is convenient also to define the notions of $(\pm 1)$-possibility and $(+1)$-possibility, in which all the non-zero multiplicities of irreducibles in $\alpha_1, \ldots, \alpha_n$ are required to be (respectively) $\pm 1$ or all $+1$.

*Theorem* 2$(+1)$. A matrix $\begin{bmatrix} a & h \\ h & b \end{bmatrix}$ is $(+1)$-possible

if and only if its entries are non-negative integers with $a \geqslant h \leqslant b$.

*Proof.* If $\alpha$ and $\beta$ are multiplicity-free characters with the given inner products, then $h$ is the number of common constituents, and so plainly $a \geqslant h \leqslant b$. On the other hand, if $a, h, b$ are non-negative integers satisfying this condition, then (in any group with at least $a+b-h$ irreducibles) we can take

$$\alpha = \chi_1 + \ldots + \chi_a, \; \beta = \chi_1 + \ldots + \chi_h + \chi_{a+1} + \ldots + \chi_{a+b-h}.$$

*Lemma.* If $\begin{bmatrix} a & h-a \\ h-a & b-2h+a \end{bmatrix}$ is $(+)$-possible, then so is $\begin{bmatrix} a & h \\ h & b \end{bmatrix}$.

*Proof.* If $\alpha$ and $\gamma$ have the first matrix, then $\alpha$ and $\gamma + \alpha$ have the second.

*Theorem* 2(+). The matrix $\begin{bmatrix} a & h \\ h & b \end{bmatrix}$ is (+)-possible

if and only if its entries are non-negative and satisfy $ab \geqslant h^2$.

*Proof.* If $\alpha, \beta$ have the given inner products, then the matrix must clearly have non-negative entries if $\alpha$ and $\beta$ are proper, and it must also be positive semi-definite, which implies $ab \geqslant h^2$.

Now suppose that $\begin{bmatrix} a & h \\ h & b \end{bmatrix}$ satisfies the given conditions.

If $h \leqslant a$ it is (+1)-possible by Theorem 2(+1), and so *a fortiori* (+)-possible. If instead $h > a$, then the matrix

$$\begin{bmatrix} a & h-a \\ h-a & b-2h+a \end{bmatrix}$$

can easily be checked to satisfy the same conditions, and may therefore inductively be supposed (+)-possible, since it has smaller entries. We now apply the Lemma.

Although it does not seem very important, we might as well append:

*Theorem* 2($\pm$). An integral matrix $\begin{bmatrix} a & h \\ h & b \end{bmatrix}$ is ($\pm$)-possible

if and only if it is positive semi-definite.

*Proof.* This condition (which can be written $a \geqslant 0$, $b \geqslant 0$, $ab-h^2 \geqslant 0$) is obviously necessary. If it holds, and also $h \geqslant 0$, we can apply Theorem 2(+) directly to $\begin{bmatrix} a & h \\ h & b \end{bmatrix}$. Otherwise we apply it to $\begin{bmatrix} a & -h \\ -h & b \end{bmatrix}$ and then change the sign of $\beta$.

## 6.1 Guy's inclusion criterion

*Theorem* 2($\leqslant$). Suppose that $\alpha$ and $\beta$ are proper characters with inner products

$$\begin{bmatrix} a & h \\ h & b \end{bmatrix}, \text{ and that } 0 < a \leqslant b.$$

Then if $ab-h^2 < b$, we must have $\alpha \leqslant \beta$. If instead $ab-h^2 \geqslant b$, then there is a group $G$ with two proper characters $\alpha_1, \beta_1$ having this inner product matrix, and for which $\alpha_1 \nleqslant \beta_1$.

*Proof.* We handle the second part first. If $ab-h^2 \geqslant b$, then the hypotheses imply that

$$\begin{bmatrix} a-1 & h \\ h & b \end{bmatrix} \text{is } (+)\text{-possible,}$$

and we can suppose it is the matrix of two proper characters $\alpha_0, \beta_0$, which do not involve some irreducible, $\chi_t$. Then $\alpha_0+\chi_t$ and $\beta_0$ are the desired two characters $\alpha_1$ and $\beta_1$.

Now for the first part. If $\alpha \nleqslant \beta$, then for some irreducible $\chi_t$, we must have $a_t > b_t$ for the corresponding multiplicities, and we shall write $a_t = b_t+c$, noting that $c > 0$. Now the matrix of inner products of $\alpha-a_t\chi_t$ and $\beta-b_t\chi_t$ must be possible, and so we must have

$$(a-a_t^{\,2})(b-b_t^{\,2})-(h-a_tb_t)^2 \geqslant 0.$$

Also, we can suppose $ab-h^2 = b-d$ for some $d > 0$, from the given condition. We shall deduce a contradiction from these inequalities.

The displayed inequality simplifies to give

$$ab-h^2 \geqslant a_t^{\,2}b+b_t^{\,2}a-2ha_tb_t \geqslant a_t^{\,2}b+b_t^{\,2}a-(a+b)a_tb_t,$$

since we have $2h \leqslant a+b$ by the arithmetic-geometric mean inequality.

But this now tells us that

$$b > b-d \geqslant (a_tb-b_ta)(a_t-b_t) = [(b-a)b_t+bc]c \geqslant bc^2 \geqslant b,$$

the desired contradiction.

The point of the Theorem is that the failure of $\alpha \leqslant \beta$ is equivalent to the possibility of $\alpha-a_t\chi_t$ and $\beta-b_t\chi_t$ for some pair of numbers with $a_t > b_t$, and that it turns out that whenever this happens, we can actually take $a_t = 1$, $b_t = 0$.

Note that when the criterion fails, we may still have $\alpha \leqslant \beta$. But there will exist a group with characters $\alpha^*$ and $\beta^*$ having the same inner products, and $\alpha^* \nleqslant \beta^*$.

We shall call this (the indicator-free form of) *Guy's Inclusion Criterion* and the early Cambridge experience shows that it is very useful indeed. Its main value is that by a trivially testable condition on the numbers it can give results that would be totally intractable by backtracking, and which enable us to reinstate a form of the Euclidean algorithm for proper characters. In the *positive Euclidean algorithm* for characters $\alpha_1, \ldots, \alpha_n$, of course, one sees if the condition of Guy's inclusion criterion holds for any pair $\alpha_i, \alpha_j$, and if so, replaces $\alpha_j$ by $\alpha_j - \alpha_i$. The simplest non-trivial case where the criterion holds is

$$\begin{bmatrix} 2 & 2 \\ 2 & 3 \end{bmatrix},$$

which could obviously have been found by backtracking, since $\alpha$ and $\beta$ can only be multiplicity-free characters which must have both constituents of $\alpha$ in common (this could plainly also be found by the "difference-is-an-irreducible" test): a larger example is

$$\begin{bmatrix} 20 & 31 \\ 31 & 50 \end{bmatrix}, \text{ for which we find } ab - h^2 = 1000 - 961 = 39 < 50,$$

and so we should replace $\beta$ by $\gamma = \beta - \alpha$,

$$\text{giving} \begin{bmatrix} 20 & 11 \\ 11 & 8 \end{bmatrix} \text{ for the new inner products,}$$

a substantial simplification.

## 6.2 The equality case of Guy's inclusion criterion

It is sometimes useful to note that we can still deduce something when $ab - h^2 = b$.

*Theorem* 2(=). Suppose that $\begin{bmatrix} a & h \\ h & b \end{bmatrix}$, with $0 < a \leqslant b$, is the matrix of two proper characters for which $\alpha \nleqslant \beta$, and that $ab - h^2 = b$. Then, there must exist integers $p$ and $q$, and characters $\gamma$ and $\chi$, with $\chi$ irreducible, for which

$$\alpha = p\gamma + \chi, \; \beta = q\gamma.$$

*Proof.* A careful analysis of the proof of Theorem 2($\leqslant$) shows that every inequality appearing there must now become an equality. This implies $b_t = 0$, $c = a_t = 1$, and the condition

$$(a - a_t^2)(b - b_t^2) = (h - a_t b_t)^2$$

now shows that $\alpha - \chi_t$ and $\beta$ must be proportional characters.

This theorem, applied at a suitable stage in the square-dancing process, occasionally produces an irreducible $\chi$ "by magic".

### 6.3 A possibility theorem with indicators

*Theorem* 2(+, ind). For the array

$$\begin{bmatrix} a & h \\ h & b \end{bmatrix}\begin{matrix} i \\ j \end{matrix} \text{ to be properly possible,}$$

it is necessary and sufficient that

$$\begin{bmatrix} a & h \\ h & b \end{bmatrix} \text{ be properly possible,}$$

and that for all integers $r, s$ we have

$$|ri+sj| \leqslant ar^2+2hrs+bs^2.$$

Moreover, given the numbers $a, b, h$ we can explicitly produce three particular pairs $(r_1, s_1)$, $(r_2, s_2)$, $(r_3, s_3)$ for which this condition suffices.

*Proof.* The extra condition amounts to the assertion

$$|\text{ind}(r\alpha+s\beta)| \leqslant \text{norm}(r\alpha+s\beta)$$

and so it is obviously necessary. Now, if $h \leqslant a \leqslant b$ and $i$ and $j$ are two numbers for which

$$|i| \leqslant a, \ |j| \leqslant b, \ |i-j| \leqslant a+b-2h$$

it is easy to check that indicator values 0 or $\pm 1$ can be attached to the irreducibles $\chi_t$ appearing in the proof of Theorem 2(+1) so as to realize

$$\begin{bmatrix} a & h \\ h & b \end{bmatrix}\begin{matrix} i \\ j \end{matrix}.$$

We now work back, as in the proof of Theorem 2(+1).

An example will illustrate the process and clarify the argument (which is quite subtle). We find the exact condition on $i$ and $j$ under which the array

$$\begin{bmatrix} 20 & 31 \\ 31 & 50 \end{bmatrix}\begin{matrix} i \\ j \end{matrix} \text{ is properly possible.}$$

If $\alpha$ and $\beta$ are any characters corresponding to these numbers, then $\alpha$ and the (possibly virtual) character $\gamma = \beta-\alpha$ will have the array

$$\begin{bmatrix} 20 & 11 \\ 11 & 8 \end{bmatrix}\begin{matrix} i \\ j-i \end{matrix}$$

and then $\delta = \alpha-\gamma$ and $\gamma$ will have the array

$$\begin{bmatrix} 6 & 3 \\ 3 & 8 \end{bmatrix}\begin{matrix} 2i-j \\ j-i \end{matrix}.$$

We now stop, since the new $h$ is smaller than the new $a$ and $b$.

Now the condition $|\text{ind}| \leqslant \text{norm}$ must certainly hold for $\gamma, \delta, \delta-\gamma$ even though these may be virtual characters, and so the numbers $i$ and $j$ must satisfy

$$|j-i| \leqslant 8, \quad |2i-j| \leqslant 6, \quad |3i-2j| \leqslant 8.$$

On the other hand, when these three conditions hold, we can actually find *proper* characters $\delta^*$ and $\gamma^*$ realizing the final array, and work back by addition to derive proper characters $\alpha^*$ and $\beta^*$ for the original one.

## 6.4 Guy's inclusion criterion, with indicators

We shall not state this as a Theorem, but rather describe briefly the algorithm by which one determines whether the information

$$\begin{bmatrix} a & h \\ h & b \end{bmatrix}\begin{matrix} i \\ j \end{matrix} \text{ for proper characters } \alpha, \beta$$

necessarily implies that $\alpha \leqslant \beta$. As usual, when the criterion fails to tell us that $\alpha \leqslant \beta$, it does not mean that the given characters fail to have $\alpha \leqslant \beta$, but that there will be some group $G$ and some characters $\alpha^*, \beta^*$ of $G$ having the given array

$$\begin{bmatrix} a & h \\ h & b \end{bmatrix}\begin{matrix} i \\ j \end{matrix},$$

but $\alpha^* \nleqslant \beta^*$.

If $\alpha \leqslant \beta$ fails, then as usual, there will be some irreducible $\chi_t$ whose multiplicities $a_t, b_t$ in $\alpha$ and $\beta$ have $a_t > b_t$, and $\chi_t$ will have an indicator $i_t = 0, 1$, or $-1$. In other words, the array

$$\begin{bmatrix} a-a_t^2 & h-a_tb_t \\ h-a_tb_t & b-b_t^2 \end{bmatrix}\begin{matrix} i-a_ti_t \\ j-b_ti_t \end{matrix}$$

will be possible for some integers $a_t > b_t \geqslant 0, \quad -1 \leqslant i_t \leqslant 1$.

This condition is exact in the sense that if such integers exist we cannot deduce that $\alpha \leqslant \beta$ merely from the given information. On the other hand, it is readily testable, since $a_t$ and $b_t$ are bounded by $\sqrt{a}$ and $\sqrt{b}$. (It is actually much easier tested than this.) Of course, it allows us to strengthen the positive Euclidean algorithm to a form taking into account the indicator information in an optimal way. However, the extra complexity involved probably means that this extension will only be worthwhile in a few cases, which should be looked for just before the square-dance starts.

## 7 Possibility Theorems in Higher Dimensions?

It is natural to ask for extensions of Theorems 2(+) and (2±) to higher dimensions. I know one such extension, but will first prove a simple lemma.

*Lemma.* Suppose that the integers $a, b, c, f, g, h$ satisfy

$$|g|+|h| \leqslant a, \quad |h|+|f| \leqslant b, \quad |f|+|g| \leqslant c.$$

Then the matrix

$$\begin{bmatrix} a & h & g \\ h & b & f \\ g & f & c \end{bmatrix} \quad \text{is } (\pm 1)\text{-possible.}$$

*Proof.* Look at the Venn diagram below.

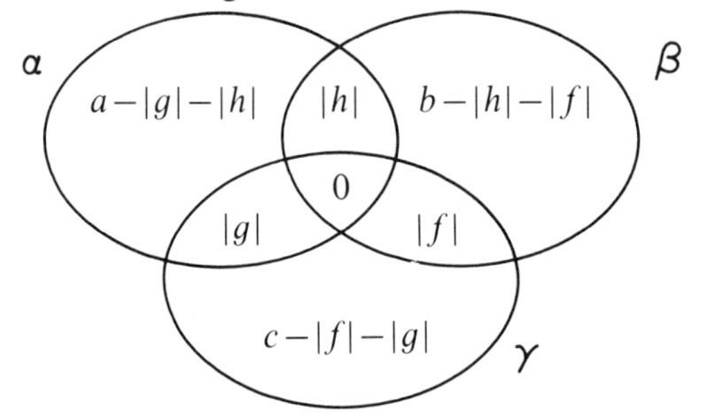


Each of the circles indicates a character $\alpha, \beta, \gamma$, and the numbers inside the regions indicate sums of irreducibles. Thus $\alpha$ involves $|h|$ irreducibles that are also involved in $\beta$, $|g|$ that are also involved in $\gamma$, and $a-|g|-|h|$ that only appear in $\alpha$. These irreducibles appear with multiplicity $\pm 1$ in all characters and the multiplicities of the irreducibles that appear in both $\alpha$ and $\beta$ are the same or opposite, according as $H$ is positive or negative (etc.).

*Theorem* 3(±). A $3 \times 3$ matrix of integers is (±)-possible if and only if it is positive semi-definite.

*Proof.* It does no harm to suppose that the matrix is Minkowski reduced. But this implies

$$2|f| \leqslant \min(b, c), \quad 2|g| \leqslant \min(c, a), \quad 2|h| \leqslant \min(a, b)$$

and these inequalities in turn imply those of the Lemma.

The 4 and 5-dimensional problems are still unsolved:

*Conjectures* 4(±) *and* 5(±). A $4 \times 4$ or $5 \times 5$ matrix of integers is (±)-possible if and only if it is positive semi-definite.

On the other hand, we have a

*Counterexample* 6(±). There is a $6 \times 6$ positive matrix of integers which is not (±)-possible, namely:

$$\begin{bmatrix} 2 & 1 & & & & \\ 1 & 2 & 1 & & & \\ & 1 & 2 & 1 & & 1 \\ & & 1 & 2 & 1 & \\ & & & 1 & 2 & \\ & & 1 & & & 2 \end{bmatrix}.$$

It is easiest to see the impossibility by looking at the associated $E_6$ diagram

$$\begin{array}{ccccccccc} & & & & \pi & & & & \\ & & & & | & & & & \\ \alpha & — & \beta & — & \gamma & — & \delta & — & \varepsilon \end{array}$$

(The result is essentially just the fact that it is impossible to take Euclidean co-ordinates so that all vectors of the $E_6$ root lattice have integer co-ordinates.) It is easy to check that essentially the only solution for $\alpha, \beta, \gamma, \delta, \varepsilon$ is

$$110000 \quad 011000 \quad 001100 \quad 000110 \quad 000011.$$

(This is not quite so obvious as it sounds, since there is another solution for $\alpha, \beta, \gamma$ namely 110, 011, $\bar{1}10$, but this yields no possibility for $\delta$.) Now for $\pi$ to have inner product 1 with $\gamma$ it must have co-ordinates ??10?? or ??01?? on these irreducibles, which extend to $1\bar{1}1000$ or $0001\bar{1}1$ using the other inner products, both of which contradict norm$(\pi) = 2$.

Note that actually all the entries of our matrix are non-negative.

It is also worthwhile to point out that for *proper* possibility there are already 3- and 4-dimensional counterexamples to the obvious conjectures:

*Counterexamples* 3(+) *and* 4(+). The matrices

$$\begin{bmatrix} 3 & 2 & \\ 2 & 3 & 2 \\ & 2 & 3 \end{bmatrix} \quad \text{and} \quad \begin{bmatrix} 2 & & & 1 \\ & 2 & & 1 \\ & & 2 & 1 \\ 1 & 1 & 1 & 2 \end{bmatrix}$$

have non-negative entries and are positive-definite (and indeed) (±)-possible, but are not (+)-possible.

The first example has the solution

$$\alpha = 1110, \quad \beta = 1101, \quad \gamma = 01\bar{1}1$$

but if proper characters had these inner products, all three would have to be made out of 3 distinct irreducibles, and $\beta$ would have 2 of these in common with each of $\alpha$ and $\gamma$, which would force $\alpha$ and $\gamma$ to meet.

The second example is associated with the $D_4$ diagram

$$\begin{array}{c} \beta \\ | \\ \alpha - \delta - \gamma \end{array}$$

and has the solution

$$\alpha = 1100, \quad \beta = 0011, \quad \gamma = 001\bar{1}, \quad \delta = 0110.$$

Were it possible for proper characters, $\delta$ would be a sum of 2 irreducibles, at least one of which must appear in any of $\alpha, \beta, \gamma$. But $\alpha, \beta, \gamma$ must also be disjoint! Note that no virtual solution to this matrix has the same amount of symmetry as the matrix itself. This places limitations on the form of a proof of Conjecture 4($\pm$).

## Acknowledgements

The entire "philosophy" of character reduction presented here is due to M. J. T. Guy, who is also responsible for many particular ideas other than those attributed to him directly in the text. I also benefited greatly from conversations during the Durham conference with M. D. Atkinson, J. Neubüser, S. P. Norton, R. A. Parker, and W. Plesken.

## References

1. Cassels, J. W. S. (1978). Rational Quadratic Forms. London Mathematical Society Monographs, No. 13, Academic Press, London, Orlando and New York.
2. Neubüser, J., Pahlings, H. and Plesken, W. (1983). CAS; Design and use of a system for the handling of characters of finite groups, (these Proceedings).

# The Computer Calculation of Modular Characters (The Meat-Axe)

R. A. PARKER

## 1 Introduction

The Brauer character table is easiest to obtain theoretically for fields whose characteristic divides the group order to a small power only. For "interesting" (sporadic and related) groups, 2 divides a large number of times, and indeed the 2-modular characters are particularly difficult to calculate.

By way of contrast, a digital computer works very efficiently with matrices over the field of two elements. This is true for several reasons. For example:

(a) Digital computers work internally in binary. The "exclusive or" instruction, which adds many elements mod 2 simultaneously, exists on most computers;
(b) A matrix entry occupies just one bit;
(c) Representations mod 2 tend to be smaller and less numerous than for other characteristics.

Thus, four years ago, it seemed likely that a computer could be used to calculate the 2-modular characters of groups, where theoretical methods had failed.

Before describing in detail how a computer can be used, here is a summary of the basic method.

At the outset, two (or more) matrices are produced which generate the required group in some representation over $GF_2$ (or some other finite field). Corresponding pairs of matrices are then made in other representations. Larger representations are made by tensoring smaller ones. Smaller representations are obtained as constituents of larger ones. Some representations prove to be irreducible—the rest are discarded. Eventually all (or most) of the irreducibles are made and these are studied. Sufficient information is collected to ensure the easy calculation of the character table.

The above simplification is necessarily inaccurate in several details, but should give the reader an idea of what is going on.

COMPUTATIONAL GROUP THEORY
ISBN 0-12-066270-1

Perhaps the heart of the process is the sentence "smaller representations are obtained as constituents of larger ones".

A representation can be broken up into two constituents once a non-trivial invariant subspace has been found. This can be readily made from any single vector in it. A null vector of a random element of the group algebra is likely to lie in a non-trivial invariant subspace.

## 2 The Five Main Programs AD, MU, NS, SP, TE

For the basic process summarized in the previous section, five programs are needed. Four are quite simple. The other—SP—is rather more complex. All are listed in Table 1.

In practice, the vector used in Split is the top row of the output of Null-space. The rest of the Null-space output (if any) is ignored by Split. Of the five main programs, only the Split program warrants further description. Below is given a fairly detailed description of this program.

First two matrices (**A** and **B**) are read, along with a single vector **w**. A space (**W**) is calculated as the smallest space containing **w** that is invariant under **A** and **B**. In the computer, a basis $\bar{\mathbf{W}}$ for **W** is built up in echelon form. To start with, this echelon form matrix ($\bar{\mathbf{W}}$) contains the single vector **w**. Vectors of $\bar{\mathbf{W}}$ are multiplied by **A** and **B**, and the resultant two vectors are included in $\bar{\mathbf{W}}$, which is again reduced to echelon form. This process is continued until the state is reached where every basis vector of $\bar{\mathbf{W}}$ has been multiplied by **A** and **B**. $\bar{\mathbf{W}}$ is now a basis for **W**.

*Table 1*

| Name | Action | Description |
|---|---|---|
| AD | Add | Two matrices are read, and one—their sum—is written out. |
| MU | Multiply | Two matrices are read, and one is written out. This is the matrix product of the input matrices. |
| NS | Null-space | One matrix is read in, and a Gaussian elimination performed. A basis for the null-space is obtained and written out. The rank, obtained as a by-product is also output. |
| SP | Split | One vector and two matrices are read in. First the smallest invariant subspace containing the input vector is found. Four matrices are then written out. Two of them describe the action of the input on the invariant subspace. The other two similarly describe the quotient space. |
| TE | Tensor | Two matrices are read, and one—their tensor product—is written out. |

Each vector of $\bar{\mathbf{W}}$ is now multiplied by **A**, and the result expressed as a linear combination in the vectors of $\bar{\mathbf{W}}$. The coefficients of this expression are written out as a row of the first output matrix. In this way the action of **A** on $\bar{\mathbf{W}}$ is calculated. This process is repeated for matrix **B**, giving the second output matrix.

$\bar{\mathbf{W}}$ is then completed to a non-singular matrix **X** by adding further rows (not in **W**). The action of the matrix **A** on the quotient of the original space by **W** can now be calculated by multiplying the new rows by **A**, and expressing the result as a linear combination of the rows of **X**. Only the coefficients of the new rows are written out. This forms the third output matrix. This process is repeated for **B**, giving the last output matrix. The user is notified of the dimension of **W**.

## 3 The Basic System

The first problem is to obtain $GF_2$ matrices generating the required group. The most convenient way to do this is to use permutations. For example, (1 2 3) and (3 4 5 6 7 8 9) clearly generate the alternating group $A_9$. Other groups can be more difficult to obtain, indeed I have yet to get 3 . McL (the Schur triple cover of McLaughlin's group). Having obtained generating matrices, the splitting and tensoring process can begin.

Splitting a representation into its constituent irreducibles is normally the most time-consuming step. The method is to multiply and add matrices more or less at random to obtain a matrix **F**. In practice $\mathbf{F} = \mathbf{A}+\mathbf{B}+\mathbf{AB}$, where **A** and **B** are the generators, is often good enough. A null vector (**w**) of **F** is then found using the Null-space program. With luck, **w** lies in a proper invariant subspace, so that running the Split program will result in two smaller representations. If **w** does not lie in a proper invariant subspace, another matrix $\mathbf{F}'$ is made, and the Null-space/Split process tried again. Soon the representation yields and a splitting is achieved. Continued application ultimately splits the representation into its irreducible constituents.

Two irreducible representations are then tensored to obtain a new representation. This is split into its constituents in the same way. Continuing in this fashion, all the irreducible representations can (in principle) be made.

The central fact in all this is that

> "A null-vector of a random element of the group algebra is *likely* to lie in an invariant sub-space."

The method has been extensively used, and has been found to work. The first vector tried lies in a proper invariant subspace more than 50% of the time. Exceptionally, ten attempts may be necessary, but I cannot remember ever needing twenty.

*Table 2*

| Nullity | Probability | Number of null vectors | |
|---|---|---|---|
| 0 | 29% | 0 | (Of course, there are no proper invariant subspaces.) |
| 1 | 57% | 1 | |
| 2 | 13% | 3 | |
| 3 | 1% | 7 | |
| $\geqslant 4$ | Negligible | $\geqslant 15$ | |

Suppose a space **V** has an invariant subspace **W**. Acting on **W**, **F** probably has a few null vectors. On **V**, **F** may have more null vectors. Nevertheless, there will not usually be too many, and some of them lie in **W**. Thus a null vector of **F** (on **V**) has a fair chance of being in **W**.

In simple cases, the probabilities can be calculated. First suppose $\rho$ is an absolutely irreducible representation over $GF_2$, whose dimension is at least 10. We obtain the distribution of nullities in the group algebra in $\rho$ as in Table 2. Now suppose $\rho$ is indecomposable, with just two constituents, distinct and absolutely irreducible over $GF_2$. Thus $\rho$ has a unique proper invariant subspace, **W**. Thus we see (in Table 2), in this case, that 34% of the time there is just one null vector and it lies in **W**.

Using tensor products and splits etc. the stage is ultimately reached where all the irreducible representations of the group have been made. Proving irreducibility can then be done using Norton's irreducibility test, or by examining fingerprints, as described in the next two sections. The final stage is to compute the modular character table. Given the ordinary character table, knowledge of the degrees of the modular irreducibles is usually sufficient to calculate the table, but ambiguities occasionally arise. These are easily resolved by computer measurement of character values.

*Table 3*

| Nullity | Probability | Number of null vectors | Number of null vectors in **W** |
|---|---|---|---|
| 0 | 8% | 0 | 0 |
| 1 | 50% | 1 | 1 (34%) 0 (16%) |
| 2 | 36% | 3 | 3 (10%) 1 (22%) 0 (4%) |
| 3 | 5% | 7 | 7 (1%) 3 (3%) 1 (1%) 0 (small) |
| $\geqslant 4$ | $\sim 1\%$ | $\geqslant 15$ | |

## 4 The Nullity Fingerprint

Although of little theoretical interest, an important practical tool for use while working on a computer is the Nullity Fingerprint.

Let $R$ be a representation generated by two matrices $\mathbf{A}$ and $\mathbf{B}$.

Define
$$\begin{aligned}
\mathbf{F}_1(R) &= \mathbf{A}+\mathbf{B}+\mathbf{AB}\\
\mathbf{F}_2(R) &= \mathbf{A}+\mathbf{B}+\mathbf{AB}+\mathbf{AB}^2\\
\mathbf{F}_3(R) &= \mathbf{A}+\mathbf{BA}+\mathbf{B}^2+\mathbf{BAB}+\mathbf{BAB}^2\\
\mathbf{F}_4(R) &= \mathbf{B}+\mathbf{A}+\mathbf{BA}+\mathbf{B}^2+\mathbf{BAB}+\mathbf{BAB}^2\\
\mathbf{F}_5(R) &= \mathbf{B}+\mathbf{A}+\mathbf{AB}+\mathbf{BA}+\mathbf{B}^2+\mathbf{BAB}+\mathbf{BAB}^2\\
\mathbf{F}_6(R) &= \mathbf{B}+\mathbf{AB}+\mathbf{BA}+\mathbf{B}^2+\mathbf{BAB}+\mathbf{BAB}^2+\mathbf{A}+\mathbf{A}
\end{aligned}$$

Define $N_i(R) = \text{nullity}\,(\mathbf{F}_i(R)) \quad i = 1,\ldots,6$

Then the set of six numbers $N_1,\ldots,N_6$ is called the (nullity) fingerprint of $R$.

Although there is nothing particularly interesting about the above choice of six particular elements of the group algebra, some care was needed in their selection.

The fingerprint has several useful properties.

(a) Two equivalent representations have the same fingerprint. This is usually used in the negative sense—two representations with different fingerprints are not equivalent. Of course it can happen that inequivalent representations have the same fingerprint, but this does not arise often. A point worth noting is that a representation and its dual (if inequivalent) have different fingerprints in general.

(b) If $S$ is a subquotient of $R$, $N_i(S) \leqslant N_i(R) \quad i = 1,\ldots,6$.

The usual application of this is when $S$ is an irreducible. In this case, fingerprints will often show that $S$ is not a constituent of $R$. This is especially likely to happen if $R$ is irreducible—if $R$ and $S$ are distinct absolutely irreducible representations over $\mathrm{GF}_2$, for example, there is an 87% chance that, for some $i$, $N_i(S) > N_i(R)$. By extending the definition of fingerprint (by defining $\mathbf{F}_7, \mathbf{F}_8 \ldots$) if necessary, this method can be used to show that none of the known irreducibles are constituents of $R$.

(c) If a representation is irreducible over $\mathrm{GF}_q$, but reducible over $\mathrm{GF}_{q^n}$, there is a minimal value $m$ of $n$ such that the constituents are absolutely irreducible. We have that

$$m \text{ divides } N_i, \quad i = 1,\ldots,6$$

This can be used to prove the absolute irreducibility (one merely needs to observe that the greatest common divisor of the $N_i$ is 1) of a representation already known to be irreducible. If the G.C.D. of the $N_i$ is greater than one,

*Table 4*

| | | |
|---|---|---|
| $R_1 = \mathbf{A}$ | $\mathbf{A}$ | |
| $R_2 = \mathbf{B}$ | $\mathbf{B}$ | |
| $R_3 = R_1 * R_2$ | $\mathbf{AB}$ | |
| $R_4 = R_1 + R_2$ | $\mathbf{A+B}$ | |
| $R_5 = R_3 + R_4$ | $\mathbf{A+B+AB}$ | $\mathbf{F}_1$ |
| $R_6 = R_3 * R_2$ | $\mathbf{AB}^2$ | |
| $R_7 = R_5 + R_6$ | $\mathbf{A+B+AB+AB}^2$ | $\mathbf{F}_2$ |
| $R_8 = R_2 * R_7$ | $\mathbf{BA+B}^2\mathbf{+BAB+BAB}^2$ | |
| $R_9 = R_1 + R_8$ | $\mathbf{A+BA+B}^2\mathbf{+BAB+BAB}^2$ | $\mathbf{F}_3$ |
| $R_{10} = R_2 + R_9$ | $\mathbf{B+A+BA+B}^2\mathbf{+BAB+BAB}^2$ | $\mathbf{F}_4$ |
| $R_{11} = R_3 + R_{10}$ | $\mathbf{AB+B+A+BA+B}^2\mathbf{+BAB+BAB}^2$ | $\mathbf{F}_5$ |
| $R_{12} = R_1 + R_{11}$ | $\mathbf{AB+B+2.A+BA+B}^2\mathbf{+BAB+BAB}^2$ | $\mathbf{F}_6$ |

this strongly suggests that a field extension would reduce the representation further. This can be readily established by finding a commuting matrix using the Standard Basis program.

(d) The distribution of nullities suggests the module structure to some extent. For example, for each representation made, one has to decide whether to try to split it, or to prove its irreducibility. Over $GF_2$ a reasonable criterion is to attempt a split if $\sum N_i \geqslant 7$.

(e) The matrices $\mathbf{F}_1 \ldots \mathbf{F}_6$ are reasonably "random", and their null vectors can be used to split representations. By measuring the fingerprint, a "good" candidate can be chosen—usually an $\mathbf{F}_i$ with the smallest non-zero nullity.

There are two reasons why care was needed in the selection of group algebra elements used in the definition of the $\mathbf{F}_i$. Firstly, although all group elements are random, it appears that some are more random than others. For example $\mathbf{A+B+AB+B}^2$ would be no good, because it factorizes as $(\mathbf{A+B})(\mathbf{I+B})$. $\mathbf{I+B}$ is likely to have a fairly large null space if it has any. Similarly $\mathbf{A+B} = \mathbf{A}(\mathbf{I+A}^{-1}\mathbf{B})$. Thus, this element should not be used. In practice, the selected elements seem to be suitably random. It may be that there is some subtle bias in their nullity statistics, but it does not seem to matter much. The second reason for care is to minimize the work needed to calculate the $\mathbf{F}_i$. It should be noted that all six can be calculated using the scheme in Table 4.

## 5 Norton's Irreducibility Test (S. P. Norton)

Although fingerprints can be used to demonstrate the irreducibility of a representation, a direct and more general method is available.

Let $\mathbf{A}_1 \ldots \mathbf{A}_n$ be square matrices of the same size. Let $\mathbf{B}$ be an element of the algebra generated (under multiplication and addition) by the $\mathbf{A}_i$.

At least one of the following occurs:

(1) $\mathbf{B}$ is non-singular;
(2) At least one non-zero null vector of $\mathbf{B}$ lies in a proper subspace invariant under $\mathbf{A}_1, \mathbf{A}_2, \ldots, \mathbf{A}_n$;
(3) Every non-zero null vector of $\mathbf{B}'$ (the transpose of $\mathbf{B}$) lies in a proper subspace invariant under $\mathbf{A}_1', \mathbf{A}_2', \ldots, \mathbf{A}_n'$;
(4) There is no proper subspace invariant under $\mathbf{A}_1, \ldots, \mathbf{A}_n$.

In computer terms, if every null vector of $\mathbf{B}$ fails to split the representation generated by $\mathbf{A}_1$ and $\mathbf{A}_2$, try splitting the group generated by the transposes of $\mathbf{A}_1$ and $\mathbf{A}_2$ with one null vector of $\mathbf{B}'$. If this also fails, the representation is irreducible.

In practice, a matrix $\mathbf{B}$ of nullity one is usually readily available. Thus proving irreducibility requires just two runs of the split program.

This also provides an indication of why null vectors split representations.

## 6 The Standard Basis Program

This rather technical program has a wide range of uses. Like Split, it reads in two matrices and one vector. Its output consists of just one matrix, the rows of which form a standard (or canonical) basis for the representation space.

To understand the uses of this versatile program, it is necessary to understand exactly what the program does. This is particularly unfortunate, since the program may not at first sight appear to achieve anything! This may be because, in most applications, two runs are needed.

The description given below is not intended to show how the program works, but what it does. An efficient implementation would do things in a different order from that suggested below.

The program reads in two matrices ($\mathbf{A}$ and $\mathbf{B}$) and one vector ($\mathbf{V}_1$). The vector $\mathbf{V}_1\mathbf{A}$ is calculated. If $\mathbf{V}_1\mathbf{A}$ and $\mathbf{V}_1$ are linearly independent, $\mathbf{V}_2$ is set to be $\mathbf{V}_1\mathbf{A}$. Otherwise, $\mathbf{V}_1\mathbf{B}$ is calculated and if independent of $\mathbf{V}_1$, $\mathbf{V}_2$ is set to be $\mathbf{V}_1\mathbf{B}$. If linear dependence occurs in both cases, the process stops; otherwise the process continues. At any stage in the process, there are $n$ vectors $\mathbf{V}_1, \mathbf{V}_2, \ldots, \mathbf{V}_n$. $\mathbf{V}_{n+1}$ is calculated (if possible) as the first vector in the list

$$\mathbf{V}_1\mathbf{A}, \mathbf{V}_1\mathbf{B}, \mathbf{V}_2\mathbf{A}, \mathbf{V}_2\mathbf{B}, \ldots, \mathbf{V}_{n-1}\mathbf{B}, \mathbf{V}_n\mathbf{A}, \mathbf{V}_n\mathbf{B}$$

such that $\mathbf{V}_1, \mathbf{V}_2, \ldots, \mathbf{V}_n, \mathbf{V}_{n+1}$ are linearly independent.

The process continues with the calculation of $\mathbf{V}_{n+2}\ldots$. If $\mathbf{V}_{n+1}$ does not exist—i.e. if every vector in the list $\mathbf{V}_1\mathbf{A}, \ldots, \mathbf{V}_n\mathbf{B}$ is in the space spanned by the

$\mathbf{V}_i$—then the process stops. A matrix is output whose $i$th row is $\mathbf{V}_i$. The program has now finished.

The most common use of this program is in equivalence testing. Suppose $\mathbf{A}_1$, $\mathbf{B}_1$ generate a representation and $\mathbf{A}_2$, $\mathbf{B}_2$ generate another of the same dimension. The standard basis program can be used to test whether these two representations are equivalent. In fact, if it exists, a matrix $\mathbf{X}$ can be calculated such that

$$\mathbf{X}\mathbf{A}_1\mathbf{X}^{-1} = \mathbf{A}_2 \quad \textit{and} \quad \mathbf{X}\mathbf{B}_1\mathbf{X}^{-1} = \mathbf{B}_2.$$

Conversely, if no such $\mathbf{X}$ exists, standard basis can prove this. The first step is to find an element in the algebra generated by $\mathbf{A}_1$ and $\mathbf{B}_1$ of nullity one (or small non-zero nullity). Let $\mathbf{V}_1$ be its null vector. Compute the corresponding algebra element in the algebra generated by $\mathbf{A}_2$ and $\mathbf{B}_2$, and let $\mathbf{V}_2$ be its null vector (the nullity must be one if the representations are equivalent).

Run the standard basis program twice, once with $\mathbf{A}_1$, $\mathbf{B}_1$ and $\mathbf{V}_1$ (call the output $\mathbf{Y}_1$) and once with $\mathbf{A}_2$, $\mathbf{B}_2$ and $\mathbf{V}_2$ (output $\mathbf{Y}_2$). Calculate $\mathbf{X} = \mathbf{Y}_2{}^{-1}\mathbf{Y}_1$.

Either $\mathbf{X}$ is the required matrix and the representations are equivalent, or the representations are not equivalent.

# Group Theory on a Micro-Computer

M. D. ATKINSON, R. A. HASSAN AND M. P. THORNE

Until recently, most group theoretic computation was done either by hand or on a large mainframe computer. With the current hardware revolution, it is now possible to adopt an intermediate course by using a micro-computer. A micro-computer allows a user more control over when and where he does his computing. The local computer centre will not switch it off overnight or at weekends and it can be taken wherever the user finds it most convenient. Another obvious advantage is cheapness. For under £2000 one can buy a reasonable configuration with 64K bytes of memory, screen, keyboard and dual floppy disk drives together with supporting software. On the other hand 64K bytes should be contrasted with a typical mainframe memory of several megabytes; moreover both the CPU and the backing storage of a micro may be several hundred times slower than their mainframe counterparts. The consequence of this for group theory is that large integrated systems such as CAYLEY [**4**], cannot be run on micros. Unfortunately, a collection of totally separate programs implementing algorithms such as coset enumeration or finding strong generators for a permutation group, may not prove to be especially useful since group theoretic investigations are not commonly of the "one-off" type: "Find the index of this subgroup" or "Find the order of this permutation group". Rather, a user may wish to build up information by executing a series of connected algorithms. In CAYLEY this would be done by writing a CAYLEY program. On a micro it has to be done by running a succession of programs which communicate by reading and writing disk files. Since floppy disk accesses are so slow, the programs have to be organized to access the files sequentially as far as possible.

Any collection of complicated computer programs can only be developed in a supportive software environment. The choice of such an environment on a micro is limited. The system of programs developed in Cardiff is written in PASCAL and runs under the UCSD operating system. This operating system has proved to be a good development tool; some of our early programs were written in PASCAL/Z running under CP/M and this was much less satisfactory. The main disadvantage of UCSD PASCAL is that it is a hybrid

COMPUTATIONAL GROUP THEORY
ISBN 0-12-066270-1

compiler/interpreter and so object programs run more slowly than with PASCAL/Z.

According to [8] the execution time ratio between UCSD PASCAL and PASCAL/Z seems to be about two to one. The UCSD system is widely available on microcomputers and we therefore hope that the Cardiff programs will be useful elsewhere. However, they were written as a pilot study and we are now planning a unified improved system which will supersede them.

At present the system contains a set of routines for character theory calculations, a coset enumeration program, and permutation group programs for finding group orders and block systems. The character theory programs allow standard calculations such as inner product, tensor product, induction and restriction to be carried out interactively. They make use of 3 different sorts of disk file. Each group has an associated *irreducible file* containing centralizer orders, powermaps and some irreducible characters. In the course of an interactive session other characters are generated and these are all stored in a *work file*. For passing information between a group and its subgroups there may also be a *fusion file* which specifies how conjugacy classes of a subgroup fuse in the group. In a typical session the user might run programs to:

(a) define an irreducible file for a given group, irreducible files for various subgroups, together with a fusion file;
(b) form tensor products, exterior squares and cubes, and place them in the work file;
(c) calculate characters induced from subgroups and place them in the work file;
(d) remove, from the characters in the work file, all the constituent characters found in the irreducible file;
(e) calculate the inner products and Schur indicators of characters in the work file; and thereby
(f) recognize further irreducible characters for placing in the irreducible file.

The final step of using inner product information to obtain irreducible constituents is certainly the hardest (see [5], [6]). We have found generalized characters to be useful. Let $\theta$, $\phi_1, \phi_2, \ldots, \phi_n$ be $n$ (possibly generalized) characters and suppose that $a_{ij} = (\phi_i, \phi_j)$ and $b_i = (\theta, \phi_i)$.

The norm of $\theta + \sum x_i \phi_i$ is

$$(\theta, \theta) + 2\sum b_i x_i + \sum a_{ij} x_i x_j = (\theta, \theta) + 2bx' + xAx'.$$

This quadratic function is positive definite and hence is minimal when its $n$ partial derivatives are zero. The minimizing vector $m$ satisfies $mA = -b$. This equation may not have an integral vector as a solution. However, we can take $p$ to be the nearest lattice point to $m$ and evaluate the norm of $\theta + \sum p_i \phi_i$, which will often be considerably less than the norm of $\theta$. We can then replace $\theta$ by $\theta+\sum p_i\phi_i$ and repeat the calculation with a different set of characters. This method of reducing norms has proved successful in the case $n = 1$ and we are currently investigating higher values of $n$. The aim of course is to obtain characters of norm 1.

The other programs in the system also operate on disk files having a common format: *permutation group files* which contain generators for a permutation group. Thus the coset enumerator offers the user the chance of saving, in such a file, the permutation representation on the cosets of the given subgroup. This file may then be read by another program to find the order of the permutation group. It may also be read by a program for finding systems of imprimitivity; this program will, if asked, generate permutation group files for the groups induced on block systems and these in turn may be examined to find their order or block systems.

None of the programs involve radical new ideas. However, we have tried to take advantage of software developments in Computer Science and this has made many of the technical problems easier to solve. We will give two examples of how Computer Science theory has helped in creating the system.

The coset enumeration program has been designed to make the input of relator and subgroup generator words relatively painless. The user inputs these much as they would be written in mathematics. Thus he types

$$[a, b * c] \wedge 2 \text{ in place of } [a, b^c]^2$$

($a, b, c$ being group generators). The program then has the job of expanding this to

$$a^{-1}\ c^{-1}\ b^{-1}\ c\ a\ c^{-1}\ b\ c\ a^{-1}\ c^{-1}\ b^{-1}\ c\ a\ c^{-1}\ b\ c.$$

This sort of textual translation is similar to (but easier than) that performed by a compiler in translating source code to object code. The methodology worked out by compiler writers has been employed to allow such textual expansion to be done easily. The input words are defined by the following extended BNF

| | | |
|---|---|---|
| *relation* | ::= | *word* \| *word* = 1 \| *word* = *word* |
| *word* | ::= | *expression* {*expression*} |
| *expression* | ::= | *term conjugator* |
| *conjugator* | ::= | **expression conjugator* \| *null* |
| *term* | ::= | *factor power* |
| *factor* | ::= | *generator* \| (*word*) \| [*word*, *word*] |
| *power* | ::= | ∧ *integer* \| *null* |

The terminal symbols *generator* and *integer* stand for single letter variables and optionally signed integer constants. The above description satisfies the

LL(1) condition [1] and so the translator can use recursive descent syntax analysis.

The second example of how Computer Science theory can be exploited is a simplification of the algorithm and its proof [3] for finding the blocks of a transitive permutation group. Let $G$ be a group generated by permutations $g_1, \ldots, g_m$ on a set $\Omega$ and suppose that it is required to find the minimal block containing the pair of symbols $\alpha, \beta$. Initially then $\Omega$ is partitioned by the subset $\{\alpha, \beta\}$ and singleton subsets $\{\gamma\}$, $\gamma \notin \{\alpha, \beta\}$, and the initial partition has to be transformed into a block system for $G$. The major operation which successively modifies the partition is replacing two distinct parts by their union. This scenario is well known in the theory of algorithms (see [2]) and is generally handled by representing the partition as a collection of trees. For every tree node $x$ we record $f[x]$, the node above $x$ in the unique path from $x$ to the root of the tree. This representation allows two parts to be united rapidly and allows us to detect when two nodes lie in the same tree.

The block finding algorithm in its simplest form defines a list $C$ of branches. Initially, $C$ contains just one branch connecting $\alpha$ to $\beta$, and this will be the only branch in the tree. It then proceeds as follows.

```
repeat
        delete a branch (x, y) from C;
        for each generator g do
        begin
                w: = xg; z: = yg;
                r:  = root of w's tree;
                s:  = root of z's tree;
                if    r < > s then
                begin
                        define a branch from r to s (i.e. f[r]: = s);
                        add this branch to C
                end
        end
until C is empty
```

Of course, since branches have the form $(x, f[x])$ it is only necessary to store the first component of each branch. To justify that this algorithm is correct, we use another idea from Computer Science—that of a loop invariant. Observe that the following statement $S$ is true initially and is preserved by every iteration of the **repeat** . . . **until** loop.

$S$: "For every tree branch $(x, y)$ one of the following holds:
(i) $(x, y) \in C$
(ii) for every generator $g$, $xg$ and $yg$ belong to the same tree."

Consequently, on termination, when $C$ is empty, $xg$ and $yg$ belong to the same tree for every branch $(x, y)$. It follows immediately that the generators (and so all group elements) permute the trees and so the node sets of the trees are a set of blocks for $G$.

The algorithm spends most of its time tracing tree paths to find a root. There are two well known techniques, the *weighting rule* and *path compression*, designed to flatten the trees and so reduce this time. The weighting rule requires that when two different trees are joined together, the smaller tree is made a subtree of the larger. This ensures that every node is distant at most $\log_2 n$ from a root and leads to an execution time of order $n \log n$.

Path compression is performed whenever a path from a node to a root is traced. The nodes $\alpha_1, \alpha_2, \ldots$ which are visited in such a path to the root $\rho$ are made children of the root (by defining $f[\alpha_i] := \rho$). It is not evident that path compression is applicable in our situation since the statement $S$ above would no longer be a loop invariant. However, it can be justified with the help of the following definition. At any point in the algorithm an ordered pair $(\sigma, \tau)$ is said to be *bonded* if $(\sigma, \tau)$ is, or previously was, a branch (i.e. if $f[\sigma]$ has been defined as $\tau$ at some stage) and if, for each generator $g$, $\sigma g$ and $\tau g$ belong to the same tree. Then we can take the following statement $T$ as an invariant.

$T$: "For every node $\sigma$ there is a sequence $\sigma_0, \sigma_1, \ldots, \sigma_p$ of nodes such that $\sigma = \sigma_0$, $\sigma_p$ is a root, and each $(\sigma_i, \sigma_{i+1})$ is either in $C$ or is bonded."

It is relatively easy to check that $T$ is preserved by the operations of joining trees and finding roots with path compression. On termination, $C$ is empty and, as above, we can argue that the node sets of the trees are a set of blocks for $G$.

Using both the weighting rule and path compression gives an execution time of order $ng(n)$ where $g(n)$ is a very slow-growing function related to a functional inverse of Ackermann's function (see [7]). However, with the modest values of $n$ that can be handled by a micro-computer (a few hundred at most), path compression does not, in practice, improve the execution time.

Our program to find the blocks of $G$ uses this algorithm (without path compression) successively with initial values $(\alpha, \beta) = (1,2), (1,3) \ldots (1, n)$. By incorporating the techniques of [3] for improving the speed, the overall execution time appears on average to be subquadratic in $n$.

One of our main objectives in producing these programs was to encourage the wider use of machine computation in group theory. In order for the programs to appeal to group theorists who are new to computing, it is crucial that they be easy to use and resistant to user errors. A mainframe operating system will usually trap user errors and terminate the program without crashing the system. On a micro, the programs themselves must be designed to trap errors. The extra responsibility thus placed on the programmer has some compensations; he is encouraged to tailor the input routines to the specific application, to allow the user to recover if he types a nonsense input, and to exploit to the full the interactive nature of a micro. For example, it is useful in a long running program for the user to be given reassurance that the program has not gone to sleep; thus in a coset enumeration, a dot can be printed (or deleted) every time a coset is defined (or deleted through coincidence); this gives the user an insight into how difficult the enumeration is.

Another important consideration for ease of use is the execution time of each program. As might be expected, the character theory programs which each carry out rather short calculations produce their results almost immediately. On the other hand, coset enumerations and permutation group calculations usually take longer. For example the 272 cosets of $\langle a \rangle$ in

$$\langle a, b \mid a^9 = b^2 = (ab)^4 = (a^2b^3)^3 = 1 \rangle = \mathrm{PSL}(2,17)$$

require about 3 minutes to generate and this is, by no means, an unfavourable enumeration.

The programs have been successfully used in Cardiff by group theorists who are not used to computing. However, the system is not intended to compete with mainframe systems such as CAYLEY. We envisage it being used for small- or medium-sized routine calculations which lie beyond the scope of hand calculation but which do not merit large resources.

## References


1. Aho, A. V. and Ullmann, J. D. (1977). "Principles of compiler design". Addison-Wesley, Reading, Mass.
2. Aho, A. V., Hopcroft, J. E. and Ullman, J. D. (1974). "The design and analysis of computer algorithms". Addison-Wesley, Reading, Mass.
3. Atkinson, M. D. (1975). An algorithm for finding the blocks of a permutation group, *Mathematics of Computation* **29**, 911–913.
4. Cannon, J. J. (1982). (Preprint). A Language for Group Theory. University of Sydney.
5. Conway, J. H. (1983b). Character Calisthenics, (these Proceedings).
6. Neubüser, J., Pahlings, H. and Plesken, W. (1983). CAS; Design and use of a system for the handling of characters of finite groups, (these Proceedings).
7. Tarjan, R. E. (1975). On the efficiency of a good but not linear set merging algorithm, *J. ACM* **22**, 215–225.
8. Woteki, T. H. and Sand, P. A. (1982). Four implementations of PASCAL, *Byte 7* (March edition), 316–352.

# Permutation Groups and Combinatorics

# On Computing Double Coset Representatives in Permutation Groups

GREGORY BUTLER

## 1 Introduction

The enumeration of single cosets [**18**] dates from the dawn of computational group theory, but the enumeration of double cosets is still very much in its infancy. To date, the literature contains the efforts of: Butler [**2**]—the straightforward approach restricted to very small groups; Laue [**12**]—an inductive approach for soluble groups based on enumerating single cosets, where the index is small, and grouping the single cosets into orbits; Holt [**11**]—a similar approach to Laue's that also needs an enumeration of single cosets; and a hand method of Conway [**8**] that is restricted by the amount of information one needs.

In this paper, we treat only one part of the problem of double coset enumeration, and we only treat the case of permutation groups. We define a canonical double coset representative and present an algorithm for determining the canonical representative, given any element in the double coset. The algorithm was developed during the study (see section 7 of [**3**]) of a similar algorithm (see [**7**], [**13**]) for single cosets. The algorithm (for double cosets) has not been extensively applied. However, from our experience with the techniques for computing with permutation groups, it should be applicable to very large groups of degree in the thousands.

The paper continues by presenting the now standard concepts of this area, and then the algorithm. This is followed by an example and a Cayley procedure that implements the algorithm [**6**], [**7**]. A conclusion discusses the performance of the algorithm, and suggests some directions for solving the problem of double coset enumeration.

## 2 Concepts

The standard concepts of permutation groups that are used in computational

COMPUTATIONAL GROUP THEORY
ISBN 0-12-066270-1

group theory are widely discussed in the literature ([**3**], [**5**], [**10**], [**16**], [**17**], [**19**]) so we will be brief.

A permutation group $G$ acts on a set of points $P$ on the right. The image of $p$ in $P$ under $g$ in $G$ is denoted $p^g$ and $(p^g)^h = p^{(gh)}$ for all $p$, $g$, and $h$. The orbit of $p$ under $G$ is the set $p^G = \{p^g \mid g \text{ in } G\}$ of images of $p$. The stabilizer of $p$ in $G$ is the group $G_p = \{g \text{ in } G \mid p^g = p\}$ of elements that fix $p$. A base for $G$ is a sequence $\boldsymbol{b} = [b_1, b_2, \ldots, b_k]$ of points such that only the identity element of $G$ fixes each point in the base. An element $g$ of $G$ is uniquely determined by its base image

$$\boldsymbol{b}^g = [b_1{}^g, b_2{}^g, \ldots, b_k{}^g].$$

Associated with a base is a chain of stabilizers $G^{(i)}$, $i = 1, 2, \ldots, k+1$, where

$$G^{(i)} = G_{b_1, b_2, \ldots, b_{i-1}}.$$

A subset $S$ of $G$ is a strong generating set of $G$ relative to $\boldsymbol{b}$ if $S$ contains a generating set for each stabilizer $G^{(i)}$ in the chain.

Given a base and a strong generating set, we can determine an element $g_i(p)$ of $G^{(i)}$ mapping $b_i$ to any point $p$ in the orbit

$$b_i{}^{G^{(i)}}.$$

We can also determine a strong generating set relative to any other base by using the base change algorithm [**1**], [**3**], [**17**].

If we define a total order on $P$, then there is an induced lexicographical order on $G$ and on the base images $\{\boldsymbol{b}^g \mid g \text{ in } G\}$. If we insist that $b_1, b_2, \ldots, b_k$ are the first $k$ points of $P$, then the order on $G$ corresponds to the order on the base images. In any case, the identity element is the smallest element of the group.

For a sequence $\bar{a} = [a_1, a_2, \ldots, a_k]$, the partial sequence $[a_1, a_2, \ldots, a_i]$ is denoted $\bar{a}(i)$.

If $H$ and $K$ are subgroups of $G$ and $g$ is in $G$, then the size of the double coset $HgK$ is $|H||K|/|K \cap H^g|$ since $|K \cap H^g|$ is the number of pairs $(h, k)$ in $H \times K$ such that $hgk = g$ (or any other fixed element of the double coset).

## 3 Algorithm

This section presents and justifies the correctness of our algorithm. It relies on an order on $P$ where the base points of $G$ are the first $k$ points, because the algorithm needs the correspondence between the order on the elements of $G$ and the order on the base images. Given:

(a) a permutation group $G$ acting on a set $P$ with a base $\boldsymbol{b} = [b_1, b_2, \ldots, b_k]$;
(b) subgroups $H$ and $K$ of $G$ given by a base and strong generating set; and
(c) an element $g$ of $G$;

the algorithm determines:

(i) the image $\bar{w} = [w_1, w_2, \ldots, w_k]$ of the base $\boldsymbol{b}$ under the first element $\bar{g}$ of the double coset $HgK$; and
(ii) the size of the double coset $HgK$.

The base of $H$ will be set to $\boldsymbol{b}$ and, as $\bar{w}$ is determined, the base for $K$ will be changed to $\bar{w}$. The algorithm considers translates of orbits of $H$ (and its stabilizers) and translates of unions of orbits of $K$ (and its stabilizers) as it incrementally determines the smallest possible base image $\bar{w}$ of an element in the double coset $HgK$. Having determined $\bar{w}(i)$, it must find the smallest point $w_{i+1}$ that is an image of $b_{i+1}$ under an element of $HgK$ that maps $\boldsymbol{b}(i)$ to $\bar{w}(i)$. This is done by keeping track of some pairs $(h, k)$ of elements in $H \times K$. With a pair $(h, k)$, is associated the base image $\bar{a} = [a_1, a_2, \ldots, a_k]$ of the element $h$. The images $\bar{a}$ are constructed incrementally as $\bar{w}$ is determined. The algorithm is a simple iteration involving the following tasks.

### 3.1 Task 1, getting started

The base of $H$ must be $\boldsymbol{b}$.

The point $w_1$ is the smallest point in $b_1{}^{HgK}$ (which is the union over $p$ in $(b_1{}^H)^g$ of the orbits $p^K$).

### 3.2 Task 2, determining $w_{i+1}$, assuming $\bar{w}(i)$ is known

The base of $K$ must begin with $\bar{w}(i)$.

Define

$$\text{ALPHA}(i) = \{\bar{a}(i) \mid \bar{a}(i) \text{ in } \boldsymbol{b}(i)^H \text{ and } \bar{a}(i)^g \text{ in } \bar{w}(i)^K\}.$$

The set ALPHA(1) is easily determined from $b_1{}^H \cap (w_1{}^K)^{g^{-1}}$. As Task 3, below, shows how to determine ALPHA$(i)$ from ALPHA$(i-1)$, we will assume ALPHA$(i)$ is known. For each $\bar{a}(i)$ in ALPHA$(i)$ there is a pair $(h(\bar{a}(i)), k(\bar{a}(i)))$ of elements in $H \times K$ such that $h(\bar{a}(i))gk(\bar{a}(i))$ maps $\boldsymbol{b}(i)$ to $\bar{w}(i)$. This pair will often be called $(h, k)$.

If $f$ is some element of $HgK$ mapping $\boldsymbol{b}(i)$ to $\bar{w}(i)$, then $f$ is in

$$H_{b_1, b_2, \ldots, b_i} hgk \, K_{w_1, w_2, \ldots, w_i}$$

for some $\bar{a}(i)$ in ALPHA$(i)$. Therefore, the images of $b_{i+1}$ under the elements of $HgK$ that map $\boldsymbol{b}(i)$ to $\bar{w}(i)$ are

$$\text{IMAGES}(i+1) = \bigcup_{\substack{\bar{a}(i) \text{ in} \\ \text{ALPHA}(i)}} ((b_{i+1}^{H^{(i+1)}})^{hgk})^{K^{(i+1)}}.$$

So $w_{i+1}$ is the smallest point in IMAGES$(i+1)$.

### 3.3 Task 3, determining ALPHA$(i+1)$ from $w_{i+1}$ and ALPHA$(i)$

For each $\bar{a}(i)$ in ALPHA$(i)$, define

$$\mathrm{NEXT}(\bar{a}(i)) = (w_{i+1}^{K^{(i+1)}})^{k^{-1}g^{-1}} \cap (b_{i+1}^{H^{(i+1)}})^h.$$

This is the set of points $a_{i+1}$ that can extend $\bar{a}(i)$ to an image in ALPHA$(i+1)$. For each $a_{i+1}$ in NEXT$(\bar{a}(i))$,

(i) the image $\bar{a}(i+1) = [a_1, a_2, \ldots, a_{i+1}]$ is in ALPHA$(i+1)$,
(ii) the element

$$h(\bar{a}(i+1)) = u_{i+1}(a_{i+1}^{h(\bar{a}(i))^{-1}})\, h(\bar{a}(i+1)),$$

where $u_{i+1}(p)$ is an element of $H^{(i+1)}$ mapping $b_{i+1}$ to $p$, and
(iii) the element

$$k(\bar{a}(i+1)) = k(\bar{a}(i))v_{i+1}(a_{i+1}^{gk(\bar{a}(i))})^{-1},$$

where $v_{i+1}(p)$ is an element of $K^{(i+1)}$ mapping $w_{i+1}$ to $p$.

In this way we construct the whole of ALPHA$(i+1)$.

### 3.4 Task 4, determining the size of $HgK$

On termination, $\bar{w}$ is the base image of the canonical representative and the images $\bar{a}$ in ALPHA$(k)$ are in one-to-one correspondence with the elements of $K \cap H^g$. Hence, the size of the double coset $HgK$ is easily calculated from the size of ALPHA$(k)$.

(The correspondence with the intersection algorithm is given as follows. Take a fixed $\bar{a}_0$ in ALPHA$(k)$. Then the tuples $\bar{a}^g$, for $\bar{a}$ in ALPHA$(k)$, are the base images of the elements of $K \cap H^g$, where $\bar{a}_0{}^g$ is the base of $H^g$ and $K$.)

## 4 Example

The group $G$ is the alternating group of degree six, and the subgroups $H$ and $K$ are the centralizers of $(1,2)(3,4)$ and $(1,2,3)$ respectively. The base $\boldsymbol{b}$ is $[1,2,3,4]$ and the element $g$ is $(1,2,3,4,5)$.

The orbit $1^H$ is $\{1,2,3,4\}$ and the orbits of $K$ are $\{1,2,3\}$ and $\{4,5,6\}$. Hence IMAGES(1) is $\{1,2,3,4,5,6\}$ and $w_1$ is 1.

The set ALPHA(1) is $\{[1], [2]\}$ with associated pairs of elements (identity, $(1,3,2)$), and ($(1,2)(3,4)$, $(1,2,3)$). The orbit $2^{H_1}$ is $\{2,4\}$ and the orbits of $K_1$ are $\{1\}$, $\{2\}$, $\{3\}$, and $\{4,5,6\}$. The set IMAGES(2) is $\{2,3,4,5,6\}$ so $w_2$ is 2.

The set ALPHA(2) is $\{[1,2]\}$ with associated pair (identity, $(1,3,2)$). The orbit $3^{H_{1,2}}$ is $\{3\}$ and the orbits of $K_{1,2}$ are the orbits of $K_1$. The set IMAGES(3) is $\{4,5,6\}$ so $w_3$ is 4.

Since there is a unique member of ALPHA(2) and since $H_{1,2,3}$ and $K_{1,2,4}$

are trivial, the sets ALPHA(3) and ALPHA(4) also only have one member. The associated pair is always (identity, (1,3,2)). The final base image $\bar{w} = [1,2,4,5]$ giving the canonical representative (3,4,5) in a double coset of size 72.

## 5 A Cayley Procedure

A direct implementation of the algorithm is very space consuming because of the possibly large size of the sets ALPHA($i$). The procedure we give enumerates ALPHA($i$) whenever it is needed to compute ALPHA($i+1$) or IMAGES(i+1). The enumeration is easily expressed recursively because ALPHA($i$) is defined in terms of ALPHA($i-1$) and NEXT. (At the moment, Cayley does not allow recursion. However, there are standard techniques [**14**], [**15**], for unfolding recursion, and we leave this as an exercise for the reader.)

```
procedure canonicalrep (G, H, K, GELT; WTILDA, LTH);

"Determine the base image WTILDA of the canonical representative of
the double coset H * GELT * K, and determine the length LTH of the
double coset."

    GBASE = base (G); WTILDA = empty;
    for    I = 1 to length (GBASE) do
       "form IMAGES (I+1) by enumerating ALPHA(I)"
       enumerate (I, GBASE, WTILDA, GELT, H, K; IMAGES,
       LTH);
       "select the smallest point as the next point in the base image"
       minpt (degree (G), IMAGES; PT);
       WTILDA = append (WTILDA, PT);
    end;
    "find the length of ALPHA(K)"
    enumerate (length (GBASE), GBASE, WTILDA, GELT, H, K;
    IMAGES, LTH);
    "and compute the length of the double coset"
    LTH = order (H) * order (K)/LTH;
end;

procedure enumerate (I, GBASE, WTILDA, GELT, H, K; IMAGES,
LTH);

"enumerate the set ALPHA(I) and return the set IMAGES(I + 1) (if it is
```

defined) in IMAGES, and return the length of ALPHA($I$) in LTH"

```
    IMAGES = [ ]; LTH = 0;
    extend (I, empty, GBASE, WTILDA, identity of H, GELT,
            identity of K, H, K; IMAGES, LTH);
end;
procedure extend (I, ATILDA, GBASE, WTILDA, HELT, GELT,
                  KELT, HSUB, KSUB; IMAGES, LTH);
```

"Extend the partial base image ATILDA in ALPHA($J$) to give all possible extensions of it in ALPHA($I$). The pair associated with ATILDA is (HELT, KELT). The base of $G$ and $H$ is GBASE, the base image of the canonical representative is WTILDA. The $(J+1)$st group in the stabilizer chains of $H$ and $K$ are HSUB and KSUB respectively"

```
    J = length (ATILDA);
    if J eq I then
      "next element ATILDA of ALPHA(I) is known so we
      increase IMAGES (I+1)"
      if I lt length (GBASE) then "IMAGES(I+1) is defined"
        for each PT in orbit (HSUB, GBASE[I+1])
         ↑(HELT * GELT * KELT) do
          if not PT in IMAGES then
            IMAGES = IMAGES join orbit (KSUB, PT);
          end;
        end;
      end;
      "count members in ALPHA(I)"
      LTH = LTH+1;
    else
      "extend ATILDA by the points in NEXT(ATILDA)"
      NEXT = (orbit (HSUB, GBASE[J+1]↑HELT) meet
      orbit (KSUB, WTILDA[J+1])↑(KELT↑-1 * GELT↑-1);
      for each PT in NEXT do
        svrep (HSUB, GBASE[J+1], PT↑(HELT↑-1); U);
        svrep (KSUB, WTILDA[J+1], PT↑(GELT * KELT); V);
        extend (I, append (ATILDA, PT), GBASE,
                WTILDA, U * HELT, GELT, KELT * V↑-1,
                stabilizer (HSUB, GBASE[J+1]),
                stabilizer (KSUB, WTILDA[J+1]);
                IMAGES, LTH);
      end;
    end;
end;
```

```
procedure svrep(GP, BASEPT, IMPT; ELT);

"Cayley does not allow access to the Schreier vectors so we use this
procedure to return an element ELT of the group GP that maps
BASEPT to IMPT"

    T = transversal (GP, stabilizer (GP, BASEPT));
    for each ELT in T do
        if BASEPT↑ELT eq IMPT then break; end;
    end;
end;

procedure minpt (DEG, S; PT);

"This procedure returns the smallest point PT
in the set S. The set is a subset of {1, 2, ..., DEG}."

    for PT = 1 to DEG do
        if PT in S then break; end;
    end;
end;
```

## 6 Conclusion

The algorithm we presented bears a marked similarity to the intersection algorithm [**4**], [**17**], for permutation groups. Although the intersection algorithm has been shown [**9**] to be exponential in the worst case, experience has shown it to be effective in groups of degree in the thousands. Therefore, we believe that our algorithm for determining a double coset representative will also be as effective.

At this time, we know of no systematic way of effectively enumerating double cosets. Randomly generating elements of $G$ may produce an element in each double coset, and our algorithm can eliminate duplications and determine whether the double cosets account for all the elements of the group. However, until we have some experience with this approach, we do not know how serious a problem it will be to find elements in very small double cosets.

## References

1. Butler, G. (1979). Computational approaches to certain problems in the Theory of Finite Groups, Ph.D. Thesis, University of Sydney.

2. Butler, G. (1981). Double cosets and searching small groups. Proc. of the 1981 ACM Symposium on Symbolic and Algebraic Computation. 182–187.
3. Butler, G. (1982a). Effective computation with group homomorphisms. Technical Report 184, University of Sydney.
4. Butler, G. (1982b). Computing in permutation and matrix groups II: backtrack algorithm, *Mathematics of Computation* **39**, 671–680.
5. Butler, G. and Cannon, J. J. (1982c). Computing in permutation and matrix groups I: normal closure, commutator subgroup, series, *Mathematics of Computation* **39**, 663–670.
6. Cannon, J. J. (1980). Software tools for group theory, *AMS Proc. Symp. Pure Math.* **37**, 495–502.
7. Cannon, J. J. (1982). The Group Theory Language Cayley, University of Sydney.
8. Conway, J. H. (1983). An algorithm for double coset enumeration? (these Proceedings).
9. Hoffman, C. M. (1980). On the complexity of intersecting permutation groups and its relationship with graph isomorphism. Technical Report 4/80. Christian Albrechts Universitaet, Kiel.
10. Hoffman, C. M. (1982). Group-theoretic algorithms and graph isomorphism, Lecture Notes in Computer Science vol. **136**, Springer-Verlag, Berlin.
11. Holt, D. F. (1983). The Calculation of the Schur Multiplier of a Permutation Group, (these Proceedings).
12. Laue, R. (1982). Computing double coset representatives for the generation of solvable groups, Lecture Notes in Computer Science vol. **144**, 65–70. Springer-Verlag, Berlin.
13. Richardson, J. S. (1973). GROUP: A computer system for group-theoretical calculations, M.Sc. Thesis, University of Sydney.
14. Rohl, J. S. (1977). Converting a class of recursive procedures into non-recursive ones, *Software Practice and Experience* **7**, 231–238.
15. Rohl, J. S. (1981). Eliminating recursion from combinatoric procedures, *Software Practice and Experience* **11**, 803–817.
16. Sims, C. C. (1971a). Determining the conjugacy classes of a permutation group. *In* "Computers in Algebra and Number Theory" (Eds. G. Birkhoff and M. Hall Jr.). *SIAM—AMS Proc.* **4**, 191–195. Amer. Math. Soc.
17. Sims, C. C. (1971b). Computation with permutation groups. *In* "Proc. of the Second Symposium on Symbolic and Algebraic Manipulation" (Ed. S. R. Petrick), 23–28. Assoc. Comput. Mach., New York.
18. Todd, J. A. and Coxeter, H. S. M. (1936). A practical method for enumerating cosets of a finite abstract group, *Proc. Edinburgh Math. Soc.* (*2*), **5**, 26–34.
19. Wielandt, H. (1964). "Finite Permutation Groups". Academic Press, London, Orlando and New York.

# An Algorithm for Computing Galois Groups

L. H. SOICHER

## 1 Introduction

This paper describes a new algorithm which is often useful in the practical computation of the Galois group over the rationals $\mathrm{Gal}(f)$ of an irreducible polynomial $f(x)$ of positive degree $n$. We regard $\mathrm{Gal}(f)$ as a (transitive) permutation group on the zeros of $f$. The algorithm may be used to determine the orbit-lengths of the action of $\mathrm{Gal}(f)$ on the $r$-subsets of $\{1,\ldots,n\}$ or on the $r$-sequences of distinct elements of $\{1,\ldots,n\}$. We assume, without loss of generality, that $f$ is monic and has integer coefficients.

To compute $\mathrm{Gal}(f)$, it is sufficient to determine various properties of $\mathrm{Gal}(f)$ and then to distinguish $\mathrm{Gal}(f)$ from amongst the transitive permutation groups of degree $n$. First, we may compute the discriminant, $\mathrm{disc}(f)$, which is an integral square if and only if $\mathrm{Gal}(f)$ is a subgroup of $A_n$. Factorizing $f$ modulo $p$ for primes $p$ not dividing $\mathrm{disc}(f)$, we determine cycle types in $\mathrm{Gal}(f)$: the degrees of the irreducible factors of $f$ mod $p$ are the lengths of the disjoint cycles of a permutation in $\mathrm{Gal}(f)$. (See, for example [4].) This usually determines the Galois group very quickly if it is $A_n$ or $S_n$; otherwise additional techniques are necessary.

One such technique is the following (see [5] and [6] for details, tables, etc.). Let $F = e_1x_1 + \ldots + e_rx_r$ be a polynomial in $\mathbb{Z}[x_1,\ldots,x_n]$, where the $e_i$ are non-zero integers. We wish to determine the orbit-lengths of the action of $\mathrm{Gal}(f)$ on the orbit $F^{S_n}$ of $F$ under the action of $S_n$ (the action is by permuting the variables $\{x_1,\ldots,x_n\}$ in the obvious way). If the $e_i$ all equal 1, the orbit-lengths on the $r$-subsets of $\{1,\ldots,n\}$ are determined; if the $e_i$ are pairwise distinct, the orbit-lengths on the $r$-sequences of distinct elements of $\{1,\ldots,n\}$ are determined. To determine the orbit-lengths of $Gal(f)$ on $F^{S_n}$ we construct (using the algorithm described later) the "linear" resolvent polynomial with respect to $F$ and $f$:

$$R(F,f) = \prod_{J\in F^{S_n}} (x - J(\alpha_1,\ldots,\alpha_n)),$$

where $\{\alpha_1,\ldots,\alpha_n\}$ are the zeros of $f$. Note that the coefficients of $R(F,f)$ are

COMPUTATIONAL GROUP THEORY
ISBN 0-12-066270-1

given by integral symmetric polynomials in the zeros of $f$ and hence are rational integers. If $R(F,f)$ has distinct zeros, then the action of Gal$(f)$ on the zeros of $R(F,f)$ is equivalent to the action of Gal$(f)$ on $F^{S_n}$; hence the degrees of the irreducible factors of $R(F,f)$ are the orbit-lengths of the action of Gal$(f)$ on $F^{S_n}$.

A method of dealing with the occurrence of multiple zeros is to apply an appropriate Tschirnhaus transformation to $f$ which leaves $f$ irreducible and hence Gal$(f)$ unchanged, and then to recompute $R(F,f)$. A Tschirnhaus transformation may be calculated by a method mentioned later.

The algorithm LRINT (Linear Resolvent over the INTegers), we describe, constructs the resolvent directly over the integers. A version of LRINT is implemented in the multilength integer language ALGEB [3] on the PDP-11/34 minicomputer. Resolvent polynomials constructed are factorized using the Hensel factorization algorithm of Zassenhaus [**8**] implemented by Ford in ALGEB. These programs were demonstrated at the Symposium.

In [**5**], we discuss a different version of LRINT which instead works over GF$(p)$ for large primes $p$ and the resolvent is constructed over the integers using the Chinese Remainder Algorithm. We have found the simpler direct algorithm, described here, to be useful in determining Gal$(f)$ for $f(x)$ with moderate-sized coefficients and degree up to about 7 or 8. The GF$(p)$ algorithm, implemented in PASCAL on the CDC Cyber, is found to be useful for polynomials of higher degrees and with large coefficients.

*Notation.* Our notation for group names is as follows:

$S_n$ and $A_n$ are respectively the symmetric and alternating groups of degree $n$;
$n$ denotes the cyclic group of order $n$;
$p^n$ denotes the elementary abelian group of order $p^n$ ($p$ prime).

It will be clear from the context whether a possibly ambiguous symbol denotes a group or an integer. For groups $A$ and $B$, a group (of shape) $A\,.\,B$ is an extension of $A$ by $B$.

## 2 The Algorithm

The resolvent algorithm revolves around the following observation of Trager [7]. Let

$$f(x) = (x-\alpha_1)\ldots(x-\alpha_n),\;\; g(x) = (x-\beta_1)\ldots(x-\beta_m)$$

be non-constant polynomials over the integers.

Then the resultant eliminating $y$,

$$\mathrm{res}_y(f(y),\; g(x-y)) = \prod_{i=1}^{n} g(x-\alpha_i),$$

is the degree $nm$ monic polynomial having zeros $\alpha_i+\beta_j$ $(i=1,\ldots,n,$ $j=1,\ldots,m)$.

We take this opportunity to note that

$$\operatorname{disc}(f)=(-1)^{n(n-1)/2}\operatorname{res}_x(f(x),\mathrm{d}f/\mathrm{d}x)$$

and the Tschirnhaus transformation of $f$ by $g$ is $\operatorname{res}_y(f(y),\ x-g(y))$. Computing resultants is discussed in [**1**] and [**2**].

The following notation is used in algorithm LRINT. Let $t$ be a non-zero integer and $f(x)$ a monic polynomial of degree $n$. Then we define

$$f_{(t)}(x)=t^n f(x/t).$$

Thus $f_{(t)}(x)$ is the monic polynomial whose zeros are $t$ times those of $f(x)$. Next, we define $\operatorname{mult}(t,F)$ to be the number of distinct terms of the (multivariate) polynomial $F$ having the coefficient $t$.

## 2.1 Algorithm LRINT

Input: A monic integral polynomial $f(x)$ of positive degree $n$, and $F=e_1x_1+\ldots+e_rx_r$, where $r\leqslant n$ and the $e_i$ are non-zero integers.

Returns: $R(F,f)$.

1 if $r=0$ $(F\equiv 0)$ then return ("$x$") and stop.

2 if $r=1$ then return $(f_{(e_1)}(x))$ and stop.

3 Permute the labelling of $x_1,\ldots,x_r$ in $F$ so that $\operatorname{mult}(e_r,F)\leqslant\operatorname{mult}(e_i,F)$ for $i=1,\ldots,r$.

(This ensures that the degree of $u(x)$ in (4) is as small as possible. Note that the symmetry allows relabelling of the variables of $F$ without changing $R(F,f)$.)

4 set $F':=e_1x_1+\ldots+e_{r-1}x_{r-1}$, and set $u(x):=R(F',f)$ (recursively).

5 Let $a_1,\ldots,a_k$ be the $k$ (say) distinct elements of $\{e_1,\ldots,e_{r-1}\}$, and set $F_i:=F'+e_rx_{i'}$ where $i'$ (not necessarily uniquely determined) is chosen so that $a_ix_{i'}$ is a term of $F'$, $(i=1,\ldots,k)$.
If any of the $F_i$ now have only $r-2$ terms (i.e. $a_i+e_r=0$), then relabel the variables of these $F_i$ with $1,2,\ldots,r-2$ to conform with the input rules for this algorithm.

6 set $v(x):=\prod_{i=1}^{k}R(F_i,f)^{c_i}$ (recursively),

where $c_i=n-r+2$ if $a_i+e_r=0$, and $c_i=\operatorname{mult}(a_i+e_r,F_i)$ otherwise.

(Observe that $\operatorname{res}_y(u(y),f_{(e_r)}(x-y))/v(x)=R(F,f)^c$, where $c=\operatorname{mult}(e_r,F)$.)

7 set $c:=\operatorname{mult}(e_r,F)$ and $m:=(n\deg(u)-\deg(v))/c+1$.
$[m-1=\deg(R(F,f))]$

Choose distinct integers $s_1, \ldots, s_m$ such that for $i = 1, \ldots, m$: $v(s_i) \neq 0$ and if $c$ is even then $|s_i| > |\theta|$ for any zero $\theta$ of $R(F, f)$.
(A bound on the magnitude of the zeros of $R(F, f)$ is calculated by bounding the magnitude of the zeros of $f$.)

8 for $i = 1, \ldots, m$: set $t_i := \text{res}_y(u(y), f_{(e_r)}(s_i - y))/v(s_i)$.

9 (For non-negative real $t$, let $t^{1/c}$ denote the non-negative real $c$th root of $t$.)
for $i = 1, \ldots, m$:
if $t_i < 0$ then set $t_i := -|t_i|^{1/c}$
else if $c$ is even and $m-1$ is odd and $s_i < 0$ then set $t_i := -t_i^{1/c}$
else set $t_i := t_i^{1/c}$.
(now $t_i = R(F, f)(s_i)$)

10 set $w(x)$ to be the unique polynomial of degree (at most) $m-1$ such that $t_i = w(s_i)$ for $i = 1, \ldots, m$.
return $(w(x))$ and stop.

## 3 Example

The transitive subgroups of $A_7$ and their orbit-lengths on the 3-subsets of $\{1, \ldots, 7\}$ are as follows: $A_7$, (35); $L_3(2)$, (7,28); 7 . 3, ($7^2$,21); 7, ($7^5$).

Consider $f(x) = x^7 - 7x^3 + 14x^2 - 7x + 1$; $f$ is irreducible and $\text{disc}(f) = 7^8 17^2$, thus Gal$(f)$ is a transitive subgroup of $A_7$. Letting $F = x_1 + x_2 + x_3$, we compute and factorize $R(F, f)$ of degree 35 to determine the orbit-lengths of the action of Gal$(f)$ on the 3-subsets of $\{1, \ldots, 7\}$. It takes 6 minutes to compute $R(F, f)$ using LRINT on the PDP-11/34. $R(F, f)$ is found to have irreducible factors of degrees 7 and 28 which proves that Gal$(f)$ is $L_3(2)$. The degree 7 factor is $x^7 - 14x^4 + 7x^3 + 14x^2 - 56x - 32$. The factorization of $R(F, f)$ takes approximately 10 minutes on the PDP-11/34.

We remark that the PDP-11/34 minicomputer is much slower than a typical large main-frame computer.

## Appendix: A Polynomial with Group $2^3 . L_3(2)$

For every transitive group $G$ of degree $\leqslant 7$ a polynomial $f(x)$ with Gal$(f) = G$ is exhibited in [**5**]. In considering the problem for degree 8, we were led to ask, at the Symposium, about constructing a polynomial whose Galois group over the rationals is the transitive degree 8 group of shape $2^3 . L_3(2)$. We have since constructed such a polynomial $u(x)$ as described, below.

*Observation.* Let $f(x)$ and $g(x)$ be (monic integral) polynomials of respective degrees $n > 1$ and $m > 1$. If $h(x) = f(g(x))$ is irreducible, then Gal$(h)$ is imprimitive, having $n$ blocks of size $m$, and $\mathrm{Gal}(f)$ is the homomorphic image of $\mathrm{Gal}(h)$ acting on these blocks. This is apparent from the factorization of $h$ in the splitting field of $f$:

$$h(x) = \prod_{i=1}^{n} (g(x)-\alpha_i),$$

where $\{\alpha_1,\ldots,\alpha_n\}$ are the zeros of $f$.

Now, let $f(x) = x^7-7x^3+14x^2-7x+1$ and recall that $\mathrm{Gal}(f)$ is $L_3(2)$. To place $\mathrm{Gal}(f)$ in a specific conjugate of $L_3(2)$ in $S_7$ we label the zeros $\{\alpha_1,\ldots,\alpha_7\}$ of $f(x)$ as follows:

$$\alpha_1 \approx -0{\cdot}401-1{\cdot}76i, \alpha_2 \approx -1{\cdot}99, \alpha_3 = \overline{\alpha_1}, \alpha_4 \approx 1{\cdot}06-0{\cdot}544i,$$

$$\alpha_5 = \overline{\alpha_4}, \alpha_6 \approx 0{\cdot}406, \text{ and } \alpha_7 \approx 0{\cdot}263.$$

With respect to this labelling,

$$\mathrm{Gal}(f) = \langle(1,2,3,4,5,6,7), (1,3)(4,5)\rangle.$$

Now consider $h(x) = f(x^2)$; $h(x)$ is irreducible (modulo 3 for example). Fix a labelling of the zeros $\{\beta_1,\ldots,\beta_{14}\}$ of $h(x)$ so that $\beta_{2i} = \alpha_i{}^{1/2}$ and $\beta_{2i-1} = -\beta_{2i}$ (i $= 1,\ldots,7$). Then, with respect to this labelling, $\mathrm{Gal}(h)$ is a subgroup of $H = \langle A,B,C\rangle$, where

$$A = (1,3,5,7,9,11,13)(2,4,6,8,10,12,14),$$

$$B = (1,5)(2,6)(7,9)(8,10), \text{ and } C = (1,2);$$

$H$ is of shape $2^7 . L_3(2)$. The factorization of $h$ modulo 59 shows that there is an element $T$ in $\mathrm{Gal}(h)$ with cycle type $(2,3^4)$. The conjugates of $T^3$ in $\mathrm{Gal}(h)$ must be

$$\{(2i-1,2i):1 \leqslant i \leqslant 7\}$$

and we conclude that $\mathrm{Gal}(h) = H$. $H$ has a subgroup $K = \langle A,B,D,E\rangle$ of shape $2^4 . L_3(2)$, where $A$, $B$ are as before,

$$D = (1,2)(3,4)(5,6)(7,8)(9,10)(11,12)(13,14), \text{ and}$$

$$E = (1,2)(7,8)(11,12)(13,14).$$

We find an element $\gamma$ in the splitting field of $h$ such that $\mathrm{stab}_H(\gamma) = K$. Such an element is

$$\gamma = \sum \beta_{2i}\beta_{2j}\beta_{2k}\beta_{2l}\ (\{i,j,k,l\} \in \{1,4,6,7\}^{\mathrm{Gal}(f)}).$$

(That is, the sum is over the complements of the Steiner triples in the $S(2,3,7)$ fixed by $\mathrm{Gal}(f)$.) The images of $\gamma$ under $H$ are

$$\gamma_i = \gamma^{(2i-1,2i)}\ (i = 1,\ldots,7), \text{ and } \gamma_8 = \gamma.$$

We form

$$t(x) = \prod_{i=1}^{8} (x-\gamma_i)$$

using high-precision approximations to the zeros of $h(x)$.

As $H$ ($= \mathrm{Gal}(h)$) permutes the zeros of $t(x)$, we know that the coefficients of $t$ are rational integers. Thus, by determining the coefficients of $t$ sufficiently accurately, they are found exactly by rounding. It turns out that

$$u(x) = t(2x)/256 = x^8 + 14x^5 + 7x^4 - 14x^3 + 4x + 14,$$

and we know by the construction that

$$\mathrm{Gal}(u) = H \Big/ \Big( \bigcap_{P \in H} P^{-1}KP \Big)$$

which is of shape $2^3 . L_3(2)$.

In a similar way, we have constructed irreducible degree 8 polynomials with groups of shape $2^3 . (7 . 3)$ and $2^3 . 7$.

## Acknowledgements

This research, except that described in the Appendix, was conducted at Concordia University, Montreal, Canada, where the author received funding from the Natural Sciences and Engineering Research Council of Canada and from the Ministry of Education of Quebec.

## References

1. Childs, L. (1979). "A Concrete Introduction to Higher Algebra". Springer-Verlag, New York.
2. Collins, G. E. (1971). The calculation of multivariate polynomial resultants, *J. Assoc. Comput. Mach.* **18**, 515–532.
3. Ford, D. J. (1978). On the computation of the maximal order in a Dedekind domain, Ph.D. Thesis, Ohio State University, Columbus. ("ALGEB", submitted to Digital Equipment Computer Users Soc.).
4. Jacobson, N. (1974). "Basic Algebra I". W. H. Freeman and Co., San Francisco.
5. Soicher, L. (1981). The computation of Galois groups, M. Comp. Sci. Thesis, Concordia University, Montreal. (Reprinted as Technical Report No. MPA V4 by University of Laval, Quebec.)
6. Soicher, L. and McKay, J. (1982). Computing Galois groups over the rationals, *J. Number Theory* (to appear).
7. Trager, B. M. (1976). Algebraic factoring and rational function integration, *In* Proc. of the 1976 ACM Symposium on Symbolic and Algebraic Comput., New York, 219–226.
8. Zassenhaus, H. (1969). On Hensel factorization, I, *J. Number Theory* **1**, 291–311.

# Practical Strategies for Computing Galois Groups

G. KOLESOVA AND J. McKAY

Given an explicit polynomial over the rationals, how does one determine its Galois group? Little help can be found in the literature since almost all the examples are either of the type that one can readily see the Galois symmetries which lead to a solvable group, or they lead to the symmetric group or the alternating group which can be determined from decomposition into irreducibles in $F_p[x]$ together with knowledge of the discriminant. These methods are inadequate to deal with most other groups.

Here strategies are described which have been developed to compute Galois groups of polynomials of small degree (say up to 11).

First, some notation:

$$f = f(x) = \prod_{1}^{n} (x - x_i), \; K = Q(x_1, x_2, \ldots, x_n), \; s_k = \sum x_i^k,$$

$a_k$ is the elementary symmetric function of weight $k$ in the roots of $f$.

Our input data is the polynomial $f$, or more precisely its coefficients which, after a trivial transformation, may be taken to be integers. Now $\mathrm{Gal}(K/Q)$ induces a permutation group on the roots of $f$; this group is determined only to within a labelling of the roots and thus to conjugacy in the symmetric group of degree $n$. For more precise information on the group, we identify each root explicitly with some $x_i$ and this needs some numerical approximation. Since our philosophy is that all our computations should be exact and the computation should provide a proof, we prefer to avoid approximations at the price of computing only the similarity class of $\mathrm{Gal}(K/Q)$.

The first step is to check irreducibility over the rationals. If reducible, each irreducible constituent can be treated separately and common subfields of the splitting fields can be found later. We shall assume that the polynomial $f$ is irreducible over the rationals. The discriminant is computed. There are two ways of doing this.

One may use a polynomial remainder sequence and compute the discriminant from the resultant of $f$ and its derivative. This has the advantage

COMPUTATIONAL GROUP THEORY
ISBN 0-12-066270-1

of yielding a Sturm sequence which tells us the number of real roots of $f$ and so the cycle type of the involution realized by complex conjugation in $\mathrm{Gal}(K/Q)$.

An alternative is to compute the discriminant as a determinant in the symmetric functions with $i, j$ entry given by

$$s_{i+j-2} \; (i, j = 1, 2, \ldots, n).$$

This follows by multiplying the van der Monde with entries $x_j^{i-1}$ by its transpose. The $s_k$ are found from the $a_k$ using Newton's relations:

$$ra_r + \sum (-1)^i s_i a_{r-i} = 0.$$

If the discriminant is a square, then $\mathrm{Gal}(K/Q) \leqslant \mathrm{Alt}(n)$. The discriminant is an invariant of the group $\mathrm{Alt}(n)$. In the general situation we shall seek some minimal set of invariants which together determine $\mathrm{Gal}(K/Q)$. These invariants are the orbit lengths of $\mathrm{Gal}(K/Q)$ acting on subsets of roots of $f$.

The subsets chosen are either unordered or ordered. In practice, the ordered sets are replaced by partial sums with distinct integer coefficients. Unordered sets are replaced by partial sums with common coefficient. As long as the numerical values of the partial sums are all distinct, the action of $\mathrm{Gal}(K/Q)$ is that on ordered and unordered sets. It may be that values are not distinct in which case a Tschirnhaus transformation is applied which preserves $\mathrm{Gal}(K/Q)$ and changes $f$.

The orbit lengths of all permutation groups of degree up to 11 are tabulated by Butler and McKay **[1]** and, with few exceptions, these are sufficient to determine the group. For the few exceptional cases, it seems necessary to use *ad hoc* methods.

We shall have to factor polynomials of large degree. It may be time-consuming to do this directly, but it is often more efficient to use the geometry of the expected group to find a likely factor with approximate numerical coefficients, and then to round them to integers and test for divisibility. One simple example of this is the dihedral group of order $2n$ as a permutation group of degree $n$. The roots may be arranged on a circle so that the orbits on unordered pairs are given by the paths $P_k$ traced by edges joining roots $x_i$ and $x_{i+k}$.

More complex is the case of the Mathieu group $M_{11}$ acting as a permutation group of degree 11. To date, no such polynomial has been found, but we can give a method for proving the result once the polynomial is provided.

The group is 4-fold transitive and so will be intransitive on 5-sets. By using the method of cycle types—for each good prime $p$ (one not dividing the discriminant), the partition of $n$ induced by the degrees of the irreducible factors of $f \bmod p$ is the cycle type of an element of $\mathrm{Gal}(f)$ and the proportion in which these cycle types occur is that in which they occur in $\mathrm{Gal}(f)$—we can show that $M_{11} \leqslant \mathrm{Gal}(f) \leqslant \mathrm{Alt}(11)$. Now $M_{11}$ is maximal in Alt(11), so all that is needed is to show that $\mathrm{Gal}(f) \neq \mathrm{Alt}(11)$.

It is known that $M_{11}$ fixes the Steiner system S(4,5,11) which consists of 66 5-ads chosen from a set of cardinal 11 with the property that any of the ${}^{11}C_4$ subsets of cardinal 4 lies in a unique 5-ad. The action of $M_{11}$ on the subsets of cardinal 5 is intransitive, having two orbits of degrees 66 and 396, respectively.

The problem now is to find the factor of degree 66 corresponding to S(4,5,11). There are, *a priori*, $[\mathrm{Sym}(11){:}M_{11}] = 5040$ possible labellings to examine.

The values of an appropriate function (say sums of products 5 at a time) can be examined for a close approximation to an integer. If $f$ has any complex roots, this calculation can be reduced further. We note that involutions in $M_{11}$ have type $2^4 1^3$, so there are three real roots if there are any complex ones. By finding elements commuting with complex conjugation we need examine only 48 choices for the relabelling. The program has been written—all that is needed is the polynomial!

The method has been developed by Soicher [**2**] who has found polynomials for all transitive groups of degree up to 7. It would be most useful to have a practical computational method which uses the composition factors of the group, even for the solvable case. Although the groundwork has been done for polynomials of degree up to 11, there remains the problem of dealing with the exceptional cases.

## References

1. Butler, G. and McKay, J. Transitive permutation groups of degree up to 11, *Comm. in Alg.* (to appear).
2. Soicher, L. H. (1983). An algorithm for computing Galois groups, (these Proceedings).

# On the Number of Certain Permutation Representations of $\langle 2, 3, n\rangle$-Groups

A. A. HUSSEIN OMAR

Let $G$ be a finite group of order $N$, with conjugate classes $C_1, \ldots, C_r$, and let $\alpha_i$ denote a fixed element in the class $C_i$. Let $\chi_1, \ldots, \chi_r$ be the distinct irreducible complex characters of $G$, and $C_{ijk}$ denote the number of ordered pairs $(X, Y)$ such that $X \in C_i$, $Y \in C_j$ and $YX = \alpha_k$. Then it is known that

$$C_{ijk} = \frac{|C_i||C_j|}{N} \sum_{t=1}^{r} \frac{\chi_t^{(\alpha_i)} \chi_t^{(\alpha_j)} \overline{\chi_t^{(\alpha_k)}}}{\chi_t^{(1)}} \tag{A}$$

Our aim is to investigate this formula, when $G$ is the symmetric group of degree $n = 6t$, and compute $C_{ijk}$ where $C_i = \{2^{n/2}\}$, $C_j = \{3^{n/3}\}$ and $C_k = \{n\}$. In other words, for the permutation representation of degree $n$, of the $\langle 2, 3, n\rangle$-groups with the elements of order 2 and 3 fixed point free and $n$ a complete cycle, we show that

$$C_{ijk} = \frac{n!}{2^{n/2}\left(\dfrac{n}{2}\right)!\,3^{n/3}\left(\dfrac{n}{3}\right)!} \sum_{r=0}^{n-1} \frac{(-1)^r a_r b_r}{\dbinom{n-1}{r}} \tag{B}$$

where $a_r$ and $b_r$, $0 \leqslant r \leqslant n-1$ are the coefficients appearing in the following two equations:

$$\frac{(x^2-1)^{n/2}}{x-1} = \sum_{r=0}^{n-1} a_r x^{n-r-1}$$

(C)

$$\frac{(x^3-1)^{n/3}}{x-1} = \sum_{r=0}^{n-1} b_r x^{n-r-1}$$

It is clear that $n$ will be a multiple of 6. Now we will give an outline of the proof of formula (B).

## Outline of the proof

In the permutation representation of $S_n$, an element of the class $\{n\}$ is

COMPUTATIONAL GROUP THEORY
ISBN 0-12-066270-1

represented by a matrix which corresponds to the action of a generator of the cyclic group of order $n$ in its regular representation. Hence, the eigenvalues of this matrix are exactly the distinct $n$th roots of unity, i.e. the roots of the equation $x^n - 1 = 0$.

The space of this representation has as factor space modulo the 1-dimensional fixed space, a space $V$ affording the $[n-1, 1]$ representation and in this an element of the class $\{n\}$ has eigenvalues the roots of the equation

$$\sum_{r=0}^{n-1} x^{n-r-1} = \frac{x^n - 1}{x - 1} = 0.$$

The $r$th exterior power of $V$, $V_r$, is an irreducible $S_n$-module and affords the $[n-r, 1^r]$ representation [1], [2], [3]. If for $\sigma \in S_n$, $\mathbf{A} = \mathbf{A}(\sigma)$ is the matrix for $\sigma$ afforded by $V$ with respect to some basis $B$, then the matrix for $\sigma$ afforded by $V_r$ with respect to the basis obtained canonically from $B$ is the $r$th compound of $\mathbf{A}$, $\mathbf{A}^{(r)}$. Hence the trace

$$\chi^{[n-r, 1^r]}(\sigma) \text{ of } \mathbf{A}$$

is given by the $r$th elementary symmetric function of the roots of the characteristic equation of $\mathbf{A}$.

The elementary symmetric functions of the roots of the characteristic equation of $A$ are given by the coefficients of the characteristic equation. Hence when $\sigma$ is an $n$ cycle,

$$\chi^{[n-r, 1^r]}(\sigma) = (-1)^r.$$

Since the order of the centralizer of the class $\{n\}$ is $n$, the equation

$$\sum_{\lambda} \chi^{\lambda}(\sigma)\overline{\chi^{\lambda}(\sigma)} = n.$$

shows that $\chi^{\lambda}(\sigma) = 0$ for all other values of $\lambda$.

Since in the permutation representation on $n$ points, the characteristic equations of $\mathbf{A}(\alpha_i)$ and $\mathbf{A}(\alpha_j)$ for $\alpha_i \in \{2^{n/2}\}$, $\alpha_j \in \{3^{n/3}\}$, are $(x^2-1)^{n/2} = 0$ and $(x^3-1)^{n/3} = 0$, respectively, the traces $\chi_r(\alpha_i)$ and $\chi_r(\alpha_j)$ are given by $(-1)^r a_r$ and $(-1)^r b_r$ respectively where $a_r$ and $b_r$ are defined in expression (C). Hence we can write expression (B) for (A).

**Analysing expression (B)**

In this section we show that

$$\lim_{n \to \infty} C_{ijk} = \infty,$$

The proof is composed of the following two lemmas.

*Lemma (1).*

$$\lim_{n\to\infty} \frac{n!}{2^{n/2}\left(\frac{n}{2}\right)!\,3^{n/3}\left(\frac{n}{3}\right)!} = \infty .$$

*Proof.* The proof follows easily by using Stirling's formula.

*Lemma (2).* For $n = 6k$ we have

$$\lim_{n\to\infty} \sum_{r=0}^{n-1} \frac{(-1)^r a_r b_r}{\binom{n-1}{r}} = \begin{cases} 2 & \text{for } k \text{ odd} \\ 0 & \text{for } k \text{ even.} \end{cases}$$

*Proof.* Let

$$W(n) = \sum_{r=0}^{6k-1} \frac{(-1)^r a_r b_r}{\binom{6k-1}{r}} \tag{1}$$

where $a_r$, $b_r$ are the coefficients in the following two expressions:

$$\frac{(x^2-1)^{3k}}{x-1} = \sum_{r=0}^{6k-1} a_r x^{6k-r-1} \tag{2}$$

$$\frac{(x^3-1)^{2k}}{x-1} = \sum_{r=0}^{6k-1} b_r x^{6k-r-1}. \tag{3}$$

Apart from sign, the coefficients in these expressions are symmetrical, more precisely

$$a_r = (-1)^{k+1} a_{6k-r-1} \tag{4}$$

$$b_r = -b_{6k-r-1}. \tag{5}$$

It follows at once from (4) and (5) that in the expression (1), if $k$ is even, terms at equal distances from the ends of the sum cancel out. Thus for even $k$ we have

$$W(n) = 0.$$

Therefore, from now on, we suppose that $k$ is odd. Terms at equal distances from the ends of the sum in (1) are equal, so

$$W(n) = 2 \sum_{r=0}^{3k-1} \frac{(-1)^r a_r b_r}{\binom{6k-1}{r}}. \tag{6}$$

Now from (2),

$$\sum_{r=0}^{6k-1} a_r x^r = (1-x)^{-1}(1-x^2)^{3k}$$
$$= (1+x+x^2+\dots)\sum_{c=0}^{3k}(-1)^c\binom{3k}{c}x^{2c}.$$

Hence by comparing the coefficients of $x^r$

$$a_r = \sum_{2c \leqslant r}(-1)^c\binom{3k}{c}.$$

Then

(7) $$a_{2m} = a_{2m+1} = \sum_{c=0}^{m}(-1)^c\binom{3k}{c} = (-1)^m\binom{3k-1}{m}.$$

This will hold for all $m$, including $m \geqslant 3k$, if we take $\binom{c}{d}$ to mean zero when $d > c$.

Similarly, from (3) we have

(8) $$b_{3m} = b_{3m+1} = b_{3m+2} = (-1)^m\binom{2k-1}{m}.$$

Now we write

$$T_r = \frac{(-1)^r a_r b_r}{\binom{6k-1}{r}},$$

and by (7), (8) we write the terms $T_{6m}$, $T_{6m+1}$, $T_{6m+2}$, $T_{6m+3}$, $T_{6m+4}$ and $T_{6m+5}$. Using these terms, it follows that, for $0 \leqslant r \leqslant 3k-2$,

(9) $$\left|\frac{T_{r+1}}{T_r}\right| \leqslant 1$$

except when $r$ is of the form $6m+5$. However, provided that $6m+5 \leqslant 3k-1$, we still have

(10) $$\left|\frac{T_{6m+6}}{T_{6m}}\right| \leqslant 1.$$

Now we can verify that, as $k \to \infty$,

$$T_1, T_2, T_3, T_4, T_5, T_6 = O(1/k)$$
$$T_7, T_8, T_9, T_{10}, T_{11}, T_{12} = O(1/k^2)$$

By (9) and (10) we have uniformly in $7 \leqslant m \leqslant 3k-1$;

$$T_m = O(1/k^2).$$

Thus, in the sum in (6), we have one term equal to 1, 6 terms which are $O(1/k)$ and $(3k-7)$ terms which are $O(1/k^2)$, hence

$$W(n) = 2+O(1/k).$$

## Acknowledgements

The author wishes to express his gratitude to Professor D. Livingstone. Further thanks are due to Professor B. Kuttner and Dr. R. List.

## References

1. Littlewood, D. E. (1940). "The theory of group characters and matrix representation of groups". Oxford.
2. Livingstone, D. (1957). Proof of a theorem discovered by Murnaghan. *Proc. Nat. Acad. Sci.* **43**, 618–619.
3. Murnaghan, F. D. (1955). On the generation of the irreducible representations of the symmetric group, *Proc. Nat. Acad. Sci.* **41**, 514–515.

# The Calculation of the Schur Multiplier of a Permutation Group

D. F. HOLT

## 1 Introduction

The object of this paper is to describe an algorithm for calculating the $p$-primary part of the Schur Multiplier of a finite permutation group, for a given prime $p$. This algorithm has been successfully implemented on a machine, and it appears to give accurate results in a reasonable time for groups of degree up to a few hundred.

Let us first give a brief description of the underlying group theory. This theory is the dual of the theory developed in Section 10, Chapter XII of [**2**] for cohomology groups of finite groups. The prime $p$ will be fixed throughout and, to simplify notation, we will use $M(G)$ to denote the $p$-primary part of the Schur Multiplier of a group $G$. Let $P \in \mathrm{Syl}_p(G)$, and let $N(P)$ denote the normalizer of $P$ in $G$. Then $M(G)$ is naturally an epimorphic image of $M(H)$ for any group $H$ that satisfies $P \subseteq H \subseteq G$. The aim is to calculate $M(P)$ first, and then to calculate the factor groups of $M(P)$ that are naturally isomorphic to $M(N(P))$ and $M(G)$, and possibly $M(H)$, for suitable intermediate subgroups $H$. We will again simplify notation, by identifying $M(N(P))$, $M(G)$ and $M(H)$ with these factor groups of $M(P)$. There is a standard algorithm for calculating $M(P)$, due to Macdonald [**6**] and Wamsley [**9**], which we shall say more about later. The action of $N(P)$ on $P$ induces an action of $N(P)$ on $M(P)$, and then we have $M(N(P)) = M(P)/[M(P), N(P)]$. To calculate $M(G)$, we must first find double coset representatives $g_i$ $(1 \leqslant i \leqslant n)$ of $N(P)$ in $G$. For each such $g_i = h$, we then calculate $Q := P \cap h^{-1}Ph$. Let $M(\bar{Q})$ denote the image of $M(Q)$ in $M(P)$ induced by the insertion map $Q \to P$. Then conjugation by $h^{-1}$ induces a map $\phi: M(\bar{Q}) \to M(h\bar{Q}h^{-1})$. Let

$$M(P, h) = \langle x^{-1}\phi(x) \mid x \in M(\bar{Q}) \rangle$$

and let

$$M(P, G) = \langle M(P, g_i) \mid 1 \leqslant i \leqslant n \rangle.$$

COMPUTATIONAL GROUP THEORY
ISBN 0-12-066270-1

Then we have $M(G) = M(N(P))/M(P, G)$. In practice, if $|G:N(P)|$ is too large, we seek one or more intermediate subgroups $H$ with $N(P) \subset H \subset G$, and then calculate $M(H)$ first and $M(G)$ as a factor group of $M(H)$. In order to achieve this, it is necessary to find double coset representatives $g_i$ of $H$ in $G$ for which $g_i^{-1}Pg_i \cap P$ is a Sylow $p$-subgroup of $g_i^{-1}Hg_i \cap H$.

We chose to apply the above theory in the case in which $G$ is a permutation group on a set $\Omega$, because there are existing efficient methods, originally due to Sims [**8**], for the manipulation of permutation groups on computers.

An interesting feature of the complete algorithm is that, in addition to its use of permutation groups, it involves the manipulation of power-commutator presentations of nilpotent groups (introduced by Macdonald [**6**]). This is another area in which there exist efficient algorithms for mechanical computation. On account of this dependence on two essentially different sets of algorithms, it was necessary to write procedures for the conversion of permutation representations of $p$-groups to power-commutator representations, and there are times when it is unclear which of these two is the more convenient.

In the existing implementation, the complete algorithm is split up into six programs, which exchange input and output, and we shall describe these, together with their principal input and output files, in the remaining sections of this paper. Several of them use existing algorithms which have been described adequately elsewhere and we shall give less details in these cases. The programs are entitled GENPERM, SYLNORM, POWERCOMM, SYLCUTS, DCOSETS and NILPQUOT. Somewhat surprisingly, the slowest part of the whole process turned out to be the determination of the double coset representatives.

All of the programs are written in Burroughs Algol, which is an extended version of Algol 60, with record-oriented input–output. They were designed for use on the Burroughs 6700 computer at Warwick University, which is medium fast, and has a fairly large amount of available storage space. They were therefore written with a view to minimizing process times rather than storage requirements. We will give some examples of performance times in the final section of this paper.

## 2 GENPERM

This algorithm is due to Sims, using theory due to Schreier, and more details can be found in [**8**]. Let $G$ be a permutation group on a finite set $\Omega$. In order to perform efficient computations with $G$, it is convenient to have a basis and a strong generating set, which are defined as follows. Let $(\alpha_1, \alpha_2, \ldots, \alpha_r)$ be an ordered subset of $\Omega$, and define

$$G^1 = G \quad \text{and} \quad G^s = G_{\alpha_1 \alpha_2 \ldots \alpha_{s-1}},$$

for $2 \leqslant s \leqslant r+1$. Then $(\alpha_1, \ldots, \alpha_r)$ is called a basis for $G$ if $G^{r+1} = 1$; in other words, if no nontrivial element of $G$ fixes each $\alpha_i$. Let $S$ be a subset of $G - \{1\}$. Then $S$ is called a strong generating set for the basis $(\alpha_1, \ldots, \alpha_r)$, if $\langle G^s \cap S \rangle = G^s$ for $1 \leqslant s \leqslant r$. Let $\Omega^s$ denote the orbit of $\alpha_s$ under $G^s$. Then we have

$$|G| = \prod_{s=1}^{r} |\Omega^s|.$$

For a given $\alpha$ in $\Omega$, it is necessary to be able to determine rapidly whether $\alpha \in \Omega^s$ and, if so, to be able to calculate a permutation $g^s$ that takes $\alpha$ to $\alpha^s$. This can be achieved by means of a Schreier Vector (see [**8**]). It is then easy to determine whether a given permutation of $\Omega$ lies in $G$ and, if so, to express it as a product of the elements of $S$.

In certain circumstances, it is convenient to store a unique permutation $g_\beta{}^s$, for each $\beta \in \Omega^s$, even though this involves the use of a large amount of storage space. Each element $g \in G$ can then be expressed uniquely in the form

$$g = g_{\beta_1}^{1} g_{\beta_2}^{2} \cdots g_{\beta_r}^{r},$$

for suitable $\beta_s \in \Omega^s$. ($gh$ means permutation $g$ followed by permutation $h$, throughout.) If we now order each set $\Omega^s$, then this gives us a natural lexicographical ordering of the elements of $G$, and so the elements of $G$ can be scanned in order rapidly. We always choose the ordering such that $\alpha^s$ is the first element of $\Omega^s$, which has the effect that the elements of $G^s$ all precede those of $G^t - G^s$, when $t < s$. This process is used in SYLNORM and DCOSETS, and we say that $G$ is stored in expanded form in these cases. On the other hand, in the programs POWERCOMM and SYLCUTS, it is merely necessary to be able to determine rapidly whether a given permutation of $\Omega$ lies in $G$, and for this purpose it is more efficient and uses less storage space, to store the strong generating set and the Schreier Vectors.

The algorithm GENPERM starts with a set $T$ of permutations on $\Omega$ and calculates the order of the group $G$ generated by $T$, a basis $B$ for $G$, and a strong generating set $S$ for $G$ (containing $T$), together with Schreier Vectors $SV$. In practice, $B$, $S$ and $SV$ seem to be produced fairly rapidly but, when $\Omega$ contains more than a few hundred points, the verification that $S$ really is a strong generating set is painfully slow. A faster method, which brings in Todd-Coxeter coset enumeration, is described in [**5**]. For our purposes, we may assume that $|G|$ is already known and so we only wish to compute $S, B$ and $SV$, which is quite quick.

Input File:

INPERM. This gives the permutations in $T$ and, if possible, the order of $G$ or, better still, a basis, together with the orders $|\Omega^s|$ $(1 \leqslant s \leqslant r)$.

Output File:

OUTPERM. This gives a basis, strong generating set and Schreier Vectors for $G$, together with the orders $|\Omega^s|$.

A small additional program EXPAND, uses OUTPERM as input and has an output file OUTEXP, which gives the permutations in the expanded form described above.

## 3 SYLNORM

This program calculates a Sylow $p$-subgroup $P$ of $G$. It can also calculate the normalizer, or the orbit-permuter, of an arbitrary subgroup $H$ of $G$, which will usually be $P$. It is very inefficient, particularly when symmetric or alternating groups of moderately large degree are involved in $G$, but it seems to be adequate in our particular context, and it does not take up a large percentage of the total computation time for the groups under consideration (for example the Mathieu groups and the Higman-Sims group). The efficient calculation of normalizers is a notoriously difficult process, but there do exist efficient algorithms for the computation of Sylow subgroups (see, for example [**1**]).

Suppose that we wish to calculate the normalizer of a subgroup $H$ of $G$. We assume that we are given a strong generating set $S(H)$, and Schreier Vectors $SV(H)$ for $H$, using the same basis $B$ as for $G$. Then, for a given $g \in G$, we can determine rapidly whether $g \in N(H)$, by testing whether $g^{-1}hg \in H$ for all $h \in S(H)$ (or better, for those $h$ in a minimal generating set for $H$). We will assume that $G$ is stored in expanded form. Roughly speaking, we search through the elements of $G$, using the lexicographical ordering described in section 2, looking for elements of $N(H)$. Of course, we do not really need to test every element of $G$. Suppose that we are considering an element

$$g_{\beta_1}^1 g_{\beta_2}^2 \cdots g_{\beta_r}^r.$$

The next element in the lexicographical ordering would result from changing $g_{\beta_r}^r$ only, but we attempt to speed up the searching process by advancing $g_{\beta_s}^s$, where $s$ is as small as possible. In this program, this is achieved in a rather unsophisticated manner, and a more efficient program would utilize more tests in order to reduce this value of $s$. The methods used here are as follows.

Firstly, if $g_{\beta_1}^1 \neq 1$ and $\beta_1$ is already known to be in the orbit of $\alpha_1$ under $N(H)$, then we can take $s = 1$ (the best possible), because we are assuming that $N(H) \cap G_{\alpha_1}$ has already been found. Otherwise, we use the fact that any $g$ in $N(H)$ must permute the orbits of $H$, and it must take an orbit to one of the same length. By considering the partial words $g_{\beta_1}^1 \ldots g_{\beta_s}^s$ for $s = 1, 2, \ldots$ in turn, we can calculate the action of $g$ on the $H$-orbits of $\alpha_1, \ldots, \alpha_s$. If we find a contradiction for a particular value of $s$, then we can change $g_{\beta_s^s}$. This process works best when $H$ has several orbits of different lengths (which, fortunately, is often the case when $H$ is a $p$-subgroup of $G$), and it is useless when $H$ acts regularly on $\Omega$. So, for example, it will take ridiculously long to calculate the normalizer of a Sylow 17-subgroup of Sym(17).

A Sylow $p$-subgroup of $G$ is computed as follows. We may assume inductively that, for some $s > 1$, $Q = P^s \in \mathrm{Syl}(G^s)$ has been found, together with $S(Q)$ and $SV(Q)$, and we aim to find $P^{s-1}$. We carry out the above searching procedure to look for a $p$-element $g \in N(Q) \cap G^{s-1}$. If we find such a $g$, then we replace $Q$ by $R = \langle Q, g \rangle$, and $S(Q) \cup \{g\}$ will automatically be a strong generating set for $R$, but $SV(R)$ must be calculated. We then repeat this process with $R$ in place of $Q$. Since $|P^{s-1}|$ and $|R|$ are known, we will know immediately when $P^{s-1}$ has been found. If $S(P) = \{g_1, \ldots, g_r\}$ and we define $P(i) = \langle g_1, \ldots, g_i \rangle$, then we have the additional property that $P(i) \trianglelefteq P(i+1)$ for all $i$.

This algorithm seems to work reasonably quickly most of the time and it only exposes its defects when it undertakes a lot of irrelevant searching. For example, when looking for a Sylow 3-subgroup of Sym(27), it found a subgroup of index 3 in $P$, having 3 orbits of length 9, very quickly, but then it took much too long to find the final generator of $P$.

Input Files:

1. INPERMG. Output from GENPERM using EXPAND, giving the expanded form of the representation of $G$.
2. INPERMH (if required). Output from GENPERM for a subgroup $H$ of $G$, where $N_G(H)$ is to be calculated. If only $P$ and $N(P)$ are required, then this file is not needed.

Output Files:

1. OUTSYL. This gives $S(P)$ and $SV(P)$.
2. OUTNORM. This gives $S(N(H))$ and $SV(N(H))$, where $H$ may be equal to $P$.

Both OUTSYL and OUTNORM have the same format as OUTPERM (see section 2), and so they can be used in place of OUTPERM if necessary.

## 4 DCOSETS

This program computes a set $D(K, H)$ of double coset representatives of the subgroups $K$ and $H$ of $G$. (So $G = \cup KgH$ for $g \in D(K, H)$.) It achieves this by first calculating a set $T$ (transversal) of left coset representatives of $H$ in $G$, and then calculating $D(K, H)$ by computing the orbits of the action of $K$ on $T$ by left multiplication. It is the second half of this computation which takes up most of the time (more than 80 %), and it is easily the slowest part of the whole algorithm. The difficulty is that the coset representative of $kgH$, for a given $k \in K$ and $g \in T$, requires some computation and, when $|G:H|$ is larger than a few thousand, the whole process becomes prohibitively slow. As far as I know, no essentially quicker method of finding double coset representatives is known.

It is most convenient to have both $G$ and $H$ stored in expanded form (with a common basis). $K$ is involved only in orbit calculations, and so we simply require a set of generators for $K$, which should always be as small as possible. In practice, $K$ is often a $p$-group, in which case a minimal generating set is readily available. The algorithm for determining $T$ is quite simple and efficient, and it is roughly the same as that described in [**1**]. For each coset $gH$, we choose as representative the least element of $gH$ in the lexicographical ordering of $G$. Then we see easily that, if

$$g_{\beta_1}^{1} g_{\beta_2}^{2} \cdots g_{\beta_r}^{r} \in T, \quad \text{then} \quad g_{\beta_s}^{s} \cdots g_{\beta_r}^{r} \in T$$

for $1 \leqslant s \leqslant r$. Furthermore, $T^s = T \cap G^s$ is a transversal for $H^s$ in $G^s$.

Assume inductively that $U = T^{s+1}$ has been found for some $s > 0$, and that we wish to find $T^s$. Then we search through the elements $g = g_{\beta_s}^{s} u$, for all $u \in U$ and $\beta_s \in \Omega^s$. For each such element, we know that either $g \in T^s$, or there exists $h \in H^s$ with

$$g_{\beta_s}^{s} uh = g_{\gamma_s}^{s} v,$$

where $v \in U$ and $\gamma_s < \beta_s$ in the ordering of $\Omega^s$. To determine whether such an $h$ exists, we merely have to check whether $(\gamma_s) g_{\beta_s}^{s} u$ lies in the orbit of $\alpha_s$ under $H^s$ for any $\gamma_s \in \Omega^s$ with $\gamma_s < \beta_s$. As $|T^s|$ is known, we will know as soon as the computation is complete.

For the second part of the computation, namely the left permutation action of $K$ on $T$, we need to be able to calculate the element $k$ of $T$ in $gH$, for a given $g \in G$. Suppose that

$$k = g_{\beta_1}^{1} \cdots g_{\beta_r}^{r},$$

and we assume inductively that

$$g_{\beta_1}^{1}, \ldots, g_{\beta_{s-1}}^{s-1}$$

have been calculated for some $s \geqslant 1$, together with an element $h$ of $H$ such that

$$x = (g_{\beta_1}^{1} \cdots g_{\beta_{s-1}}^{s-1})^{-1} gh \in G^s.$$

($h = 1$ when $s = 1$.) Then we search through the elements $\beta_s \in \Omega^s$, until we find the least such element for which $(\beta_s)x$ lies in the image of $\alpha_s$ under $H^s$. Then there exists $h' \in H^s$ with $(\beta_s)xh' \in G^{s+1}$, and replacing $x$ by $(g_{\beta_s}^{s})^{-1} xh'$ completes the induction.

Input Files:

INPERMG, INPERMH. Output from GENPERM (using EXPAND), giving the expanded form of the presentations of $G$ and $H$.
INPERMK. A minimal generating set for $K$.

Output Files:

1. DCREPS. The elements in $D(K, H)$.
2. COSETREPS (if required). The elements in $T$.

## 5 POWERCOMM

This program uses the output from SYLNORM as its input. Thus, we assume that we are given a strong generating set $S = \{g_1, \ldots, g_m\}$ and basis $\{\alpha_1, \ldots, \alpha_r\}$ for a $p$-group $P$, together with Schreier Vectors $SV$. $S$ will have the additional property that $P(i) \trianglelefteq P(i+1)$ for $1 \leqslant i \leqslant m$, where $P(i) = \langle g_1, \ldots, g_i \rangle$. Furthermore, for each $s$, there will be an $i$ such that $P(i) \in \mathrm{Syl}_p(G^s)$. The purpose of the program is to calculate a power-commutator presentation (PCP) for $P$. This is achieved in two stages. In the first stage, $S$ is replaced by a new strong generating set $S^* = \{h_1, \ldots, h_n\}$ (where $|P| = p^n$), which has the additional property that $P^*(i+1)/P^*(i) \subseteq Z(P/P^*(i))$ for $0 \leqslant i < n$, where $P^*(i) = \langle h_1, \ldots, h_i \rangle$ and $P^*(0) = 1$. In other words, we have calculated a maximal central series for $P$. It is now possible to express each $g \in P$ uniquely in the form $h_n^{a_n} \ldots h_1^{a_1}$ where $0 \leqslant a_i < p$, and we will call this the normal form of $g$. In the second stage, we express the commutators $[h_i, h_j]$ $(i < j)$ and the powers $h_i^p$ in normal form. The generators $h_i$, together with these equations, constitute the PCP of $P$. A number of further properties of this presentation, which will be needed in NILPQUOT, are also calculated at this stage. For example, it is important to know which of the $h_i$ lie in the commutator subgroup $[P, P]$ of $P$, and which of the equations in the PCP can be regarded as definitions of these generators in $[P, P]$.

We now have two distinct representations of $P$, a permutation representation (together with strong generating set, etc.) and a PCP. The former will be required again in SYLCUTS, but the latter, which will be used in NILPQUOT, no longer depends on the permutation representations of the $h_i$. A final section of POWERCOMM calculates the permutations $g^{-1}h_ig$ for generators $g \in N(P)$, and puts these elements in normal form. These will be required in the calculation of $M(N(P))$ as a quotient of $M(P)$. We will now give more details pertaining to the two stages described, above.

Let us assume inductively that, for some $i$, $\{g_1, \ldots, g_{i-1}\}$ has been replaced by $\{h_1, \ldots, h_k\}$, where $Q := P(i-1) = \langle h_1, \ldots, h_k \rangle$, $|Q| = p^k$, and the series $1 = Q(0) \subset Q(1) \subset \ldots \subset Q(k) = Q$ is a central series for $Q$, where

$Q(j) = \langle h_1, \ldots, h_j \rangle$. We now want to extend this construction to $P(i)$. Put $g = g_i$. Then $g \in N(Q)$. First of all, if $g^p \notin Q$, then we can replace $\{g\}$ by the set

$$\{g^{p^{a-1}}, g^{p^{a-2}}, \ldots, g^p, g\},$$

where $a$ is minimal such that $g^{p^a} \in Q$. We may therefore assume that $g^p \in Q$. Let us now assume inductively that, for some $l > 0$, $g$ centralizes $Q(j)/Q(j-1)$ for $0 < j < l$. Then we wish to alter the set $\{h_l, \ldots, h_k\}$ in order to force $g$ to centralize $Q(l)/Q(l-1)$. Put $x = h_l^{-1} g^{-1} h g$. Then, since $g \in N(Q)$, we have $x \in Q$. We can now use $SV$ to express $x$ as a product of the generators in $S$. From the way that $P$ was originally constructed in SYLNORM, it will be the case that all of the elements of $S$ that are involved in this expression for $x$ will lie in the set $\{h_1, \ldots, h_k\}$. Furthermore, the highest value of $m$ for which $h_m$ is involved in the expression will be equal to the least value of $m$ for which $x \in Q(m)$. If $m < l$, then $g$ centralizes $Q(l)/Q(l-1)$ already, whereas $m = l$ cannot occur. If $m > l$, then the basic idea is to replace $h_l$ by $x$ and $h_m$ by $h_l$, and then to repeat this procedure until $g$ centralizes $Q(l)/Q(l-1)$. However, it is essential that $S$ should remain a strong generating set for $P$ and furthermore, there are certain technical properties of $S$ and $SV$ (such as the property of $h_m$ used above) which have to be preserved in the transition from $S$ to $S^*$. A slightly more complicated procedure is therefore necessary, but we shall give no further details here. Although this part of the algorithm is fairly intricate, it seems to take up very little computing time.

The basic problem in the second stage of the program is to express a given permutation $x$ in normal form. To do this, we first express $x$ as a product of elements of $S^*$, using $SV$ as usual. We now locate the greatest value of $m$ for which $h_m$ is involved in this expression, and the number $l$ of occurrences of $h_m$. Then, as stated above, $m$ is minimal such that $x \in P^*(m)$. Now let $l = l'$ (mod $p$) with $1 \leqslant l' < p$, and put $y = h_m^{-l'} x$. Then $y \in P^*(m-1)$ and, by repeating this procedure with $y$ in place of $x$, we will eventually arrive at the normal form of $x$.

Input Files:

OUTSYL, OUTNORM. Output from SYLNORM.

Output Files:

1. OUTCENT. This gives the set $S^*$ as permutations on $\Omega$, together with $SV$, in the same format as OUTPERM.
2. OUTPRES. This gives the PCP for $P$ in terms of the generating set $S^*$, without giving the permutation representations of the generators.
3. GENIMAGES. This gives the images of the generators in $S^*$ under conjugation by elements of $N(P)$, in normal form (not as permutations).

## 6 SYLCUTS

This program uses the outputs of POWERCOMM and DCOSETS as input. The idea is, for each double coset representative $g$ of $N(P)$ in $G$, to calculate generators of a central series of $Q := g^{-1}Pg \cap P$ (in normal form, using the generators of $P$), and then to calculate a PCP of $Q$ using these generators. Instead of $N(P)$, an intermediate subgroup $H$ with $N(P) \subset H \subset G$ may be used. In this case, it is necessary to replace $g$ by $h$, where

$$h^{-1}Ph \cap P \in \mathrm{Syl}_p(h^{-1}Hh \cap H).$$

This is accomplished as follows. We assume that we start with double coset representatives $D(N(P), H)$ rather than $D(H, H)$, as this seems to be more convenient at this stage. Let $g \in D(N(P), H)$, and then calculate $R = g^{-1}Pg \cap H$. We now search through the left coset representatives $g_i$ of $N(P)$ in $H$ (which are available as output from DCOSETS), until we find a $g_i$ with $g_i{}^{-1}Rg_i \subseteq P$. We then replace $g$ by $h = gg_i$, which has the required property.

Much of this program consists of procedures almost identical to those used in POWERCOMM, the only new feature being the calculation of the intersection $Q = g^{-1}Pg \cap P$. There were two possible approaches to this problem, one of which is a standard algorithm for calculating intersections of permutation groups and is mentioned in [1]. This would be the more efficient if we only required the intersection but, since we also require a central series for $Q$, it seemed to be easier to use the cruder second approach, which is simply to search through the elements of $g^{-1}Pg$ in their normal forms, testing each element for membership of $P$. This automatically produces the required generators of $Q$ in normal form and it is speeded up by the fact that we actually know $|Q|$ in advance (from information available from DCOSETS). It is possible that, if $P$ were very large, then the first approach would become faster, but in this case it would be necessary to re-apply POWERCOMM to $Q$, for all $g$, in order to get a central series. In any case, this part of the algorithm takes up considerably less process time than DCOSETS and so this argument is probably unimportant.

Input Files:

DCREPS, COSETREPS (output from DCOSETS). COSETREPS is only needed if an intermediate subgroup $H$ is being used.
OUTCENT (output from POWERCOMM).

Output File:

ADDRELS. For each $g$, generators of $Q$ and of $gQg^{-1}$ are listed in normal form, followed by the PCP for $Q$ in terms of these generators.

## 7 NILPQUOT

This program has three parts. In the first, $M(P)$ is calculated using the PCP of $P$ derived in POWERCOMM. In the second, $M(N(P))$ is calculated using the images of the generators of $P$ under $N(P)$, derived in the final part of POWERCOMM. Finally, in the third, $M(G)$ is calculated using the information available from SYLCUTS.

The algorithm used in the first part is identical to that described in [**9**], so we shall give only a brief description here. A more recent generalized version of this nilpotent quotient algorithm, which can also be used for the investigation of questions related to the Burnside Problem, is described in [**3**]. Let $h_1, \ldots, h_n$ be the generators in the PCP, and suppose, for simplicity, that $h_i \in [P, P]$ if and only if $i \leqslant m$, for some fixed $m$. Rewrite the relations of the PCP in the form of relators $R_j = 1$. Then $m$ of these relations can be regarded as definitions of $h_i$ $(i \leqslant m)$, and $n-m$ of them give the $p$th powers of $h_i$ for $i > m$. (All of this information is in fact available from the output of POWERCOMM.) Each of the remaining relations is replaced by $R_j = x_j$, where $x_j$ is a new generator, which is assumed to commute with all other generators. We then solve various equations (for example, $(h_i h_j)h_k = h_i(h_j h_k)$), which force the resulting group $P^* = \langle h_i, x_j \rangle$ to be associative and which have the effect of introducing relations amongst the $x_j$. Then $P^*$ is a covering group of $P$, where

$$\langle x_j \rangle = M(P) \text{ and } P^*/\langle x_j \rangle \cong P.$$

Furthermore, we end up with a PCP for $P^*$. The core of this algorithm is the procedure for putting an arbitrary word in the generators of $P^*$ into normal form, by using the relations in the PCP.

In the second stage of NILPQUOT, we are given a list of generators of $N(P)$ and for each of these, we are given the images $g^{-1}h_i g$ of the generators of $P$ under conjugation by $g$, in normal form. We wish to calculate the induced action of $g$ on $[P^*, P^*]$ and on $M(P)$. Now $[P^*, P^*]$ is generated by the elements $h_i$ $(i \leqslant m)$, together with the new generators $x_j$. The $h_i$ $(i \leqslant m)$ are defined in $P^*$ by the same relations as in $P$, and so we can use our knowledge of $g^{-1}h_i g$ $(i > m)$ to calculate each $g^{-1}h_i g$ $(i \leqslant m)$ in turn, as an element of $P^*$. Of course, the normal form of $g^{-1}h_i g$ in $P^*$ may involve the $x_j$ as well as the $h_j$. (In an earlier version of this program, the author made the mistake of overlooking this fact, which resulted in some wrong answers.) Since the $x_j$ themselves are defined in $P^*$ by relations involving the $h_i$, we can now calculate $g^{-1}x_j g$ in the same way. We then introduce the new relation $x_j = g^{-1}x_j g$, for each $x_j$ and each $g \in N(P)$. In the resulting quotient group $P^{**}$ of $P^*$, the remaining generators $x_j$ will generate $M(N(P))$, where $P^{**}/\langle x_j \rangle \cong P$.

The third stage is similar, but slightly more complicated. Here, we are given a list of double coset representatives $g$. For each such $g$, we are given generators $g_1, \ldots, g_l$ of $Q = P \cap g^{-1}Pg$. Furthermore, we are given a PCP of $Q$ in terms of these generators. Let $Q^{**}$ be the inverse image of $Q$ in $P^{**}$. Let

$$M(\bar{Q}) = [Q^{**}, Q^{**}] \cap M(N(P))$$

be the image of $M(Q)$ under the natural map

$$M(Q) \rightarrow M(P) \rightarrow M(N(P)).$$

Then, conjugation by $g^{-1}$ induces a map

$$M(\bar{Q}) \rightarrow M(g\bar{Q}g^{-1})$$

and we aim to factor out this action of $g$ in $M(P)$. First of all, we must calculate $g_i$ (and $gg_ig^{-1}$) as elements of $Q^{**}$ for those $g_i$ which lie in $[Q, Q]$, which can be done by using the definitions in the PCP of $Q$. For the relations $R_j = 1$ of $Q$ which are neither definitions nor $p$th powers of generators outside $[Q, Q]$, we calculate $R_j$ as an element of $Q^{**}$, which gives us a relation $R_j = y_j$ in $Q^{**}$. Then the elements $y_j$ will generate $M(\bar{Q})$. By replacing the generators $g_i$ in $R_j$ by their images $gg_ig^{-1}$, we can now calculate $gy_jg^{-1}$, and we introduce the new relation $y_j = gy_jg^{-1}$ in $M(N(P))$. By doing this for each $g$, we end up with a quotient group $P^{***}$ of $P^{**}$ which is a $p$-covering group of $G$, and the remaining generators $x_j$ generate $M(G)$.

Whenever we introduce a new relation amongst the $x_i$, in any of the three stages, we either eliminate or reduce the order of some particular $x_i$, or we change generators in such a way that the $x_i$ will continue to form a basis of the abelian group that they generate. The structure of the group $\langle x_i \rangle$ will then be apparent at each stage. In order to provide an initial order for these generators, we use a result of Schur that, if $|P| = p^n$, then $M(P)$ has exponent at most $p^{n/2}$. Another possible approach to this problem, which is used in the Havas-Newman program, is to calculate $M(P)$ in several stages, assuming exponent $p$ at each stage.

Input Files:

OUTPRES, GENIMAGES. Output from POWERCOMM.

ADDRELS. Output from SYLCUTS.

Output File:

OUTCOVER. This gives the PCP for the covering group of $P, N(P)$ or $G$, together with the orders of the $x_i$.

## 8 Results

In this final section, we will list some process times for successful calculations of $M(G)$ in three cases. In the second and third cases, some work was also required by the operator in order to find generators of a suitable intermediate subgroup $H$ with $N(P) \subset H \subset G$. In the third case, we will give the complete distribution of the process times amongst the different programs used in the calculation. The prime is 2 in each case.

1. $G = \mathrm{PSL}(3,4)$ $(= M_{21})$ acting in its natural 2-transitive representation on 21 points.
$M(G) = \mathbb{Z}_4 \times \mathbb{Z}_4$; Process Time = 32 seconds.

2. $G = M_{22}$ (Mathieu Group) acting 3-transitively on 22 points.
$M(G) = \mathbb{Z}_4$; Process Time = 65 seconds.
For a long time, it was thought that $M(G)$ was $\mathbb{Z}_2$. This error was corrected by Griess and Mazet as recently as 1979 (see [**7**]).

3. $G = \mathrm{HS}$. The Higman-Sims Group acting on 100 points. Generators were taken from [**4**].
$|G| = 2^9 . 3^2 . 5^3 . 7 . 11$. $|P| = 2^9$. $N(P) = P$, but there is an element $t$ of $P$ such that $P \subset H \subset G$, where $H = C(t)$, $|H:P| = 15$, and $|G:H| = 5775$. $M(G) = \mathbb{Z}_2$. Process times are as follows.

| | | | |
|---|---|---|---|
| (i) | GENPERM | To generate $B, S$ and $SV$ for $G$ from the two given generators. | 38 seconds |
| (ii) | EXPAND | Applied to $G$. | 15 seconds |
| (iii) | SYLNORM | To calculate $P$ | 14 seconds |
| (iv) | SYLNORM | To calculate $H = C(t) = N(\langle t \rangle)$. | 27 seconds |
| (v) | POWERCOMM | Applied to $P$. | 7 seconds |
| (vi) | NILPQUOT | To calculate $M(P)$. | 6 seconds |
| (vii) | EXPAND | Applied to $P$ and to $H$. | 7 seconds |
| (viii) | DCOSETS | Applied to $H:P$. | 8 seconds |
| (ix) | SYLCUTS | Applied to $H:P$. | 9 seconds |
| (x) | NILPQUOT | To calculate $M(H) = \mathbb{Z}_2 \times \mathbb{Z}_2$. | 9 seconds |
| (xi) | DCOSETS | Applied to $G:H$. | 329 seconds |
| (xii) | SYLCUTS | Applied to $G:H$. | 65 seconds |
| (xiii) | NILPQUOT | To calculate $M(G) = \mathbb{Z}_2$. | 24 seconds |
| Total Process Time. | | | 558 seconds |

In step (xi), the 5775 coset representatives of $H$ in $G$ were produced in about 60 seconds, and the remainder of the time was used in calculating the 33 double coset representatives $D(P, H)$. It seems clear that, in order to improve the efficiency of the whole algorithm, the one vital step would be to find a better method of calculating double coset representatives.

## References

1. Cannon, J. J. (1974). Computing local structure of large finite groups. *In* "Computers in Algebra and Number Theory". *SIAM—AMS Proc.* **IV**, 161–176.
2. Cartan, H. and Eilenberg, S. (1956). *In* "Homological Algebra". Princeton University Press.
3. Havas, G. and Newman, M. F. (1980). Applications of computers to questions like those of Burnside. *In* "Burnside Groups" (Ed. J. L. Mennicke). Lecture Notes in Mathematics, vol. **806**, 211–230. Springer-Verlag, Berlin.
4. Higman, D. G. and Sims, C. C. (1968). A simple group of order 44 352 000, *Math. Zeitschr.* **105**, 110–113.
5. Leon, J. S. (1980). On an algorithm for finding a base and strong generating set for a group given by generating permutations, *Mathematics of Computation* **35**, 941–974.
6. Macdonald, I. D. (1974). A computer application to finite $p$-groups, *J. Austral. Math. Soc.* **17**, 102–112.
7. Mazet, P. (1979). Sur le multiplicateur de Schur du groupe de Mathieu $M_{22}$, *C.R. Acad. Sci. Paris*, **289**, 659–661.
8. Sims, C. C. (1970). Computational methods in the study of permutation groups. *In* "Computational Problems in Abstract Algebra" (Ed. J. Leech), 169–183. Pergamon Press, Oxford.
9. Wamsley, J. W. (1974). "Computation in nilpotent groups (theory)". Proc. Second Internat. Conf. Theory of Groups (Canberra 1973), Lecture Notes in Mathematics vol. **372**, 691–700. Springer-Verlag, Berlin.

# Computing Automorphism Groups of Combinatorial Objects

JEFFREY S. LEON‡

## 1 Introduction

Given a combinatorial object $X$, how can we effectively compute the automorphism group of $X$? In this paper I shall discuss an approach employing backtrack search and the concepts of base and strong generating set introduced by Sims [**11**], [**12**]. The term "combinatorial object" encompasses a wide variety of structures—block designs, graphs, codes, matrices over $\mathrm{GF}(q)$, latin squares, etc. Even the precise definition of the automorphism group $\mathrm{AUT}(X)$ will vary according to the type of object; in general, $\mathrm{AUT}(X)$ will be the group of permutations or monomial permutations of some set $\Omega$ associated with $X$ which leave $X$ invariant. Any efficient automorphism group algorithm will have to deal with the specific properties of the particular type of object under study, and some of its details will be highly dependent upon those properties. Nonetheless, as Butler and Lam [**1**] point out, major portions of an automorphism group algorithm can be constructed in a type-independent manner. I shall present a general type-independent approach and then discuss some specifics in implementing this approach for one type of object (Hadamard matrices).

As mentioned above, this approach rests on the concepts of base and strong generating set introduced by Sims to facilitate computation in permutation groups of high degree. In the last few years, I have developed algorithms for computing automorphism groups of Hadamard matrices [**3**], [**6**], codes [**5**], and block designs (as a special case of codes) based on those concepts. Recently Butler and Lam [**1**] studied the general problem of constructing automorphism group algorithms and identified the type-dependent and type-independent portions of such algorithms. Earlier, B. McKay [**8**] developed a graph automorphism algorithm using an approach similar in some respects

‡ Work partially supported by National Science Foundation grant MCS 82-01311.

COMPUTATIONAL GROUP THEORY
ISBN 0-12-066270-1

to that employed here, though McKay's algorithm does not explicitly employ the base and strong generating set concepts.

Two related problems, often considered together with the automorphism group problem, are the following:

(1) Determine if two combinatorial objects $X_1$ and $X_2$ are isomorphic;
(2) Given a combinatorial object $X$, find a canonical representative canon($X$) for its isomorphism class (that is, canon($X$) is isomorphic to $X$, and $X_1$ isomorphic to $X_2$ implies canon($X_1$) = canon($X_2$)).

Problem (1) may be solved with relatively minor modifications to the automorphism group algorithm. Problem (2) also may be solved by a variation of the automorphism group algorithm; I shall discuss this approach in section 4.

Although exact definitions and notation will vary according to the type of object, we will always have a class $\Gamma$ consisting of all combinatorial objects of a given size satisfying certain properties. There will be a set $\Omega$ associated with the class $\Gamma$, and a subgroup $S_\Gamma(\Omega)$ of the symmetric group $S(\Omega)$ which acts on $\Gamma$. (Each permutation $s$ in $S_\Gamma(\Omega)$ maps any object $X$ in $\Gamma$ to another object $X^s$ in $\Gamma$, such that $(X^s)^t = X^{(st)}$). Then AUT($X$) is defined to be $\{s \in S_\Gamma(\Omega) \mid X^s = X\}$.

*Example 1* (block designs). Let $\Gamma$ consist of all block designs with point set $P$. In this case,

$$\Omega = P, S_\Gamma(\Omega) = S(\Omega), \text{ and if } X = (P, B) \in \Gamma, X^s = (P, \{b^s \mid s \in B\}).$$

*Example 2* (matrices over GF$(q)$). Let $\Gamma$ consist of all $m$ by $n$ matrices over GF$(q)$. Let

$$\begin{aligned} \Omega_R &= \{\lambda i_R \mid 1 \leqslant i \leqslant m, \lambda \in \mathrm{GF}(q) - \{0\}\}, \\ \Omega_C &= \{\lambda j_C \mid 1 \leqslant j \leqslant n, \lambda \in \mathrm{GF}(q) - \{0\}\}, \\ \Omega &= \Omega_R \cup \Omega_C, \\ S_\Gamma(\Omega) &= \{s \in S(\Omega) \mid (\lambda i_R)^s = \lambda(i_R{}^s) \in \Omega_R \text{ and} \\ &\qquad (\lambda j_C)^s = \lambda(j_C{}^s) \in \Omega_C \text{ for all } \lambda, i, j\}. \end{aligned}$$

For $s \in S_\Gamma(\Omega)$, write $i_R{}^s = \mu(i_R) i_R{}^t$ and $j_C{}^s = \mu(j_C) j_C{}^t$, where $\mu(i_R)$, $\mu(j_C) \in \mathrm{GF}(q)$ and $t$ is a permutation of $\{1_R, \ldots, m_R, 1_C, \ldots, n_C\}$. Then, if $X = (x_{ij}) \in \Gamma$,

$$X^s = (y_{ij}), \text{ with } y_{ij} = \mu(i'_R)\mu(j'_C) x_{i'j'}, \text{ where } i'_R = i_R{}^{t^{-1}} \text{ and } j'_C = j_C{}^{t^{-1}}.$$

Here, $i_R$ and $j_C$ correspond to row $i$ and column $j$, respectively. $S_\Gamma(\Omega)$ corresponds to the set of monomial permutations of the rows and columns, and $X^s$ is obtained from $X$ by multiplying row $i$ by $\mu(i)$ and moving the result to row position $i^t$ ($1 \leqslant i \leqslant m$), and likewise for columns. Example 2 actually includes Example 1 as a subcase (Let $q = 2$, and let $X$ be the incidence matrix of the design); it also includes Hadamard matrices, which may be viewed as a special type of matrix over GF(3).

Before proceeding with the algorithm, it may be worthwhile to mention some applications that make automorphism group algorithms so useful in combinatorics. For a particular class $\Gamma$ of combinatorial structures, two questions that arise naturally are: (1) Do there exist objects in $\Gamma$? (2) If so, determine all such objects up to isomorphism. In constructing an object $X$ of size $n$, we may start with a smaller object $Y$ and attempt to extend $Y$. Knowledge of the automorphism group of $Y$ (if nontrivial) may be used to limit the number of nonequivalent extensions that must be considered. For example, if $Y = (P, B)$ is a block design, any new block to be added may be selected from a set containing one representative from each orbit of $\mathrm{AUT}(Y)$ on $k$-element subsets of $P$ ($k$ = block size).

In classifying combinatorial objects up to isomorphism, the automorphism group provides an invariant useful in distinguishing nonisomorphic objects. Moreover, in some fortunate cases, the total number $T$ of objects in the class $\Gamma$ may be computed (this has been done for certain codes; see, for example [**2**]). Since the number of distinct objects isomorphic to $X$ is $|S_\Gamma(\Omega)|/|\mathrm{AUT}(X)|$, we obtain the formula

$$T/|S_\Gamma(\Omega)| = \sum_{X \in \Gamma^*} (1/|\mathrm{AUT}(X)|)$$

where $\Gamma^*$ contains one representative from each isomorphism class of $\Gamma$. Thus one can determine immediately whether a given set $\{X_1, \ldots, X_k\}$ of non-isomorphic objects from $\Gamma$ is a complete set of representatives of the isomorphism classes of $\Gamma$.

Automorphism groups are also useful in finding minimum weights of codes [**7**]; the groups of certain smaller matrices may be used to reduce the number of vectors that must be examined.

## 2 The Automorphism Group Algorithm

In outlining the automorphism group algorithm, I shall describe four successively more sophisticated approaches.

(i) *Brute force.* For simplicity assume $S_\Gamma(\Omega) = S(\Omega)$. We generate each of the $n!$ permutations of $S(\Omega)$ and test each to see if it leaves $X$ invariant. Very efficient permutation generation algorithms are known [**10**]; each permutation is obtained from its predecessor in constant time by transposing two elements. Nonetheless, this method fails for any but the smallest $n$ (say $n \leqslant 10$) because of the extremely rapid growth of $n!$.

(ii) *Backtrack search.* It is easy to discover the major source of inefficiency in the brute force approach. Suppose, for example, that $X = (P, B)$ is a block design with $\{1,2\} \in B$ but $\{1,3\} \notin B$. Obviously no $s \in S(\Omega)$ satisfying $1^s = 1$ and $2^s = 3$ can lie in $\mathrm{AUT}(X)$, yet each of the $(n-2)!$ permutations with this property is constructed and tested separately. This suggests that the permutation $s$ be built up, step by step, from a series of partial permutations defined on $\{1\}$, $\{1,2\}$, $\{1,2,3\}$ etc. If, for some $d$, we obtain a partial permutation $s^*$ defined on $\{1,\ldots,d\}$ for which it can be determined that no automorphism $s$ can agree with $s^*$ on $\{1,\ldots,d\}$, then none of the $(n-d)!$ extensions of $s^*$ need be considered. This corresponds to backtrack search in a tree such as in Fig. 1.

Here, each node at depth $d$ corresponds to a partial permutation defined on $\{1,\ldots,d\}$; leaves correspond to permutations. For example, the starred node, which shall be referred to as node $[3,1]$, represents a partial permutation mapping 1 to 3 and 2 to 1. During the backtrack search, if the partial permutation corresponding to the current node fails to preserve the combinatorial structure, the entire subtree rooted at that node is pruned (i.e. removed from consideration).

In the general case, $\Omega = \{\alpha_1, \ldots, \alpha_n\}$, we substitute $\alpha_i$ for $i$ in the tree in Fig. 1. A node $[\beta_1 \ldots \beta_d]$ at depth $d$ corresponds to a partial permutation mapping $\alpha_j$ to $\beta_j$, $1 \leqslant j \leqslant d$; we will also use $[\beta_1 \ldots \beta_d]$ to denote the partial permutation itself. Note that, even in the case $\Omega = \{1,\ldots,n\}$, $\alpha_1, \ldots, \alpha_n$ can be any permutation of $1,\ldots,n$. If $S_\Gamma(\Omega) \neq S(\Omega)$, let $S_{\Gamma,d}(\Omega)$ denote the set of all partial permutations defined on $\{\alpha_1, \ldots, \alpha_d\}$ which can be extended to elements of $S_\Gamma(\Omega)$; usually it is easy to check whether $[\beta_1 \ldots \beta_d] \in S_{\Gamma,d}(\Omega)$. Only nodes corresponding to partial permutations in $\bigcup_d S_{\Gamma,d}(\Omega)$ will appear in the search tree. Let us assume that we have a circular linked list of $\Omega \cup \{\infty\}$ given by $\mathrm{succ}(\infty) = \alpha_1$, $\mathrm{succ}(\alpha_i) = \alpha_{i+1}$, and $\mathrm{succ}(\alpha_n) = \infty$.

We will need a function $\mathrm{TEST}([\beta_1 \ldots \beta_d])$ which returns false if it can be

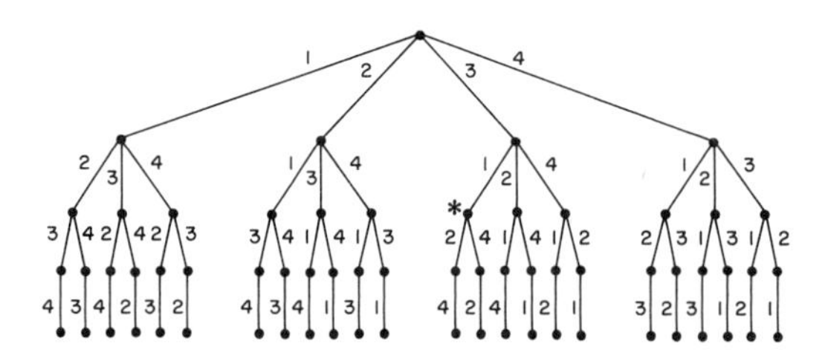


*Fig. 1*

determined that the partial permutation $[\beta_1 \ldots \beta_d]$ cannot be extended to an automorphism, and true otherwise. In computing TEST$([\beta_1 \ldots \beta_d])$, it may be assumed that TEST$([\beta_1 \ldots \beta_{d-1}])$ has already been invoked and has returned true.

The outline of the backtrack search algorithm, which generates a list $S$ of elements of AUT$(X)$, is as follows:

```
d := 1;  β1 := ∞;  S := empty set;
while d > 0 do
    begin
    repeat βd := succ(βd) until [β1 ... βd] ∈ S_{Γ,d}(Ω) or βd = ∞;
    if βd ≠ ∞ then
      if d < n and TEST([β1 ... βd]) then
        begin d := d+1;  βd := ∞ end
      else if d = n and s: αi → βi (1 ≤ i ≤ n) is an
              automorphism then S := S ∪ {s}
      else
    else d := d − 1
    end
```

The efficiency of this approach will be determined primarily by how quickly TEST can be computed and how effective it is in rejecting partial permutations $[\beta_1 \ldots \beta_d]$ for small values of $d$. Naturally the tests to be performed will depend on the particular type of combinatorial structure under consideration. Devising tests which can be computed quickly and yet which are effective, may be the most difficult task in designing the algorithm. One approach that has proven useful is to define functions $F_i : \Omega^i \to T$, where $T$ is some set, such that $F_i(\delta_1, \ldots, \delta_i) = F_i(\gamma_1, \ldots, \gamma_i)$ is a necessary condition for existence of an automorphism mapping $(\gamma_1, \ldots, \gamma_i)$ to $(\delta_1, \ldots, \delta_i)$. For example, for an $m$ by $n$ matrix over GF(2), we might define $F_i(\gamma_1, \ldots, \gamma_i)$ to be the multiset in $\mathrm{GF}(2)^i$ formed by the vectors

$$(x_{\gamma_1 j}, \ldots, x_{\gamma_i j}),\ 1 \leqslant j \leqslant n. \quad \text{(For } q > 2\text{, the situation is slightly more complex.)}$$

Since it is more convenient to store integer-valued functions, it might be preferable to consider each vector, above, as a binary integer between 0 and $2^i - 1$ and to define $F_i(\gamma_1, \ldots, \gamma_i)$, instead, to be some weighted sum of these integers.

One alternative is to precompute $F_i$ for all possible arguments; obviously this will be feasible only for a few small values of $i$. In this case, TEST$([\beta_1 \ldots \beta_d])$ simply checks that

$$F_i(\beta_{j_1}, \ldots, \beta_{j_{i-1}}, \beta_d) = F_i(\alpha_{j_1}, \ldots, \alpha_{j_{i-1}}, \alpha_d)$$

for each $F_i$ which has been precomputed and each $(i-1)$-element subset $\{j_1,\ldots,j_{i-1}\}$ of $\{1,\ldots,d-1\}$. (Some subsets might be omitted if $n$ is large.) Later we will see several advantages to this alternative.

Another alternative is for TEST to compute $F_d(\beta_1,\ldots,\beta_d)$ at the time that node $[\beta_1\ldots\beta_d]$ of the search tree is reached and to compare it to the (presumably precomputed) value of $F_d(\alpha_1,\ldots,\alpha_d)$. This computation is facilitated by the fact that $F_d(\beta_1,\ldots,\beta_{d-1})$ is already available; moreover, often it can be determined, at an early stage of the computation, that there is no possibility of $F_d(\beta_1,\ldots,\beta_d)$ equalling $F_d(\alpha_1,\ldots,\alpha_d)$. For an example of this approach, see **[5]**.

For a given selection of the combinatorial tests, the efficiency of the algorithm may vary tremendously, depending on the ordering $\alpha_1,\ldots,\alpha_n$ of $\Omega$. Consider, for example, the block design $X = (P, B)$, where $P = \{1,2,\ldots,2n\}$ and

$$B = \{\{1,n+1\},\{2,n+2\},\ldots,\{n,2n\},\{n+1,n+2,\ldots,2n\},\\ \{n+2,\ldots,2n\},\ldots,\{2n\}\}.$$

It is easy to see that $|\mathrm{AUT}(X)| = 1$. Suppose that we use the functions $F_1$ and $F_2$ given in the example above, applied to the incidence matrix of the design, considered as a matrix over GF(2) with points corresponding to rows. If we order $\Omega$ as $2n, 2n-1,\ldots,1$, we discover almost immediately that $|\mathrm{AUT}(X)| = 1$; the running time (excluding computation of $F_1$ and $F_2$) is linear in $n$. If, on the other hand, we order $\Omega$ as $1,2,\ldots,2n$, each of the $n!$ partial permutations mapping $\{1,\ldots,n\}$ into itself must be generated, and the running time is proportional at least to $n!$. Less extreme examples occur frequently in practice. In the case of 24-dimensional Hadamard matrices, running times have varied by a factor of more than fifty, depending on the ordering of the rows. In summary, the ordering $\alpha_1,\ldots,\alpha_n$ should be chosen so that $\alpha_1,\ldots,\alpha_i$ has "unusual" combinatorial properties for small values of $i$. Probably no other factor will have as much influence on the efficiency of the algorithm. One advantage of precomputing functions $F_i$ for small $i$, is that $F_i$ may be used in ordering $\Omega$.

(iii) *A small base.* A modest improvement over the method given in (ii) usually can be obtained by making use of the concept of a base introduced by Sims **[11]**, **[12]**. A base for a group $G$ on $\Omega$ is a subset $\{\gamma_1,\ldots,\gamma_k\}$ of $\Omega$ such that

$$|G_{\gamma_1\ldots\gamma_k}| = 1.$$

Note that any element of $G$ is determined uniquely by the image of a base. This concept is most useful when $k$ is much smaller than $|\Omega|$. Not every permutation group has such a base, but a great many interesting groups do.

In our case, we shall be interested in an ordered base $(\alpha_1, \ldots, \alpha_k)$ for AUT$(X)$ on $\Omega$ ($\Omega = \{\alpha_1, \ldots, \alpha_k, \ldots, \alpha_n\}$ will always be ordered so that the base comes first). For such a base to be useful, we will need an efficient procedure EXTEND($[\beta_1 \ldots \beta_k]$, $s$, flag) which computes the unique extension $s$ of $[\beta_1 \ldots \beta_k]$ to an automorphism, if one exists. EXTEND sets flag to true if the extension exists and false otherwise. I shall refer to a base for which an efficient procedure EXTEND is known as an effective base. For example, if we have a readily computable function $F_{k+1}$ of the type described in (ii), and if the multiset

$$\{F_{k+1}(\alpha_1, \ldots, \alpha_k, \gamma) \mid \gamma \in \Omega - \{\alpha_1, \ldots, \alpha_k\}\}$$

has all distinct elements, then $(\alpha_1, \ldots, \alpha_k)$ is an effective base, and the procedure EXTEND is straightforward to implement.

If a small effective base $(\alpha_1, \ldots, \alpha_k)$ is known, we can limit the depth of the backtrack search tree to $k$; when a leaf node $[\beta_1 \ldots \beta_k]$ is reached, we apply EXTEND to determine if there is a (necessarily unique) extension $s$ of the partial permutation $[\beta_1, \ldots, \beta_k]$. This may be considerably faster than applying backtrack search through another $n-k$ levels of the tree, especially if EXTEND uses different criteria than TEST, as in [**3**], [**6**]. In view of the remark at the end of (ii), it is important that the base be chosen so that short initial segments have unusual combinatorial properties.

(iv) *A strong generating set.* If AUT$(X)$ is trivial, approaches (ii) and (iii) should work quite efficiently, provided that TEST($[\beta_1 \ldots \beta_d]$) is sufficiently effective in rejecting partial permutations $[\beta_1 \ldots \beta_d]$ for small $d$. However, if AUT$(X)$ is large, these methods will be very slow because each permutation in AUT$(X)$ must be generated and saved. Moreover, the output will consist of the complete list of elements of AUT$(X)$, perhaps too massive to store and probably not useful for further computation. It would be preferable to produce only a small set of permutations which generate AUT$(X)$. At any time during the algorithm, let $H$ denote the subgroup of AUT$(X)$ generated by the permutations selected up to that time. A partial permutation $[\beta_1 \ldots \beta_d]$ should be rejected if it can be determined that any extension $s$ of $[\beta_1 \ldots \beta_d]$ to an automorphism will lie in $H$. More generally, $[\beta_1 \ldots \beta_d]$ may be rejected if it can be determined that $s$ must lie in the double coset $Hs'H$ for some element $s'$ already considered (either added to $H$ or removed from the search tree during pruning). Such a procedure will guarantee a relatively small generating set for AUT$(X)$ since each generator added multiplies $|H|$ by at least 2. We would like, of course, to obtain a generating set which facilitates further computation in AUT$(X)$ (such as determining its order).

The concept of a strong generating set, [**11**], [**12**], proves very useful here. A strong generating set for a group $G$ on $\Omega$ relative to an ordered base

$(\alpha_1, \ldots, \alpha_k)$ is a generating set $S$ such that

$$G_{\alpha_1 \ldots \alpha_{i-1}} \text{ is generated by } S \cap G_{\alpha_1 \ldots \alpha_{i-1}} \text{ for } i = 1, 2, \ldots, k.$$

Thus a strong generating set provides generators for each group in the stabilizer chain

$$G = G^{(1)} \supseteq G^{(2)} \supseteq \ldots \supseteq G^{(k)} \supseteq G^{(k+1)} = 1, \text{ where } G^{(i)} \text{ denotes } G_{\alpha_1 \ldots \alpha_{i-1}}.$$

Let $\Delta_i$ be the $G^{(i)}$-orbit of $\alpha_i$ (called the $i$th basic orbit); the $\Delta_i$ are readily computable, and $|G| = \Pi |\Delta_i|$. Given a strong generating set, many computations in $G$ can be performed easily even if $G$ is relatively large.

It is useful to define the ordering $\ll$ of $\Omega$ by $\alpha_1 \ll \alpha_2 \ll \ldots \ll \alpha_n$ (note that the base comes first). This induces an ordering on $S(\Omega)$ defined by $s \ll t$ if $(\alpha_1{}^s, \ldots, \alpha_n{}^s)$ lexicographically precedes $(\alpha_1{}^t, \ldots, \alpha_n{}^t)$. Given this ordering, we obtain a canonical generating set $S = \{s_1, \ldots, s_p\}$ for AUT($X$) as follows (assume the empty set generates the identity group):

$$s_{m+1} = \text{first element of AUT}(X) - \langle s_1, \ldots, s_m \rangle, \quad m = 0, \ldots, p-1.$$

It is easy to show that $S$ is a strong generating set (see, for example [**3**]).

The automorphism group algorithm will construct the canonical generating set $\{s_1, \ldots, s_p\}$ for AUT($X$). At some time, it will have found $s_1, \ldots, s_m$, and it will be searching, via the backtrack procedure, for the next canonical generator $s = s_{m+1}$. The search tree will be traversed from left to right, so that permutations are encountered in increasing order (relative to $\ll$). Since $s$ must be first in $HsH$, any partial permutation $[\beta_1 \ldots \beta_d]$ for which it can be determined that no extension $s$ can be first in $HsH$ will be rejected. We must consider how such a determination can be made.

Unfortunately, no useful criterion for an element to be first in its double coset is known. But Sims [**11, 12**] has proven the following necessary and sufficient conditions for an element to be first in its left and right coset.

(a) $s$ is first in $Hs$ if and only if $\alpha_j{}^s$ is first in $\Delta_j{}^s$, $1 \leqslant j \leqslant k$.
(b) $s$ is first in $sH$ if and only if $\alpha_j{}^s$ is first in its $G_{\beta_1 \ldots \beta_{j-1}}$ orbit, $1 \leqslant j \leqslant k$, where $\beta_l$ denotes $\alpha_l{}^s$.

The difficulty in applying (a) in our situation is that we know the action of $s$ only on small subsets of the $\Delta_j$, namely $\Delta_j \cap \{\alpha_1, \ldots, \alpha_d\}$. We obtain a lower bound for $\beta_d$ by

$$\beta_d > \max\{\beta_j \mid 1 \leqslant j \leqslant d-1, \ \alpha_d \in \Delta_j\}$$

(otherwise $\alpha_d{}^s$ preceeds $\alpha_j{}^s$ in $\Delta_j{}^s$). We also obtain an upper bound for $\beta_d$ by

$$\beta_d \leqslant |\Omega| - |\Delta_d| + 1.$$

It may be possible to strengthen the upper bound by using the combinatorial structure to show that not all points of $\{\gamma \in \Omega \mid \gamma > \beta_d\}$ could lie in $\Delta_d{}^s$.

Suppose we have computed invariant functions $F_i$ as in (ii); for simplicity, assume $i = 2$. Let $1 \leqslant j \leqslant d-1$. For any $\delta^s \in \Delta_d^s$,

$$F_2(\beta_j, \delta^s) = F_2(\alpha_j, \delta) = F_2(\alpha_j, \alpha_d),$$

the last equality holding because there is an automorphism fixing $\alpha_j$ and mapping $\alpha_d$ to $\delta$. Let $f_{jd} = F_2(\alpha_j, \alpha_d)$. Let $n_d = |\Delta_d|$. In order for $\beta_d$ to be first in $\Delta_d^s$, we must have

$$\beta_d \leqslant n_d\text{th largest element of } \{\gamma \in \Omega \mid F_2(\beta_j, \gamma) = f_{jd} \text{ for } j = 1, \ldots, d-1\}.$$

This idea has not been implemented, to my knowledge, but it seems quite promising.

The problem with applying (b), above, is that we do not automatically have a generating set for $H_{\beta_1 \ldots \beta_{d-1}}$ unless $(\beta_1, \ldots, \beta_{d-1}) = (\alpha_1, \ldots, \alpha_{d-1})$, that is, $\alpha_d$ is the first point moved by $s$, in which case

$$H_{\beta_1 \ldots \beta_{d-1}} = H_{\alpha_1 \ldots \alpha_{d-1}} = H.$$

So we obtain the criterion: if $d$ is minimal such that $\beta_d \neq \alpha_d$, then $\beta_d$ must be first in its $H$ orbit.

Given a strong generating set for a group $G$ relative to some base, a relatively fast procedure has been developed [**11**], [**12**], for obtaining a strong generating set relative to some other base. This process is referred to as changing the base. It is accomplished by a sequence of steps, each of which involves changing some intermediate base $\gamma_1, \ldots, \gamma_j$ to one of the following:

(1) $\gamma_1, \ldots, \gamma_j, \delta$ for some $\delta$;
(2) $\gamma_1, \ldots, \gamma_{j-1}$ provided this is a base;
(3) $\gamma_1, \ldots, \gamma_{i+1}, \gamma_i, \ldots, \gamma_j$ for some $i < j$.

Butler and Lam [**1**] were the first to employ base change in automorphism group algorithms. During the backtrack search, if $[\beta_1 \ldots \beta_d]$ is the current node, a base for $H$ of the form $\beta_1, \ldots, \beta_{d-1}, *$ will already be available (an asterisk denotes any sequence of points). If $d < k$ and the partial permutation $[\beta_1 \ldots \beta_d]$ is not rejected, then the base is changed to $\beta_1, \ldots, \beta_d, *$, using one application of (1), repeated application of (3) and possibly application of (2), to reduce the base size. This ensures that a generating set for $H_{\beta_1 \ldots \beta_{d-1}}$ is always available and makes it feasible to check criterion (b). We shall assume a procedure CHANGEBASE($[\beta_1 \ldots \beta_d]$) which changes the base from $\beta_1, \ldots, \beta_{d-1}, *$ to $\beta_1, \ldots, \beta_d, *$ and computes an array orbrep$_{d+1}$ such that orbrep$_{d+1}[\gamma]$ is the first point in the $H_{\beta_1 \ldots \beta_d}$ orbit of $\gamma$ for any $\gamma \in \Omega$. We shall also assume a function JOIN$(P, s)$, which returns the join of the partition $P$ of $\Omega$ and the partition induced by the cycles of a permutation $s$ (see Alg. 1 of [**3**]).

Incorporating all of these ideas, gives the following automorphism group algorithm. During the algorithm, $r$ is the smallest integer with $\beta_r \neq \alpha_r$, ($r = k+1$ if no such integer exists), max$_d$ will be an upper bound for $\beta_d$, and

$\Delta_1, \ldots, \Delta_k$ will denote the basic orbits of $H$ relative to the original base $(\alpha_1, \ldots, \alpha_k)$.

```
find an effective base (α_1, ..., α_k) for AUT(X);
d := 1; β_1 := ∞; r := k+1; max_1 := n; S := empty set;
for i := 1 to k do Δ_i := {α_i};
for i := 1 to k do
for j := 1 to n do orbrep_i[j] := j;
while d > 0 do
    begin
    repeat β_d := succ(β_d) until β_d > max_d or
    [β_1 ... β_d] ∈ S_{Γ,d}(Ω);
    if d < r and β_d ≠ α_d then
      begin
      if r ≤ k then for i := r−1 down to d do
        orbrep_i := orbrep_r;
      r := d
      end;
    if β_d ≤ max_d then
      if orbrep_d[β_d] = β_d and TEST([β_1 ... β_d]) then
        if d < k then
          begin
          CHANGEBASE([β_1 ... β_d]);
          d := d+1;
          max_d := n−|Δ_d|+1 (or compute max_d as above);
          β_d := max{β_j | 1 ≤ j ≤ d−1, α_d ∈ Δ_j} (∞ for
          empty set);
          end
        else
          begin
          EXTEND([β_1, ..., β_k], s, flag);
          if flag and r ≠ k+1 then
            begin
            S := S ∪ {s};
            orbrep_r := JOIN(orbrep_r, s);
            Δ_r := {γ ∈ Ω | orbrep_r[γ] = α_r};
            d := r
            end
          end
      else
    else d := d−1
    end
```

*Proposition.* $|S| \leqslant n-h$, where $h$ is the number of orbits of $\mathrm{AUT}(X)$ on $\Omega$.

*Proof.* Each generator $s_m$ which is selected must map $\alpha_r$ (the first point that it moves) to a point $\beta_r$ that is first in its $\langle s_1, \ldots, s_{m-1} \rangle$-orbit. Since $s_1, \ldots, s_{m-1}$, fix $\alpha_1, \ldots, \alpha_{r-1}$ ($r$ never increases during the algorithm), $\alpha_r$ is first in its $\langle s_1, \ldots, s_{m-1} \rangle$-orbit. Since $\beta_r \neq \alpha_r, \beta_r$ cannot lie in the $\langle s_1, \ldots, s_{m-1} \rangle$-orbit of $\alpha_r$. Thus $\langle s_1, \ldots, s_m \rangle$ has fewer orbits on $\Omega$ than $\langle s_1, \ldots, s_{m-1} \rangle$, and the result follows easily.

If $\mathrm{AUT}(X) = S(\Omega)$, the upper bound, given above, is attained. However, in most cases the actual number of generators is far smaller.

## 3 Automorphism Groups of Hadamard Matrices

The type-dependent parts of the above algorithm consist of constructing the base and devising the procedures EXTEND and TEST. I will discuss briefly how these parts can be implemented for Hadamard matrices; details appear in [**3**]. Another approach to the Hadamard matrix group problem, involving transformation to graph automorphism groups, was developed by McKay [**9**].

Let $\mathbf{H} = (h_{ij})$ be an $n$ by $n$ Hadamard matrix. Then $\mathbf{H}$ may be viewed as a matrix over GF(3), and the notation of Example 2 will be used. We define a row base of $\mathbf{H}$ to be a set $\{i_1, \ldots, i_k\}$ such that the $k$ by $n$ submatrix of $\mathbf{H}$ formed by selecting only rows $i_1, \ldots, i_k$ has distinct columns, with no column the negative of any other. It is easy to show that, if $\{i_1, \ldots, i_k\}$ is a row base, then $\{(i_1)_R, \ldots, (i_k)_R\}$ is a base for $\mathrm{AUT}(X)$, and it is straightforward to implement a procedure EXTEND for this base (Algs. 3 and 4 of [**3**]). The following result (Prop. 4.2 of [**3**]), guarantees that a small row base always exists and can be constructed quickly.

*Proposition.* $\mathbf{H}$ has a row base with at most $2(\log_2 n) - 1$ elements. Moreover, such a base may be constructed as follows:

> Choose $i_1$ arbitrarily; normalize $\mathbf{H}$ to make row $i_1$ all ones; set $Z = \{(u, v) \mid 1 \leqslant u < v \leqslant n\}$. Once $i_1, \ldots, i_j$ have been chosen, for each $i \in \{1, \ldots, n\} - \{i_1, \ldots, i_j\}$ compute $c_i = |\{(u, v) \in Z \mid h_{iu} = h_{iv}\}|$. Choose $i$ to minimize $c_i$ and set $i_{j+1} = i$. If $c_i = 0$, then $\{i_1, \ldots, i_{j+1}\}$ is a row base; otherwise replace $Z$ by $\{(u, v) \in Z \mid h_{iu} = h_{iv}\}$ and continue.

In the above, $Z$ consists of pairs of columns that are indistinguishable in rows selected thus far; the next row is selected to reduce the size of $Z$ as much as possible. When $Z$ becomes empty, we are done. The fact that such a small row base can be found makes it feasible, for all practical values of $n$, to define an array of dimension $2^k$; this greatly speeds up the procedure EXTEND (Alg. 4 of [**3**]).

A major difficulty with Hadamard matrices is that any two row triples $\{i_1, i_2, i_3\}$ and $\{j_1, j_2, j_3\}$ may be normalized to the same form, namely

$$\begin{matrix} 11\ldots11 & 11\ldots11 & 11\ldots11 & 11\ldots11 \\ 11\ldots11 & 11\ldots11 & --\ldots-- & --\ldots-- \\ 11\ldots11 & --\ldots-- & 11\ldots11 & --\ldots-- \end{matrix}$$

To obtain useful invariant functions, it will be necessary to look at $m$-tuples of rows for $m$ at least 4. Let $\{u, v, w, x\}$ be a 4-tuple; define

$$g(u, v, w, x) = |\{j \mid 1 \leqslant j \leqslant n, h_{uj} = h_{vj} = h_{wj} = h_{xj}\}|$$

and define $f(u, v, w, x) = \min(g(u, v, w, x), (n/4) - g(u, v, w, x))$.

It is easy to show that two row 4-tuples $\{i_1, i_2, i_3, i_4\}$ and $\{j_1, j_2, j_3, j_4\}$ can be normalized to the same form if and only if

$$f(i_1, i_2, i_3, i_4) = f(j_1, j_2, j_3, j_4), \text{ and } f(i_1, i_2, i_3, i_4) = f(j_1, j_2, j_3, j_4)$$

if there is an automorphism, mapping row $i_r$ to $\pm$(row $j_r$), $r = 1, 2, 3, 4$.

For large practical values of $n$, it is not feasible to store the values of $f$ for all 4-tuples. However, if $f_2$ is defined by

$$f_2(u, v) = (r_1, \ldots, r_t), \ t = \text{floor}(n/8), \text{ where}$$
$$r_j = |\{(w, x) \mid 1 \leqslant w < x \leqslant n, f(u, v, w, x) = j\}|$$

and $f_1(u)$ and $f_3(u, v, w)$ are defined analogously, then $f_1$, $f_2$, and possibly $f_3$, may be precomputed and stored. Letting $F_2(\lambda_1 u_R, \lambda_2 v_R) = f_2(u, v)$ gives an invariant function on $\Omega_R$ of the type discussed earlier. Invariant functions $F_1$ and $F_3$ can be defined analogously. To compute $f_1$, $f_2$, and $f_3$, it is necessary to compute $f(u, v, w, x)$ for all 4-tuples, but not to store all of the resulting values.

The algorithm outlined, above, and described in [**3**] has been implemented in ANSI standard Fortran [**6**] and used on matrices of size up to one hundred. Typical execution times (IBM 4341) range from several seconds for matrices of size 20–40 to several minutes for matrices of size 80–100. If bit string operations were used in computing $f(u, v, w, x)$, the algorithm should be able to handle matrices considerably larger than 100.

It should be noted that Hadamard matrices are an atypical case in two respects:

(1) in general, there is no guarantee of a small base, although one usually turns out to exist;
(2) in most cases, it is not necessary to look at 4-tuples from $\Omega$ to obtain a useful invariant.

## 4 Canonical Representatives

Isomorphism of combinatorial objects can be decided by a minor variation of the automorphism group algorithm (see section 7 of [**3**]). A somewhat more difficult problem is to find a canonical representative canon$(X)$ for the isomorphism class of $X$ in $\Gamma$, as well as a permutation $s$ in $S_\Gamma(\Omega)$ with canon$(X)^s = X$; this problem is discussed in [**8**] (for graphs) and [**1**]. Given an ordering $<'$ of $\Gamma$, it is natural to define canon$(X)$ to be $\min\{X^t \mid t \in S_\Gamma(\Omega)\}$. For example, if $\Gamma$ consists of the $m$ by $n$ matrices over GF$(q)$, we might first order GF$(q)$ and then order $\Gamma$ by $X <' Y$ if $(x_{11}, \ldots, x_{1n}, x_{21}, \ldots, x_{mn})$ lexicographically precedes $(y_{11}, \ldots, y_{1n}, y_{21}, \ldots, y_{mn})$.

Neither the concept of a base nor the combinatorial tests embodied in the procedure TEST are directly relevant here. Since $\{t \in S_\Gamma(\Omega) \mid \text{canon}(X)^t = X\}$ forms a left coset of AUT$(X)$ in $S_\Gamma(\Omega)$, we may assume that $s$ is first in its left coset $s$AUT$(X)$. Moreover, if (as in [**1**]) we define a partial order

$$<'' \text{ on } \bigcup_d S_{\Gamma,d}(\Omega)$$

with the properties that $[\gamma_1 \ldots \gamma_i] <'' [\delta_1 \ldots \delta_j]$, implies

$$X^{t^{-1}} <' X^{u^{-1}}$$

for any extensions $t, u \in S_\Gamma(\Omega)$ of $[\gamma_1 \ldots \gamma_i]$ and $[\delta_1 \ldots \delta_j]$, respectively, and that the reverse implication holds if $i = j = n$, then any partial permutation $[\beta_1 \ldots \beta_d]$ may be rejected if there is known to be another partial permutation $[\gamma_1 \ldots \gamma_e]$ with $[\gamma_1 \ldots \gamma_e] <'' [\beta_1 \ldots \beta_d]$. For example, if $\Gamma$ consists of $m$ by $n$ matrices over GF(2), ordered by $<'$ as above, the partial order $<''$ could be defined as follows:

> Given $[\gamma_1 \ldots \gamma_i]$ and $[\delta_1 \ldots \delta_j]$, let $k = \min(i, j)$, and let $X_\gamma$ and $X_\delta$ be the $k$ by $n$ matrices formed by selecting rows $\gamma_1, \ldots, \gamma_k$ (in that order) and $\delta_1, \ldots, \delta_k$ of $X$, respectively. Permute the columns of $X_\gamma$ and $X_\delta$ so that the column vectors of each matrix appear in lexicographically increasing order. Then $[\gamma_1 \ldots \gamma_i] <'' [\delta_1 \ldots \delta_j]$ if $X_\gamma <' X_\delta$.

As in section 2, part (ii), a search tree of depth $n$ in which nodes at depth $d$ correspond to partial permutations $[\beta_1 \ldots \beta_d]$, will be used to search for the permutation $s$. One alternative, which I have implemented to a limited degree for Hadamard matrices, is first to compute AUT$(X)$, using the algorithm of section 2, and then to apply breadth-first search in the tree. This has the advantage that a node $[\beta_1 \ldots \beta_d]$ can be rejected quickly if there is another node $[\gamma_1 \ldots \gamma_d]$ at the same depth with $[\gamma_1 \ldots \gamma_d] <'' [\beta_1 \ldots \beta_d]$; with depth-first (backtrack) search, a large portion of the subtree of $[\beta_1 \ldots \beta_d]$ might be searched if $[\gamma_1 \ldots \gamma_d]$ lies to the right of $[\beta_1 \ldots \beta_d]$. (By contrast, depth-first search is more appropriate in the automorphism group algorithm because it permits some automorphisms to be discovered early and used to prune the search tree.) On the other hand, breadth-first search does require saving more intermediate data. Assume that, at some stage, we have processed all nodes at depth less than $d$, and we have produced a list $L_d$ of nodes at depth $d$ that have not been pruned from the tree. This list is scanned, and any node $[\beta_1 \ldots \beta_d]$ for which there is another node $[\gamma_1 \ldots \gamma_d]$ on $L_d$ with $[\gamma_1 \ldots \gamma_d] <'' [\beta_1 \ldots \beta_d]$ is removed. If $d < n$, then for each node $[\beta_1 \ldots \beta_d]$ which remains on $L_d$, the nodes $[\beta_1 \ldots \beta_d, \gamma]$ with $\gamma$ first in its $\mathrm{AUT}(X)_{\beta_1 \ldots \beta_d}$ orbit are added to $L_{d+1}$. If $d = n$, the (necessarily unique) permutation which remains is $s$, and

$$\mathrm{canon}(X) = X^{s^{-1}}.$$

The method of partitions employed for graphs [**8**] could be used to limit the depth of the search tree to less than $n$.

Another alternative, described in [**1**], involves constructing the automorphism group simultaneously with searching for canon$(X)$ and $s$, using backtrack search. This eliminates the necessity of invoking two separate procedures to compute canon$(X)$; however, the criteria for rejecting partial permutations are considerably weaker.

## Acknowledgements

Computing services used in this research were provided by the Computer Center of the University of Illinois at Chicago. Their assistance is gratefully acknowledged.

## References

1. Butler, G. and Lam, C. W. H. A general backtrack algorithm for the isomorphism problem of combinatorial objects, (to appear).
2. Conway, J. H. and Pless, V. (1980). On the enumeration of self-dual codes, *J. Combinatorial Theory (A)* **28**, 26–53.
3. Leon, J. S. (1979). An algorithm for computing the automorphism group of a Hadamard matrix, *J. Combinatorial Theory (A)* **27**, 289–306.
4. Leon, J. S. (1980). On an algorithm for finding a base and a strong generating set for a group given by generating permutations, *Mathematics of Computation* **35**, 941–974.
5. Leon, J. S. (1982). Computing automorphism groups of error correcting codes, *IEEE Trans. on Infor. Theory* **IT-28**, 496–511.
6. Leon, J. S. Automorphism groups and equivalence of Hadamard matrices, (to appear).
7 Leon, J. S. A probabilistic algorithm for finding minimum weights of large error-correcting codes, (to appear).
8. McKay, B. D. (1977). "Computing automorphisms and canonical labellings of graphs." Proc. Internat. Conf. on Combinatorial Theory, Lecture Notes in Mathematics, vol. **686**, Springer-Verlag, Berlin.
9. McKay, B. D. (1979). Hadamard equivalence via graph isomorphism, *Discrete Math.* **27**, 213–214.

10. Nijenhuis, A. and Wilf, H. S. (1978). "Combinatorial Algorithms", 2nd ed. Academic Press, London and New York.
11. Sims, C. C. (1971a). Determining the conjugacy classes of a permutation group. *In* "Computers in Algebra and Number Theory" (Eds. G. Birkhoff and M. Hall Jr.). *SIAM—AMS Proc.* **4**, 191–195, Amer. Math. Soc.
12. Sims, C. C. (1971b). Computation with permutation groups. *In* "Proc. of the Second Symposium on Symbolic and Algebraic Manipulation" (Ed. S. R. Petrick), 23–28. Assoc. Comput. Mach., New York.

# Simple 6-(33, 8, 36) Designs from $P\Gamma L_2(32)$

SPYROS S. MAGLIVERAS‡ AND DAVID W. LEAVITT

## 1 Introduction

A $t$-design, or $t$-$(v, k, \lambda)$ design, is a pair $(X, \mathscr{B})$ where $\mathscr{B}$ is a system of $k$-sets (called blocks) from a $v$-set $X$ such that each $t$-subset of $X$ is in exactly $\lambda$ blocks of $\mathscr{B}$. A $t$-design is called *simple* if no blocks are repeated, and *trivial* if every $k$-subset of $X$ is a block and occurs the same number, $m$, of times in the design.

A necessary condition for the existence of a $t$-$(v, k, \lambda)$ design is that

$$\lambda\binom{v-i}{t-i} \equiv 0\left(\bmod\binom{k-i}{t-i}\right) \quad \text{for} \quad i = 0, 1, 2, \ldots, t.$$

In fact, Wilson has shown [**20**] that given $v, k, t$ with $0 < t < k < v$, there is a constant $N(t, k, v)$ such that $t$-$(v, k, \lambda)$ designs exist for all $\lambda > N(t, k, v)$, where $\lambda$ satisfies the above necessary conditions. A major problem is then to find the minimum value for $N(t, k, v)$ and to determine when simple, nontrivial $t$-designs exist. Kramer, Magliveras and Mesner [**11**] showed that $N(11, 12, 24) \leqslant 23$ and constructed a nontrivial, non-simple 11-(24,12,6) design. Kramer has constructed a non-trivial, 6-(17,7,2) design with repeated blocks (see [**9**]).

In sharp contrast to Wilson's result, the problem of finding simple, non-trivial $t$-designs for arbitrarily large $t$ seems rather difficult. Only in recent years have infinite families of simple, non-trivial 4- and 5-designs been constructed (see [**1**], [**2**], also [**5**]). Note, furthermore, that only finitely many Steiner systems are presently known for $t = 4$ or 5. Nevertheless, there is growing evidence that simple, non-trivial $t$-designs will exist in profusion for large values of $t$ even for small $v$, (see [**9**]). The authors believe it is remarkable that a simple, non-trivial 6-design was not found until now.

In this paper, we give a brief exposition of our construction of six non-

‡ This research was supported in part by a University of Nebraska Research Council grant to the first author.

COMPUTATIONAL GROUP THEORY
ISBN 0-12-066270-1

trivial, simple, pairwise non-isomorphic 6-(33,8,36) designs. We mention, in passing, that we have also constructed over 500 000 new, simple, nontrivial 5-designs, with $v = 33$. With the exception of two 5-designs discovered by Kramer [**9**], [**11**], these are the only known 5-designs on an odd number of points. We construct the above designs by using the group $P\Gamma L_2(32)$ in its action on the 33 points of the projective line $GF(32) \cup \{\infty\}$. The group $P\Gamma L_2(32)$ is exceptional in that it is the only permutation group which is 4-homogeneous, not 4-transitive and not set-transitive. It is, therefore, a group which achieves the highest degree of transitivity after the Mathieu groups. Extensive details of this work will appear elsewhere.

## 2 Preliminaries

A group action $G|X$ induces an action of $G$ on the collection $X_k$ of $k$-subsets of $X$ for each $k \leqslant v = |X|$. Let $\rho = (\rho(0), \rho(1), \ldots, \rho(v))$ be the vector whose $k$th entry is the number of $G$-orbits on $X_k$. The entries $\rho(k)$ are easily given by the Frobenious-Cauchy-Burnside theorem; that is,

$$\rho(k) = [\text{number of } G\text{-orbits on } X_k] = |G|^{-1} \cdot \sum_{g \in G} \theta_k(g) = (1, \theta_k)$$

where $\theta_k(g)$ is the number of $k$-subsets of $X$ fixed by $g \in G$. Note that $\theta_k$ is the character of the action $G|X_k$, and is computable by

$$\theta_k = 1_{S_k \times S_{v-k}} \uparrow^{S_v} \downarrow_G, \quad \text{where} \quad \uparrow^{S_v}, \downarrow_G$$

indicate induction to the symmetric group $S_v$ and restriction to $G$ respectively. $\theta_k$ is easily evaluated on each conjugacy class of $G$ if the cycle type $1^{\lambda_1}2^{\lambda_2} \ldots n^{\lambda_n}$ of the class elements is known. Clearly, a $k$-subset $K$ of $X$ is fixed by an element $g$ of type $1^{\lambda_1}2^{\lambda_2} \ldots n^{\lambda_n}$ if and only if $K$ is the union of cycles of $g$. Hence,

$$\theta_k(1^{\lambda_1}2^{\lambda_2} \ldots n^{\lambda_n}) = \sum_{(a_1, \ldots, a_n)} \prod_{i=1}^{n} \binom{\lambda_i}{a_i}$$

where the sum is taken over all non-negative integer vectors $(a_1, \ldots, a_n)$ such that

$$\sum_{i=1}^{n} i \, . \, a_i = k.$$

For $1 \leqslant t < k < v = |X|$, let

$$\{\Delta_i^{(t)} : i = 1, \ldots, \rho(t)\}, \{\Delta_j^{(k)} : j = 1, \ldots, \rho(k)\}$$

be the collections of orbits of $G$ on $X_t$ and $X_k$, respectively. For a fixed member $T$ of $\Delta_i^{(t)}$ let $a_{ij}(T)$ be the number of members $K \in \Delta_j^{(k)}$ such that $T \subset K$. It follows easily from the transitivity of $G$ on $\Delta_i^{(t)}$ and $\Delta_j^{(k)}$ that $a_{ij}(T)$ is independent of the choice of $T \in \Delta_i^{(t)}$; hence we may write $a_{ij} = a_{ij}(T)$. We define the $\rho(t)$ by $\rho(k)$ matrix $A_{t,k} = A(t, k; G)$ by: $A_{t,k} = (a_{ij})$. Dually, for a fixed member $K$ of $\Delta_j^{(k)}$, the number $b_{ij}(K)$ of members $T \in \Delta_i^{(t)}$ such that $T \subset K$ is independent of the choice of $K$ in $\Delta_j^{(k)}$, and we define the matrix $B_{t,k} = B(t, k; G)$ by $B_{t,k} = (b_{ij})$. For $k = 1, \ldots, v$, let $L_k = (L_k(1), \ldots, L_k(i), \ldots, L_k(\rho(k)))$ be the vector of orbit lengths of $G$ on $X_k$, that is $L_k(i) = |\Delta_i^{(k)}|$. For the pair of orbits $\Delta_i^{(t)}$ and $\Delta_j^{(k)}$, the entries $a_{ij}$ and $b_{ij}$ can be thought of as the degrees of a regular bipartite graph with vertex set $\Delta_i^{(t)} \cup \Delta_j^{(k)}$ where $T \in \Delta_i^{(t)}$ is joined to $K \in \Delta_j^{(k)}$ if and only if $T \subset K$ (see Fig. 1).

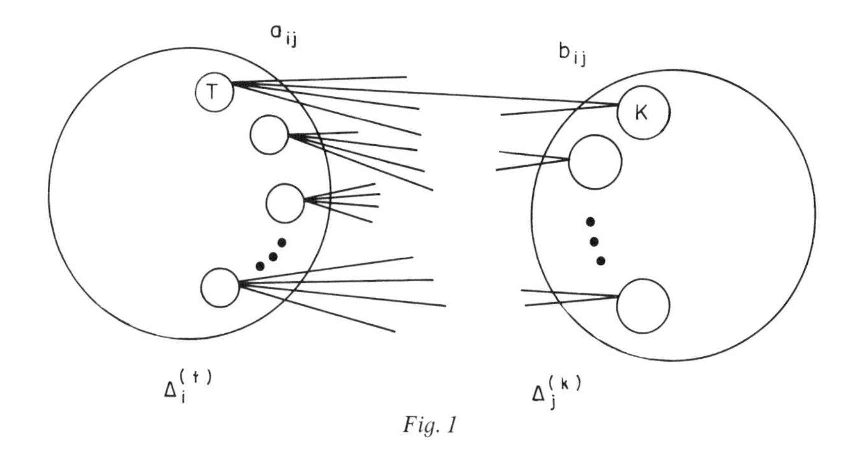


*Fig. 1*

We mention here, without proof, some of the properties of the matrices $A_{t,k}$, $B_{t,k}$ and the orbit length vectors $L_t$.

*Lemma 1.* Let $A_{t,k}$, $B_{t,k}$, $L_t$ be as defined, above.

(i) If $t \leqslant s \leqslant k$ then $A_{t,k} = \binom{k-t}{k-s}^{-1} A_{t,s} A_{s,k}$

(ii) $A_{t,k}$ has constant row sums $\binom{v}{k}\binom{k}{t} \Big/ \binom{v}{t}$

(iii) $L_t(i) A_{t,k}(i,j) = L_k(j) B_{t,k}(i,j)$

(iv) $\binom{k}{t} L_k = L_t A_{t,k}$

We remark in passing that the above lemma allows us to compute

$\{A_{t,k} : t < k\}$ and $\{B_{t,k} : t < k\}$ from $L_1$ and $\{A_{i,i+1} : i = 1, \ldots, [(v+1)/2]\}$

Let $A_{t,k}$ be as defined above for some pair $t, k, 1 \leqslant t < k < v$. Suppose furthermore that there exists a collection of columns $j_1, \ldots, j_q$ of $A_{t,k}$, corresponding to the $G$-orbits of $k$-sets $\Delta_{j_1}^{(k)}, \ldots, \Delta_{j_q}^{(k)}$, whose sum is the vector $(\lambda, \lambda, \ldots, \lambda)^T$. This simply means that the union $\mathscr{B}$ of the orbits $\Delta_{j_1}^{(k)}, \ldots, \Delta_{j_q}^{(k)}$ is a collection of $k$-subsets of $X$ with the property that any $t$-subset $T$ of $X$ occurs in exactly $\lambda$ members of $\mathscr{B}$. Hence, $\mathscr{B}$ is a $G$-invariant $t$-$(v, k, \lambda)$ design. Furthermore, if the columns $j_1, \ldots, j_q$ are distinct, no $k$-sets repeat, and consequently $\mathscr{B}$ is a simple design. The converse is easily seen to hold, so that we have the following result.

*Theorem 2.* There exists a $t$-$(v, k, \lambda)$ design with the underlying point set $X, |X| = v$, and with $G$ a group of automorphisms if and only if there exists a solution $\mathbf{u}$ to the matrix equation $\mathbf{Au} = \lambda\mathbf{J}$, where $\mathbf{A} = A_{t,k}$, $\mathbf{u}$ is a $\rho(k)$-dimensional vector of non-negative integral entries, $\mathbf{J}$ is the $\rho(t)$-dimensional vector of all 1s, and $\lambda$ a positive integer. The $t$-design is simple if and only if $\mathbf{u}$ is a 0–1 vector. (See [**10**], [**11**]).

## 3 Invariant Functions

Brute force calculation of the matrices $A_{k,k+1}$ by means of actually computing orbits $\{\Delta_i^{(k)}\}$ and $\{\Delta_j^{(k+1)}\}$ is very inefficient and should be avoided even when the sizes of $G$ and $X$ are relatively small. We will shortly describe an efficient algorithm for computing these matrices which depends on the notion of a discriminator function. Let $G|X$ be a group action and let $R^{X_k}$ denote the collection of all functions from $X_k$ into a set $R$. We extend the action to $R^{X_k}$ by $f^g(A) = f(A^g)$ for $A \in X_k, g \in G$. A function $f \in R^{X_k}$ fixed by all elements of $G$ is called $G$-invariant, or simply *invariant*. Suppressing $G$ and $X$, we denote the collection of all invariant functions in $R$ by $\Phi_k(R)$. Note that when $R$ is a ring, $\Phi_k(R)$ is a free $R$-module of rank $\rho(k)$. If $f \in \Phi_k(R)$, then the *rank* of $f$, $r(f)$, is the number of distinct values taken by $f$ in $R$, thus, $r(f) = |f(X_k)|$. An invariant function $f \in \Phi_k(R)$ is called a *discriminator* if $r(f) = \rho(k)$.

If $f \in R_1{}^{X_k}$, $g \in R_2{}^{X_k}$ are invariant, then the cartesian product of $f$ and $g$, $f \times g$ defined by $(f \times g)(A) = (f(A), g(A))$ is invariant, belonging to $\Phi_k(R_1 \times R_2)$, and we easily see that $r(f \times g) \geqslant \max\{r(f), r(g)\}$. This last condition on the rank of cartesian products allows us to increase the rank of an invariant function by iteratively taking its cartesian product with other invariant functions. This operation is used almost exclusively in creating a discriminator function from invariant functions of rank less than $\rho(k)$. The *efficiency* of an invariant function $f \in \Phi_k(R)$ is defined to be the ratio $\eta(f) = r(f)/\rho(k)$, thus, an invariant function is a discriminator if and only if it has efficiency 1. We proceed to discuss three types of invariant functions which we use to build discriminators by further taking cartesian products.

### 3.1 Anchor sets

Let $A$ be a fixed subset of $X$ which we shall call an *anchor set*. We describe an invariant function $f_A \in R^{X_k}$ as follows: Begin by calculating the orbit $\Delta = A^G = \{A_1, \ldots, A_s\}$. Now, for any $B \in X_k$, we define the *frequency vector* of $B$ relative to the anchor set $A$ to be $f_A(B) = (f_0, f_1, \ldots, f_k)$ where $f_i$ is the number of members $A_j$ of $\Delta$ intersecting $B$ in exactly $i$ points. We normally relabel the distinct frequency vectors that arise as $f_A(B)$, $B \in X_k$, by the non-negative integers $1, 2, \ldots, \mathrm{r}(f_A)$. Thus, we may think of $f_A$ as taking values in $\mathbb{Z}$. One disadvantage of invariant functions arising from anchor sets, is that the complete orbit $\Delta$ must be stored or generated in the machine. Furthermore, these invariant functions $f_A$ appear to be of low efficiency, especially as $k$ increases to $[v/2]$. For additional details on frequency vectors see [**11**].

### 3.2 Taxonomy 1

Suppose that $G$ contains an element $\pi$ which is represented on $X$ as a regular permutation of type $s^m$. Then the cyclic group $\langle \pi \rangle$ has a system $\gamma = \{C_1, \ldots, C_m\}$ of orbits on $X$, each of size $s$; that is, $\gamma$ is a regular partition of $X$. Let $F = \gamma^G$ be the orbit under $G$ of the partition $\gamma$; Thus, $F$ contains all partitions of type $\gamma^g = \{C_1{}^g, \ldots, C_m{}^g\}$, $g \in G$. Now let $B$ be any member of $X_k$. If $\delta = \{D_1, \ldots, D_m\} \in F$, we compute the frequency vector of $B$ relative to the partition $\delta$ by:

$$f(\delta, B) = (f_0, f_1, \ldots, f_q)$$

where $f_i$ is the number of blocks of $\delta$ intersecting $B$ in exactly $i$ points, $q = \min\{s, k\}$. As $\delta$ runs through the orbit of partitions in $F$, $f(\delta, B)$ runs through a specific set of distinct frequency vectors. We tabulate the frequencies with which the distinct frequency vectors appear, and obtain a frequency vector of frequency vectors $\mu_\pi(B)$. The function $\mu_\pi$ is clearly invariant, apparently of high efficiency, and it appears that the efficiency in discriminating $G$-orbits on $X$ increases with $k$ in $[1, v/2]$. In the particular case of $\mathrm{P\Gamma L}_2(32)$ acting on 33 points, the Sylow-3 subgroup is cyclic $\langle \pi \rangle$, of order 3, has type $3^{11}$ and normalizer of order 330. Hence, the taxonomy 1 orbit $F$ has length 496. This is the second shortest transitive representation of $\mathrm{P\Gamma L}_2(32)$ after the one of degree 33. Here, $\mu_\pi$ is very close to being a discriminator on 8-sets with an efficiency of 96/97. The procedure described as taxonomy 1 can clearly be generalized to an orbit $F$ of partitions which are not regular.

### 3.3 Taxonomy 2

The next procedure in computing invariant functions on $G$-orbits of $X_k$ is motivated by the matrices $B_{t,k}$. Suppose that for some $t < k$ we have been successful in obtaining a discriminator function $\phi_t$. Let $\{\Delta_i^{(t)}: i = 1, \ldots, \rho(t)\}$ be the orbits of $G$ on $X_t$, and let $B$ be an arbitrary $k$-subset of $X$. Now, consider the vector $v_t(B) = (f_1, f_2, \ldots, f_{\rho(t)})$ where $f_i$ is the number of $t$-subsets of $B$ which belong to $\Delta_i^{(t)}$. To compute $v_t(B)$, we run through the

$$\binom{k}{t} \text{ } t\text{-subsets } T \text{ of } B,$$

each time determining the orbit $\Delta_i^{(t)}$ in which $T$ falls by computing $\phi_t(T)$. It is clear that $v_t$ is an invariant function, and it appears that $v_t$ has high efficiency $\eta$. If $t = k-1$, then taxonomy 2 is also computationally efficient since in computing $v_t(B)$ there are only $k$ $(k-1)$-subsets of $B$ to be typed by $\phi_t$.

### 4.1 Algorithms for computing $A_{t,k}$

We proceed to describe how one could compute efficiently the matrices $A_{t,k}$. In the first algorithm, which we call *Diagram 1*, we assume that we have representatives of each of the $\rho(k-1)$ orbits of $G$ on $(k-1)$-subsets, say $T(k-1,1), \ldots, T(k-1, \rho(k-1))$. We also assume that we have the vector of orbit lengths $L_{k-1}$ and a discriminator function $\phi_k$ for $k$-subsets. We then proceed recursively to compute $A_{k-1,k}$, representatives of each of the $\rho(k)$ orbits of $G$ on $X_k$, and the vector of orbit lengths $L_k$.

*Diagram 1*

1 Initialize $A$ to be a $\rho(k-1) \times \rho(k)$ zero-matrix
2 Initialize $N$ to a $\rho(k)$-dimensional zero-vector
3 For $i = 1$ to $\rho(k-1)$, step $= 1$
4 Compute the complement $Y_i = X \backslash T(k-1, i)$
5 For $j = 1$ to $(v-k+1)$, step $= 1$
6 Set $q_j = j$th element of $Y_i$
7 Compute $T^+ = T(k-1, i) \cup \{q_j\}$
8 Compute $s = \phi_k(T^+)$ (Note: $s$ is the "type" of $T^+$)
9 If $N(s) = 0$ then set $T(k, s) = T^+$
10 Set $N(s) = 1$
11 Set $A(i, s) = A(i, s) + 1$
12 Next $j$
13 Next $i$
14 Output $A$ (note that $A$ is the desired $A_{k-1,k}$)
15 Stop

$L_k$ can be computed from $A_{k-1,k}$ and $L_{k-1}$ by using Lemma 1.

In practice, a discriminator function is not available at the beginning of execution of the inductive step from $(k-1)$ to $k$ in *Diagram 1*. However, a list

$\{f_1, f_2, \ldots, f_n\}$ of invariant functions can be made accessible to the program. A new algorithm, *Diagram 2* is presently described.

*Diagram 2*

1 Initialize $F$ to an $n \times (\rho(k-1) . (v-k+1))$ zero-matrix
2 For $m = 1$ to $n$, step $= 1$
3 Set function $f$ equivalent to function $f_m$
4 Set indx $= 0$
5 For $i = 1$ to $\rho(k-1)$, step $= 1$
6 Compute the complement $Y_i = X \backslash T(k-1, i)$
7 For $j = 1$ to $(v-k+1)$, step $= 1$
8 Set indx $=$ indx $+1$
9 Set $q_j = j$th element of $Y_i$
10 Compute $T^+ = T(k-1, i) \cup \{q_j\}$
11 Set $F(m,\text{indx}) = f(T^+)$
12 Next $j$
13 Next $i$
14 Compute $R =$ [the number of distinct columns of $F$]
15 If $R = \rho(k)$ then go to *Step 19*
16 Next $m$
17 Store $F$ on mass-storage device for later use
Print "Discrimination was not achieved. Increase Pool of invariant functions..."
18 Stop
19 Convert information in $F$ to $A_{k-1,k}$
20 Print $A_{k-1,k}$
21 Stop

**4.2**

The algorithm which was used to find solutions to the integer problem $\mathbf{Au} = \lambda \mathbf{J}$ will be described fully in a later publication. Here, we merely indicate that number-theoretic constraints obtainable from the original equations $\mathbf{Au} = \lambda \mathbf{J}$ are converted to new linearly independent equations which together with the previous equations, result in a system of smaller nullity. Solutions of the new system are then essentially obtained by backtrack.

## 5 Data

Tables 1–5, which follow, display the data which exhibits the construction of the 6-designs.

*Table 1*

---

**Group Generators . . .**

α = (1 2 4 8 16)(3 6 12 24 17)(5 10 20 9 18)(7 14 28 25 19)
(11 22 13 26 21)(15 30 29 27 23)(31)(32)(33)

β = (1 18 30)(2 21 12)(3 10 28)(4 31 32)(5 24 14)(6 7 17)(8 25 27)
(9 19 20)(11 15 13)(16 23 29)(22 33 26)

| Orbit Representative | Length | Orbit Representative | Length |
|---|---|---|---|
| 1 | 33 | 1 orbit on 1-sets. | |
| 1 2 | 528 | 1 orbit on 2-sets. | |
| 1 2 3 | 5456 | 1 orbit on 3-sets. | |
| 1 2 3 4 | 40920 | 1 orbit on 4-sets. | |
| 1 2 3 4 5 | 40920 | | |
| 1 2 3 4 6 | 163680 | | |
| 1 2 3 4 10 | 32736 | 3 orbits on 5-sets. | |
| 1 2 3 4 5 6 | 81840 | 1 2 3 4 5 7 | 81840 |
| 1 2 3 4 5 8 | 163680 | 1 2 3 4 5 9 | 163680 |
| 1 2 3 4 5 11 | 81840 | 1 2 3 4 5 12 | 163680 |
| 1 2 3 4 6 9 | 81840 | 1 2 3 4 6 10 | 81840 |
| 1 2 3 4 6 11 | 81840 | 1 2 3 4 6 22 | 27280 |
| 1 2 3 4 6 28 | 32736 | 1 2 3 4 6 32 | 32736 |
| 1 2 3 4 6 33 | 32736 | | |
| | | 13 orbits on 6-sets. | |
| 1 2 3 4 5 6 7 | 81840 | 1 2 3 4 5 6 8 | 163680 |
| 1 2 3 4 5 6 9 | 163680 | 1 2 3 4 5 6 10 | 163680 |
| 1 2 3 4 5 6 11 | 163680 | 1 2 3 4 5 6 12 | 163680 |
| 1 2 3 4 5 6 13 | 163680 | 1 2 3 4 5 6 14 | 81840 |
| 1 2 3 4 5 6 15 | 81840 | 1 2 3 4 5 6 16 | 163680 |
| 1 2 3 4 5 6 17 | 163680 | 1 2 3 4 5 6 19 | 81840 |
| 1 2 3 4 5 6 32 | 163680 | 1 2 3 4 5 7 9 | 163680 |
| 1 2 3 4 5 7 10 | 163680 | 1 2 3 4 5 7 12 | 163680 |
| 1 2 3 4 5 7 13 | 163680 | 1 2 3 4 5 7 15 | 81840 |
| 1 2 3 4 5 7 20 | 163680 | 1 2 3 4 5 7 24 | 81840 |
| 1 2 3 4 5 8 10 | 163680 | 1 2 3 4 5 8 11 | 163680 |
| 1 2 3 4 5 8 12 | 163680 | 1 2 3 4 5 8 13 | 163680 |
| 1 2 3 4 5 8 17 | 81840 | 1 2 3 4 5 8 24 | 163680 |
| 1 2 3 4 5 8 26 | 163680 | 1 2 3 4 5 9 11 | 163680 |
| 1 2 3 4 5 9 12 | 163680 | 1 2 3 4 6 9 17 | 32736 |
| 1 2 3 4 6 10 12 | 32736 | 1 2 3 4 6 11 16 | 32736 |
| | | 32 orbits on 7-sets. | |

*Table 1 (cont.)*

| Orbit Representative | Length | Orbit Representative | Length |
|---|---|---|---|
| 1 2 3 4 5 6 7 8 | 81840 | 1 2 3 4 5 6 7 9 | 163680 |
| 1 2 3 4 5 6 7 10 | 163680 | 1 2 3 4 5 6 7 11 | 163680 |
| 1 2 3 4 5 6 7 12 | 163680 | 1 2 3 4 5 6 7 13 | 163680 |
| 1 2 3 4 5 6 7 14 | 163680 | 1 2 3 4 5 6 7 15 | 163680 |
| 1 2 3 4 5 6 7 16 | 163680 | 1 2 3 4 5 6 7 17 | 163680 |
| 1 2 3 4 5 6 7 18 | 81840 | 1 2 3 4 5 6 7 19 | 163680 |
| 1 2 3 4 5 6 7 32 | 163680 | 1 2 3 4 5 6 8 9 | 163680 |
| 1 2 3 4 5 6 8 10 | 163680 | 1 2 3 4 5 6 8 12 | 163680 |
| 1 2 3 4 5 6 8 13 | 163680 | 1 2 3 4 5 6 8 14 | 163680 |
| 1 2 3 4 5 6 8 15 | 163680 | 1 2 3 4 5 6 8 16 | 163680 |
| 1 2 3 4 5 6 8 17 | 163680 | 1 2 3 4 5 6 8 19 | 163680 |
| 1 2 3 4 5 6 8 20 | 163680 | 1 2 3 4 5 6 8 21 | 163680 |
| 1 2 3 4 5 6 8 23 | 163680 | 1 2 3 4 5 6 8 24 | 163680 |
| 1 2 3 4 5 6 8 26 | 163680 | 1 2 3 4 5 6 8 27 | 163680 |
| 1 2 3 4 5 6 8 30 | 81840 | 1 2 3 4 5 6 8 32 | 163680 |
| 1 2 3 4 5 6 8 33 | 163680 | 1 2 3 4 5 6 9 10 | 163680 |
| 1 2 3 4 5 6 9 11 | 163680 | 1 2 3 4 5 6 9 12 | 163680 |
| 1 2 3 4 5 6 9 13 | 81840 | 1 2 3 4 5 6 9 14 | 163680 |
| 1 2 3 4 5 6 9 15 | 163680 | 1 2 3 4 5 6 9 17 | 163680 |
| 1 2 3 4 5 6 9 18 | 163680 | 1 2 3 4 5 6 9 19 | 163680 |
| 1 2 3 4 5 6 9 22 | 81840 | 1 2 3 4 5 6 9 23 | 81840 |
| 1 2 3 4 5 6 9 24 | 163680 | 1 2 3 4 5 6 9 26 | 163680 |
| 1 2 3 4 5 6 9 27 | 163680 | 1 2 3 4 5 6 9 29 | 81840 |
| 1 2 3 4 5 6 9 33 | 163680 | 1 2 3 4 5 6 10 11 | 163680 |
| 1 2 3 4 5 6 10 12 | 163680 | 1 2 3 4 5 6 10 13 | 163680 |
| 1 2 3 4 5 6 10 15 | 163680 | 1 2 3 4 5 6 10 18 | 163680 |
| 1 2 3 4 5 6 10 19 | 163680 | 1 2 3 4 5 6 10 20 | 163680 |
| 1 2 3 4 5 6 10 22 | 81840 | 1 2 3 4 5 6 10 24 | 163680 |
| 1 2 3 4 5 6 10 25 | 163680 | 1 2 3 4 5 6 10 26 | 163680 |
| 1 2 3 4 5 6 10 28 | 81840 | 1 2 3 4 5 6 10 32 | 81840 |
| 1 2 3 4 5 6 11 12 | 81840 | 1 2 3 4 5 6 11 14 | 163680 |
| 1 2 3 4 5 6 11 16 | 163680 | 1 2 3 4 5 6 11 20 | 81840 |
| 1 2 3 4 5 6 11 21 | 163680 | 1 2 3 4 5 6 11 22 | 163680 |
| 1 2 3 4 5 6 11 23 | 163680 | 1 2 3 4 5 6 11 25 | 163680 |
| 1 2 3 4 5 6 11 26 | 163680 | 1 2 3 4 5 6 11 27 | 81840 |
| 1 2 3 4 5 6 11 33 | 163680 | 1 2 3 4 5 6 12 13 | 163680 |
| 1 2 3 4 5 6 12 15 | 81840 | 1 2 3 4 5 6 12 17 | 163680 |
| 1 2 3 4 5 6 12 20 | 163680 | 1 2 3 4 5 6 12 24 | 163680 |
| 1 2 3 4 5 6 12 26 | 81840 | 1 2 3 4 5 6 12 32 | 163680 |
| 1 2 3 4 5 6 13 16 | 163680 | 1 2 3 4 5 6 14 24 | 81840 |
| 1 2 3 4 5 6 16 17 | 163680 | 1 2 3 4 5 6 16 22 | 20460 |
| 1 2 3 4 5 6 16 33 | 163680 | 1 2 3 4 5 6 17 19 | 163680 |
| 1 2 3 4 5 6 17 33 | 163680 | 1 2 3 4 5 7 9 12 | 163680 |
| 1 2 3 4 5 7 9 17 | 163680 | 1 2 3 4 5 7 9 32 | 163680 |
| 1 2 3 4 5 7 10 20 | 81840 | 1 2 3 4 5 7 10 32 | 81840 |
| 1 2 3 4 5 7 12 15 | 163680 | 1 2 3 4 5 7 12 17 | 81840 |
| 1 2 3 4 5 7 12 24 | 81840 | 1 2 3 4 5 7 13 26 | 163680 |
| 1 2 3 4 5 8 10 15 | 163680 | 1 2 3 4 5 8 13 19 | 81840 |
| 1 2 3 4 5 9 12 24 | 32736 | | |

97 orbits on 8-sets.

*Table 2*

```
        AA45
5 20 4

        AA56
4 4 4 4 4 8 0 0 0 0 0 0 0
2 2 5 4 1 3 3 2 2 1 1 1 1
0 0 0 5 5 5 0 5 5 0 1 1 1

        AA67
2 2 2 2 2 2 4 2 2 2 2 1 2 0 0 0 0 0 0 0 0 0 0 0 0 0 0 0 0 0 0 0
2 4 0 0 0 0 0 1 0 2 2 0 2 2 2 2 2 2 2 2 0 0 0 0 0 0 0 0 0 0 0 0
1 2 2 0 1 0 2 0 0 0 2 1 0 2 0 1 2 0 1 0 1 3 1 2 1 1 1 0 0 0 0 0
0 2 2 2 1 2 0 0 1 0 0 1 0 2 2 1 0 0 1 1 1 1 1 0 0 1 1 2 2 0 0 0
0 0 0 0 2 2 0 0 2 2 0 0 2 0 0 0 0 3 2 0 0 2 0 0 2 2 2 4 0 0 0 0
0 0 0 0 0 2 2 2 0 0 1 1 2 1 1 1 1 1 1 1 0 0 1 1 1 1 3 1 2 0 0 0
1 0 2 0 0 0 2 0 0 2 2 0 0 0 2 2 0 0 2 0 4 2 2 0 0 0 0 0 2 2 0 0
0 0 0 2 2 2 0 0 0 2 0 0 0 0 2 0 2 0 2 0 2 0 0 2 1 4 0 0 2 0 2 0
0 0 0 2 2 0 0 0 1 2 0 0 0 0 0 4 2 0 0 0 2 0 0 2 0 2 2 2 2 0 0 2
0 0 6 0 0 0 0 0 0 6 0 0 0 0 0 0 0 0 0 3 0 0 6 6 0 0 0 0 0 0 0 0
0 0 0 5 0 0 0 0 0 0 5 0 0 0 0 0 5 0 0 0 0 0 5 0 0 0 0 5 0 1 1 0
0 0 0 5 0 0 0 0 0 0 0 0 5 0 5 0 0 0 0 0 0 5 0 5 0 0 0 0 0 1 0 1
0 0 0 0 5 0 0 0 0 0 0 0 5 5 0 0 0 0 0 0 5 0 5 0 0 0 0 0 0 0 1 1
```

Table 2 (cont.)

```
     AA78
2222222222222.....................................................................................
21.111.......211112111111111111...................................................................
111......1...1..2..........1...111112112111111....................................................
..11..........21..1............1...11......1...1213111111111......................................
...12................1.....1..111........1..1..1..11...1....11211111111...........................
....11.........1......1...2......1..1.....11..111......1.1..1......11..2111111....................
.....13...11.1..1......1.1..1..1...1...2...1...1..1.1.1..1.......1...1...1..1...1.................
......42.........2....2..22.........2......2.............2.....2..............2...2...............
.......22.........2.....2.....2.....2....2.........2...........2....2....22.2.....................
...1...121.....11..1......1.1.........1...1...1.1..1....1........1..111.....1.1..1.111............
......2..12...1..11.11.1....11.......11.1...11..1.....1...........1........1....1.1..11...........
.......2...2.2........22...2....2......22.......2....2.......................2.........2..........
.........1.12..........1.1....22................111......11..1.11....1..1..1...1.2..1.1...........
.2..........221..2....1.1.....1111.........1..................1...11..1.1........1.....1.111......
..1.1.......1..1..12..1.......1...1.....2.........1..1.11.11...1..............1.1.........1.211...
....11..11....1.1....1.2.1................1.1......1.1........1.1..1..........1.........1.211..111
..11.2.........1..1........1..1.......1...1............1.1....1.....1...11....1...1.1.212.....1..1
.......22...........24...2.........................................2..22........22.2..........2...
.....1..1..11....1.11..1..11..111....1........1.....1.1...1.........1.......1.11...1........1...11
.2.....2.....2.....2.....2...2...2.........2.2...............2...............2...2................2
...1.....1.1..11...............1....11...1...1....1....1.....1....11......1..........2..1..31.1.1..1
..1.1.......11...1.2..1..1..........1...11....1.11..1..11...........2.....1.............1.....1.....3
.2..............1.....1...........1.22.1.....11............1..1..1..11......1...11.1..21....1.....1
..11..............1....1.......1.1...11.1.......1...1.......1.1.......2.1......1.1.......1.1..2...1..12
.......2.............2..2..................2........2........2...........2..2.22..............2...2....2
..1.................11....1.1..............1.......1..1.1....2..1.....11.111..11..........111.......211
...........1.....1.....11.1.............1...1.2.......2..11....2.......1..1.....111...1..11...1..1.1
......................1..1.1...1.11....1.....1....1.1.1.11......1...1.1...2.1..11.............3..11
........1.......1..........1.1...11.1...1..1....111..1...1.......1..1...11...1..........111..1.11..1
.........5............................5............5...5.....................................5......1
..........................................5........5........5...5............5..................1
..............................................5............5....................5..5.....5....1
```

*Table 3*

```
     RES
1.........1......1.................1..............1.......1...1......1......1........11.11...1
.....1...1...........1.........1.......1..............1..1.1..1.............1.........1.......1
......1............1...1.......11............1.1..........1...1.........................1.......1
.........1...........1...1......1............1.1..........1.1...............1....1........1.......1
...1......1....1.....1...1.....................1.....1.................1......1..................1
.1.......11..................1.............1............1..........1.11......1.....1..1.......1

     AA68
544666E866664644646224826466248422348264246441664464242642122643224621263464124421222...........
6846686886448864486A84666844486224.222.42.42412222.4.4222232142.6224412..6243645414444242I4142...
466556632446575276624745533533454434458522354453.41445.232212222752342 53.342115.4.56634431221473.
364353.332.3475644653454444642446672454362356423563646334452234323252414543542 53.2.141555225134411
..2242.882.44..22246C2268444.46442..444.23244264426262 3824.136616488A26874646845214.422.4.A.1862.
.3212486322544232614537555822543342522641.835145431363176322272233356147255647552.445544216136231
244444626C4642466244242422262222664 44C46334443622A.62A2.44.2.2444242..42.424124.812624863923326.2
..6644.242.22.4622264222.446222444.4422432222246A46644344E3142432A664966144 8222.A142646632222 8442
..4644.264.2..64422224.62624222442.44222434241 2446C66234443326A344644344326412 4.41A44C86.18936442
3C66...666...6.6C..6.6...6.6.6.6669CC6666366C66..6....3.6.63..6..C66.....C.6.66.63666..6.3.33.66.
.555.55..55...A5.55.55.5.5..55.55.5A5A.5...A5..5A5A55A555555..5.555.55..555..5A.A.5AA....5A..55.1
..A55....5.5A.555.A5.55.5....A55.5.55555...5.555AAA55A55A555.55.55.5..5..5.5.5.5..A.55.A55.5..A51
.A.55....5.5A555.5...55.5..5.5A5A555.5...5.55.555.55...555.555F55A5555A..5...555A.5A55A5.5.5..5.1

564822234.....    6644261232..    E22421614..    6442161264..    664446222.    4664111634...
6482.12.52114.    86422231.62.    6A62212322.    64821231252.    644482.26.    8642131..542.
4464425..3121.    64752222213.    6255443223.    473443224.3.    542735451.    64454223..33.
3.456324.22131    324624232521    .5466252321    244625235.21    3.64436451    62.622154.521
..24634754..1.    222428.3184.    .664424.64.    224424.3654.    2.2246668.    .2.42.287524.
.2626232521131    423317222721    84533152221    235315225521    1233546471    3223121725521
24242.2..39332    4C423...4232    6442632.432    C42632..2.32    4464262422    4C423..2..432
..2444A1.32222    422434343232    .6244263432    224426344.32    6.62444622    .2.423961.432
..246243..1932    4444443232.2    .2644143A.2    446414326..2    6.44626422    .4.413343.C.2
3..C..C...333.    .6666.6..6..    .6.666.66..    66666.6.....    6.6666..6.    C6.666.......
.5555.55..5..1    5555.55..5.1    5.555.555.1    5555.55.5..1    55555.5.51    5555.55.5...1
...55.5.5555.1    .555.55..551    .5.5.555551    55..555..551    5.55.55551    .5.555...5551
..5..5A.5.55.1    .555.5.555.1    ...5A.5.F.1    55.A.5.5.5.1    5.55.5.A51    A5.5..5..55.1
```

Table 1 exhibits the generators of the specific PΓL$_2$(32) used, and representatives and lengths of the orbits on $k$-subsets for $k = 1, 2, \ldots, 8$.

Table 2 exhibits the matrices $A_{4,5}$, $A_{5,6}$, $A_{6,7}$, $A_{7,8}$ from which $A_{t,k}$ for $4 \leqslant t < k \leqslant 8$ can be computed.

Table 3 exhibits $A_{6,8}$ and the six 6-design solutions, in hexadecimal notation.

Tables 4 and 5 exhibit the intersection numbers for the orbit constituents of the 6-designs. Here $x_i$ is the number of blocks of the design which intersect a given block of a constituent orbit in exactly $i$ points.

*Table 4*

| ORBIT REPRESENTATIVE | ORBIT LENGTH | $X_0$ | $X_1$ | $X_2$ | $X_3$ | $X_4$ | $X_5$ | $X_6$ | $X_7$ | $X_8$ |
|---|---|---|---|---|---|---|---|---|---|---|
| 1 2 3 4 5 6 7 8 | 81840 | 110927 | 394446 | 508522 | 305270 | 90720 | 13258 | 854 | 18 | 1 |
| 1 2 3 4 5 6 7 18 | 81840 | 110927 | 394446 | 508522 | 305270 | 90720 | 13258 | 854 | 18 | 1 |
| 1 2 3 4 5 6 8 14 | 163680 | 110927 | 394446 | 508522 | 305270 | 90720 | 13258 | 854 | 18 | 1 |
| 1 2 3 4 5 6 9 15 | 163680 | 110928 | 394439 | 508543 | 305235 | 90755 | 13237 | 861 | 17 | 1 |
| 1 2 3 4 5 6 10 19 | 163680 | 110933 | 394404 | 508648 | 305060 | 90930 | 13132 | 896 | 12 | 1 |
| 1 2 3 4 5 6 11 12 | 81840 | 110931 | 394418 | 508606 | 305130 | 90860 | 13174 | 882 | 14 | 1 |
| 1 2 3 4 5 6 11 22 | 163680 | 110925 | 394460 | 508480 | 305340 | 90650 | 13300 | 840 | 20 | 1 |
| 1 2 3 4 5 6 12 15 | 81840 | 110927 | 394446 | 508522 | 305270 | 90720 | 13258 | 854 | 18 | 1 |
| 1 2 3 4 5 6 14 24 | 81840 | 110933 | 394404 | 508648 | 305060 | 90930 | 13132 | 896 | 12 | 1 |
| 1 2 3 4 5 7 10 20 | 81840 | 110931 | 394418 | 508606 | 305130 | 90860 | 13174 | 882 | 14 | 1 |
| 1 2 3 4 5 7 10 32 | 81840 | 110929 | 394432 | 508564 | 305200 | 90790 | 13216 | 868 | 16 | 1 |
| 1 2 3 4 5 7 12 17 | 81840 | 110925 | 394460 | 508480 | 305340 | 90650 | 13300 | 840 | 20 | 1 |
| 1 2 3 4 5 7 12 24 | 81840 | 110933 | 394404 | 508648 | 305060 | 90930 | 13132 | 896 | 12 | 1 |
| 1 2 3 4 5 9 12 24 | 32736 | 110920 | 394495 | 508375 | 305515 | 90475 | 13405 | 805 | 25 | 1 |
| | | | | | | | | | | |
| 1 2 3 4 5 6 7 13 | 163680 | 110928 | 394439 | 508543 | 305235 | 90755 | 13237 | 861 | 17 | 1 |
| 1 2 3 4 5 6 7 17 | 163680 | 110927 | 394446 | 508522 | 305270 | 90720 | 13258 | 854 | 18 | 1 |
| 1 2 3 4 5 6 8 19 | 163680 | 110929 | 394432 | 508564 | 305200 | 90790 | 13216 | 868 | 16 | 1 |
| 1 2 3 4 5 6 9 10 | 163680 | 110927 | 394446 | 508522 | 305270 | 90720 | 13258 | 854 | 18 | 1 |
| 1 2 3 4 5 6 9 22 | 81840 | 110931 | 394418 | 508606 | 305130 | 90860 | 13174 | 882 | 14 | 1 |
| 1 2 3 4 5 6 10 24 | 163680 | 110927 | 394446 | 508522 | 305270 | 90720 | 13258 | 854 | 18 | 1 |
| 1 2 3 4 5 6 10 28 | 81840 | 110927 | 394446 | 508522 | 305270 | 90720 | 13258 | 854 | 18 | 1 |
| 1 2 3 4 5 6 11 12 | 81840 | 110927 | 394446 | 508522 | 305270 | 90720 | 13258 | 854 | 18 | 1 |
| 1 2 3 4 5 6 11 20 | 81840 | 110931 | 394418 | 508606 | 305130 | 90860 | 13174 | 882 | 14 | 1 |
| 1 2 3 4 5 6 12 32 | 163680 | 110931 | 394418 | 508606 | 305130 | 90860 | 13174 | 882 | 14 | 1 |
| 1 2 3 4 5 7 10 20 | 81840 | 110931 | 394418 | 508606 | 305130 | 90860 | 13174 | 882 | 14 | 1 |
| 1 2 3 4 5 9 12 24 | 32736 | 110935 | 394390 | 508690 | 304990 | 91000 | 13090 | 910 | 10 | 1 |
| | | | | | | | | | | |
| 1 2 3 4 5 6 7 14 | 163680 | 110926 | 394453 | 508501 | 305305 | 90685 | 13279 | 847 | 19 | 1 |
| 1 2 3 4 5 6 8 16 | 163680 | 110929 | 394432 | 508564 | 305200 | 90790 | 13216 | 868 | 16 | 1 |
| 1 2 3 4 5 6 8 21 | 163680 | 110928 | 394439 | 508543 | 305235 | 90755 | 13237 | 861 | 17 | 1 |
| 1 2 3 4 5 6 9 10 | 163680 | 110923 | 394474 | 508438 | 305410 | 90580 | 13342 | 826 | 22 | 1 |
| 1 2 3 4 5 6 9 11 | 163680 | 110926 | 394453 | 508501 | 305305 | 90685 | 13279 | 847 | 19 | 1 |
| 1 2 3 4 5 6 9 29 | 81840 | 110925 | 394460 | 508480 | 305340 | 90650 | 13300 | 840 | 20 | 1 |
| 1 2 3 4 5 6 10 11 | 163680 | 110929 | 394432 | 508564 | 305200 | 90790 | 13216 | 868 | 16 | 1 |
| 1 2 3 4 5 6 10 28 | 81840 | 110931 | 394418 | 508606 | 305130 | 90860 | 13174 | 882 | 14 | 1 |
| 1 2 3 4 5 6 11 16 | 163680 | 110925 | 394460 | 508480 | 305340 | 90650 | 13300 | 840 | 20 | 1 |
| 1 2 3 4 5 7 10 20 | 81840 | 110923 | 394474 | 508438 | 305410 | 90580 | 13342 | 826 | 22 | 1 |
| 1 2 3 4 5 9 12 24 | 32736 | 110930 | 394425 | 508585 | 305165 | 90825 | 13195 | 875 | 15 | 1 |

*Table 5*

| ORBIT REPRESENTATIVE | ORBIT LENGTH | $X_0$ | $X_1$ | $X_2$ | $X_3$ | $X_4$ | $X_5$ | $X_6$ | $X_7$ | $X_8$ |
|---|---|---|---|---|---|---|---|---|---|---|
| 1 2 3 4 5 6 7 17 | 163680 | 110927 | 394446 | 508522 | 305270 | 90720 | 13258 | 854 | 18 | 1 |
| 1 2 3 4 5 6 8 19 | 163680 | 110926 | 394453 | 508501 | 305305 | 90685 | 13279 | 847 | 19 | 1 |
| 1 2 3 4 5 6 8 24 | 163680 | 110927 | 394446 | 508522 | 305270 | 90720 | 13258 | 854 | 18 | 1 |
| 1 2 3 4 5 6 9 11 | 163680 | 110931 | 394418 | 508606 | 305130 | 90860 | 13174 | 882 | 14 | 1 |
| 1 2 3 4 5 6 9 29 | 81840 | 110929 | 394432 | 508564 | 305200 | 90790 | 13216 | 868 | 16 | 1 |
| 1 2 3 4 5 6 10 11 | 163680 | 110924 | 394467 | 508459 | 305375 | 90615 | 13321 | 833 | 21 | 1 |
| 1 2 3 4 5 6 10 28 | 81840 | 110929 | 394432 | 508564 | 305200 | 90790 | 13216 | 868 | 16 | 1 |
| 1 2 3 4 5 6 11 12 | 81840 | 110927 | 394446 | 508522 | 305270 | 90720 | 13258 | 854 | 18 | 1 |
| 1 2 3 4 5 6 12 20 | 163680 | 110927 | 394446 | 508522 | 305270 | 90720 | 13258 | 854 | 18 | 1 |
| 1 2 3 4 5 6 14 24 | 81840 | 110925 | 394460 | 508480 | 305340 | 90650 | 13300 | 840 | 20 | 1 |
| 1 2 3 4 5 7 10 20 | 81840 | 110935 | 394390 | 508690 | 304990 | 91000 | 13090 | 910 | 10 | 1 |
| 1 2 3 4 5 9 12 24 | 32736 | 110930 | 394425 | 508585 | 305165 | 90825 | 13195 | 875 | 15 | 1 |

| ORBIT REPRESENTATIVE | ORBIT LENGTH | $X_0$ | $X_1$ | $X_2$ | $X_3$ | $X_4$ | $X_5$ | $X_6$ | $X_7$ | $X_8$ |
|---|---|---|---|---|---|---|---|---|---|---|
| 1 2 3 4 5 6 7 11 | 163680 | 110927 | 394446 | 508522 | 305270 | 90720 | 13258 | 854 | 18 | 1 |
| 1 2 3 4 5 6 7 18 | 81840 | 110929 | 394432 | 508564 | 305200 | 90790 | 13216 | 868 | 16 | 1 |
| 1 2 3 4 5 6 8 12 | 163680 | 110926 | 394453 | 508501 | 305305 | 90685 | 13279 | 847 | 19 | 1 |
| 1 2 3 4 5 6 8 19 | 163680 | 110928 | 394439 | 508543 | 305235 | 90755 | 13237 | 861 | 17 | 1 |
| 1 2 3 4 5 6 8 24 | 163680 | 110929 | 394432 | 508564 | 305200 | 90790 | 13216 | 868 | 16 | 1 |
| 1 2 3 4 5 6 9 33 | 163680 | 110928 | 394439 | 508543 | 305235 | 90755 | 13237 | 861 | 17 | 1 |
| 1 2 3 4 5 6 10 19 | 163680 | 110926 | 394453 | 508501 | 305305 | 90685 | 13279 | 847 | 19 | 1 |
| 1 2 3 4 5 6 11 33 | 163680 | 110927 | 394446 | 508522 | 305270 | 90720 | 13258 | 854 | 18 | 1 |
| 1 2 3 4 5 6 12 32 | 163680 | 110930 | 394425 | 508585 | 305165 | 90825 | 13195 | 875 | 15 | 1 |
| 1 2 3 4 5 9 12 24 | 32736 | 110930 | 394425 | 508585 | 305165 | 90825 | 13195 | 875 | 15 | 1 |

| ORBIT REPRESENTATIVE | ORBIT LENGTH | $X_0$ | $X_1$ | $X_2$ | $X_3$ | $X_4$ | $X_5$ | $X_6$ | $X_7$ | $X_8$ |
|---|---|---|---|---|---|---|---|---|---|---|
| 1 2 3 4 5 6 7 9 | 163680 | 110926 | 394453 | 508501 | 305305 | 90685 | 13279 | 847 | 19 | 1 |
| 1 2 3 4 5 6 7 17 | 163680 | 110926 | 394453 | 508501 | 305305 | 90685 | 13279 | 847 | 19 | 1 |
| 1 2 3 4 5 6 7 18 | 81840 | 110925 | 394460 | 508480 | 305340 | 90650 | 13300 | 840 | 20 | 1 |
| 1 2 3 4 5 6 9 10 | 163680 | 110930 | 394425 | 508585 | 305165 | 90825 | 13195 | 875 | 15 | 1 |
| 1 2 3 4 5 6 9 29 | 81840 | 110929 | 394432 | 508564 | 305200 | 90790 | 13216 | 868 | 16 | 1 |
| 1 2 3 4 5 6 10 28 | 81840 | 110931 | 394418 | 508606 | 305130 | 90860 | 13174 | 882 | 14 | 1 |
| 1 2 3 4 5 6 11 27 | 81840 | 110927 | 394446 | 508522 | 305270 | 90720 | 13258 | 854 | 18 | 1 |
| 1 2 3 4 5 6 12 13 | 163680 | 110929 | 394432 | 508564 | 305200 | 90790 | 13216 | 868 | 16 | 1 |
| 1 2 3 4 5 6 12 15 | 81840 | 110927 | 394446 | 508522 | 305270 | 90720 | 13258 | 854 | 18 | 1 |
| 1 2 3 4 5 6 14 24 | 81840 | 110933 | 394404 | 508648 | 305060 | 90930 | 13132 | 896 | 12 | 1 |
| 1 2 3 4 5 7 9 12 | 163680 | 110927 | 394446 | 508522 | 305270 | 90720 | 13258 | 854 | 18 | 1 |
| 1 2 3 4 5 7 10 20 | 81840 | 110931 | 394418 | 508606 | 305130 | 90860 | 13174 | 882 | 14 | 1 |
| 1 2 3 4 5 9 12 24 | 32736 | 110920 | 394495 | 508375 | 305515 | 90475 | 13405 | 805 | 25 | 1 |

The reader can easily obtain 108 5-designs with $\lambda = 12$, and $\lambda = 16$ from $A_{5,6}$. We remark that there are 317 5-designs with $\lambda = 42$ and 518 401 5-designs with $\lambda = 126$ from $A_{5,7}$.

## 6 Closing Remarks

As mentioned earlier, extensive details of this work will appear in a later publication. We intend to investigate questions of extendability of the designs found, and completeness of the solution-sets for several parameter triples $(t, k, \lambda)$. We are currently investigating the question of existence of simple 7-designs with $P\Gamma L_2(32)$ as a group of automorphisms. Finally, we should like to know these geometries more intimately and perhaps gain deeper understanding of their existence.

## Acknowledgements

The authors wish to thank their colleagues Professor E. S. Kramer, Professor D. M. Mesner, Mr. D. L. Kreher, and particularly Professor L. G. Chouinard for many useful and substantial discussions. The authors also wish to thank the University of Nebraska for support of this research.

## References

1. Alltop, W. O. (1969). An infinite class of 4-designs, *J. Combinatorial Theory* **6**, 320–322.
2. Alltop, W. O. (1972). An infinite class of 5-designs, *J. Combinatorial Theory* (*A*) **12**, 390–395.
3. Dembowski, P. (1968). Finite Geometries. *In* "Ergebnisse der Mathematik und ihrer Grenzgebiete" **44**. Springer-Verlag, Berlin.
4. Denniston, R. H. F. (1976). Some new 5-designs, *Bull. London Math. Soc.* **8**, 263–267.
5. Hedayat, A. and Kageyama, S. (1980). The Family of $t$-Designs—Part I, *J. of Statistical Planning and Inf.* **4**, 173–212.
6. Hubaut, X. (1974). Two new families of 4-designs, *Discrete Math.* **9**, 247–249.
7. Hughes, D. R. (1965). On $t$-designs and groups, *Amer. J. Math.* **87**, 761–778.
8. Kantor, W. (1968). 4-homogeneous groups, *Math. Zeitschr.* **103**, 67–68.
9. Kramer, E. S. (1975). "Some $t$-designs for $t \geqslant 4$ and $v = 17, 18$", Proc. 6th South-Eastern Conf. Cominatorics, Graph Theory and Computing, Congressus Numerantium **XIV**, 443–460. Utilitas Math, Publ., Winnipeg.
10. Kramer, E. S. and Mesner, D. M. (1976). $t$-designs on hypergraphs, *Discrete Math.* **15**, 263–296.
11. Kramer, E. S., Magliveras, S. S. and Mesner, D. M. (1981). $t$-designs from the large Mathieu groups, *Discrete Math.* **36**, 171–189.
12. Lane, R. N. (1971). $t$-designs and $t$-ply homogeneous groups, *J. Combinatorial Theory* (*A*) **10**, 106–118.
13. Lindner, C. C. and Rosa, A. (1981). Topics on Steiner systems, *In* "Annals of Discrete Math." **7**, North–Holland, Amsterdam.
14. Livingstone, D. and Wagner, A. (1965). Transitivity of finite permutation groups on unordered sets, *Math. Zeitschr.* **90**, 393–403.
15. Luneburg, H. (1969). Transitive Erweiterungen endlicher Permutationsgruppen, Lecture Notes in Mathematics, vol. 84, Springer-Verlag, Berlin.
16. Mills, W. H. (1978). A new 5-design, *Ars Combinatoria* **6**, 193–195.
17. Nagao, H. (1974). "Groups and Designs". Iwanami Shoten, Tokyo.

18. Pless, V. (1972). Symmetry codes over GF(3) and new five-designs, *J. Combinatorial Theory* (*A*) **12**, 119–142.
19. Ray-Chaudhuri, D. K. and Wilson, R. M. (1975), On $t$-designs, *Osaka J. Math.* **12**, 737–744.
20. Wilson, R. M. (1973). The necessary conditions for $t$-designs are sufficient for something, *Utilitas Math.* **4**, 207–215.
21. Witt, E. (1938). Über Steinersche systeme, *Abh. Math. Sem. Hamb.* **12**, 265–275.

# The Steiner System S(5,6,12), the Mathieu Group $M_{12}$ and the "Kitten"

R. T. CURTIS

## 1 Introduction

In this article we describe the interesting properties of a "pocket calculator for $M_{12}$" which was developed by John Conway and the author; since the device claims to do for $M_{12}$ what the author's Miracle Octad Generator (MOG) does for $M_{24}$, it is known as the *Kitten.*

## 2 The Steiner System S(5,6,12) and $M_{12}$

$M_{12}$ is obtained in the usual manner as a permutation group acting on the 12 points of the projective line

$$\mathrm{PG}(1,11) = \{\infty, 0, 1, 2, 3, 4, 5, 6, 7, 8, 9, \mathrm{X}\}$$

(it is convenient to write X for 10).

as follows.

We let $Q = \{0,1,3,9,5,4\}$ be the subset consisting of the quadratic residues together with 0, and let $L$ be the linear fractional group consisting of all permutations of the form

$$y \longrightarrow \frac{ay+b}{cy+d}, \; ad-bc = 1.$$

Note: $\qquad L = \langle \alpha, \gamma;\ \alpha : y \longrightarrow y+1,\ \gamma : y \longrightarrow -1/y \rangle \cong \mathrm{PSL}_2(11).$

Now $\mathscr{S} = \{Q^x;\ x \in L\}$ consists of 132 subsets of size 6, known as *hexads*, which have the property that any five of the 12 points occur together in just one of them. That is, the hexads of $\mathscr{S}$ form a Steiner system S(5,6,12). We can now define:

$$\mathrm{M}_{12} = \langle g \in \mathrm{S}_{12};\ \mathscr{S}^g = \mathscr{S} \rangle.$$

Thus $M_{12}$ consists of all permutations preserving the Steiner system.

COMPUTATIONAL GROUP THEORY
ISBN 0-12-066270-1

## 3 The Construction of the Kitten

It was our intention to produce a simple device from which the hexad containing a given five points could be easily read off. The construction is based on the fact that the triples which complete a given three points, $\{\infty, 0, 1\}$ say, to hexads form a Steiner system $S(2,3,9)$, i.e. an affine plane. Now there is, of course, only one such plane and its lines may conveniently be taken as the rows, columns and generalized diagonals of a "noughts-and-crosses" or "tic-tac-toe" board (Fig. 1).

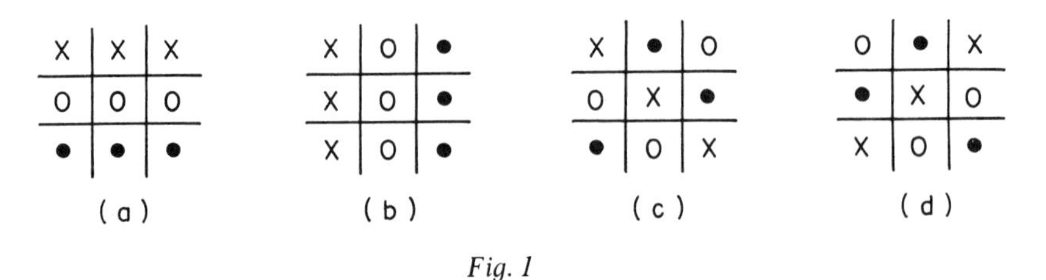


*Fig. 1*

These

$$\binom{9}{2} \Big/ \binom{3}{2} = 12 \text{ lines}$$

visibly partition into 4 sets of three parallel lines and we define (quite naturally), rows and columns to be *perpendicular* to one another, as are diagonals of types (c) and (d).

Now, unions of two perpendicular lines we term *crosses*. Examples of the $12.3/2 = 18$ crosses are shown in Fig. 2.

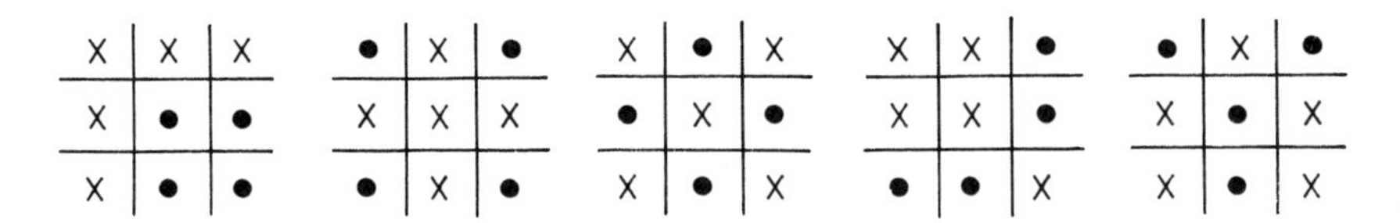

*Fig. 2*

Complements of crosses, seen in Fig. 3, are *squares*.

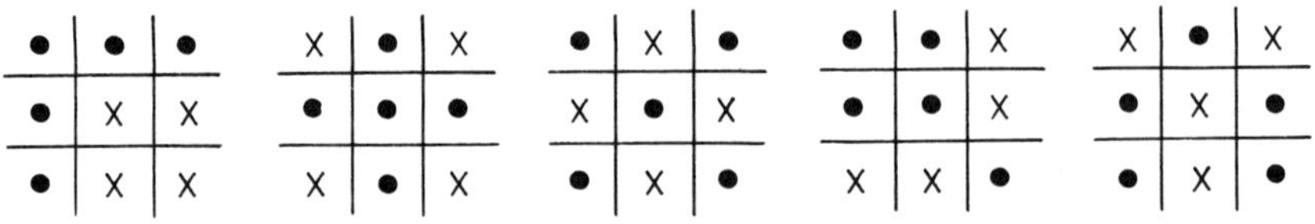

*Fig. 3*

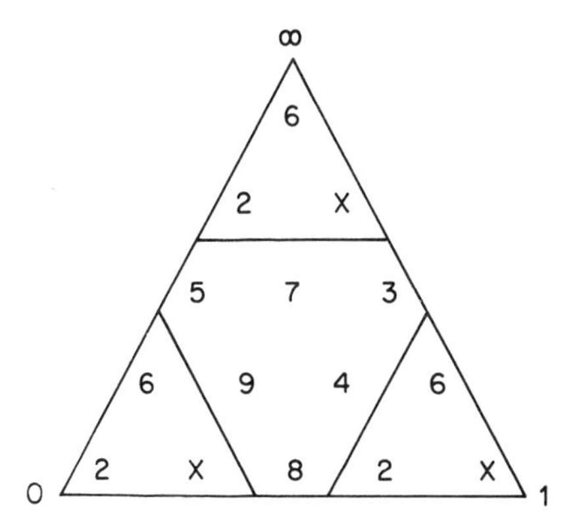


*Fig. 4 The Kitten*

The Kitten, shown in Fig. 4, consists of the three *points-at-*$\infty$ $\infty, 0, 1$, with an arrangement of the points of the plane corresponding to each of them. This correspondence is, of course, as shown in Fig. 5.

| | | |
|---|---|---|
| 6 | X | 3 |
| 2 | 7 | 4 |
| 5 | 9 | 8 |

∞ - picture

| | | |
|---|---|---|
| 5 | 7 | 3 |
| 6 | 9 | 4 |
| 2 | X | 8 |

0 - picture

| | | |
|---|---|---|
| 5 | 7 | 3 |
| 9 | 4 | 6 |
| 8 | 2 | X |

1 - picture

*Fig. 5*

The hexads are then:

I (i) $\{\infty, 0, 1\} \cup \{\text{any line}\}$ — 12 such
(ii) the union of two parallel lines — 12 such

II (i) a point-at-$\infty$ together with a cross in the corresponding picture — $3 \times 18 = 54$ such
(ii) two points-at-$\infty$ and a square in the picture corresponding to the omitted point-at-$\infty$. — 54 such.

Thus both parts of Fig. 6 are hexads.

*Examples*

1 Find the hexad containing $\{1,2,3,4,5\}$.

We look in the 1-picture and find that 7 completes the cross. Thus $\{1,2,3,4,5,7\}$ is the hexad.

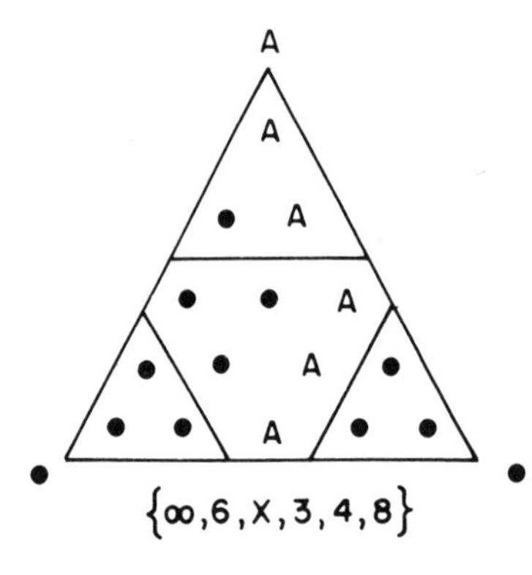


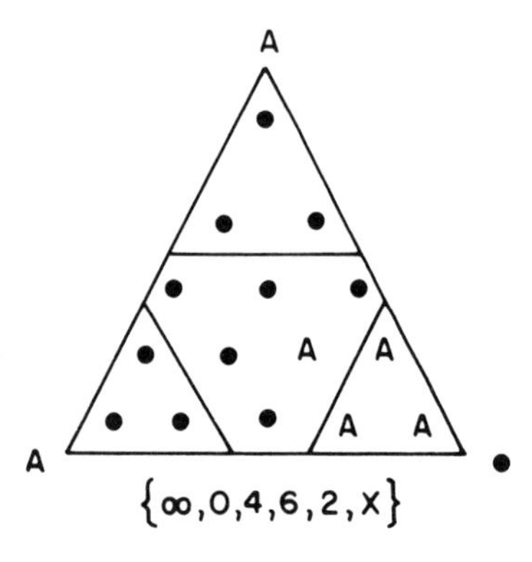


*Fig. 6*

2 Find the hexad containing $\{\infty, 1, 5, 9, X\}$.

In the 0-picture we find that 3 completes the square. Thus $\{\infty, 1, 3, 5, 9, X\}$ is the hexad.

3 Find the involution of $M_{12}$ which fixes 2, 3, 5 and 7.

We find the 4 pairs of points which when adjoined to the fixed points give hexads. As above, they are $\{\infty, 9\}$, $\{0, 6\}$, $\{1, 4\}$, $\{8, X\}$. Thus $(2)(3)(5)(7)(\infty\ \ 9)(0\ \ 6)(1\ \ 4)(8\ \ X)$ is the involution.

### 3.1 Mnemonic for construction

1 Draw the triangle with vertices labelled $\infty, 0, 1$ and write $6, 7, 8$ down the $\infty$-altitude (see Fig. 7).

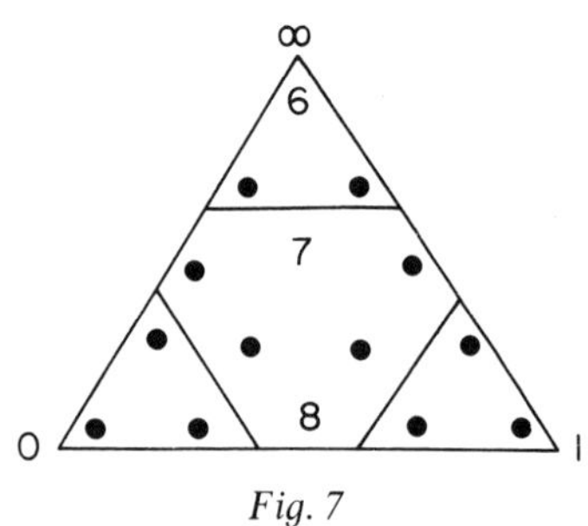


*Fig. 7*

2 Now let the anticlockwise rotation through 120° correspond to the linear fractional permutation

$$R : y \to \frac{1}{1-y} \equiv (\infty\ \ 0\ \ 1)(2\ \ X\ \ 6)(3\ \ 5\ \ 8)(4\ \ 7\ \ 9)$$

on both the large and small triangles.

### 3.2 Bonuses

From the symmetry of the construction it is clear that the three reflections in the altitudes, $f_\infty$, $f_0$ and $f_1$ must be members of $M_{12}$. We have:

$$f_\infty \equiv (0\ \ 1)(2\ \ X)(9\ \ 4)(5\ \ 3)$$
$$f_0 \equiv (1\ \ \infty)(X\ \ 6)(4\ \ 7)(8\ \ 5)$$
$$f_1 \equiv (\infty\ \ 0)(6\ \ 2)(7\ \ 9)(3\ \ 8).$$

We find:

$$\langle \alpha, R\rangle \cong \mathrm{PSL}_2(11)$$
$$\langle \alpha, f_\infty\rangle \cong M_{11} \text{ fixing } \infty$$
$$\langle \alpha, f_0\rangle \cong M_{11} \text{ transitive on the 12 points}$$

and
$$\langle \alpha, f_1\rangle \cong M_{12}.$$

Further, if we write the three altitudes as a $4 \times 3$ array, see Fig. 8(a), we find that any permutation of the columns fixing the rows is in $M_{12}$, and an *even* permutation of the rows fixing the columns is in $M_{12}$. That is, we can read off all the permutations of the maximal subgroup $A_4 \times S_3$.

Figure 8(b) is obtained from 8(a) by rotating the top row about $\infty$. The four triples thus obtained form the blocks for a further maximal subgroup of $M_{12}$ of shape $3^2 . S_4$ ($\cong M_9 . S_3$).

| ∞ | 0 | 1 |
|---|---|---|
| 6 | 2 | X |
| 7 | 9 | 4 |
| 8 | 3 | 5 |

(a)

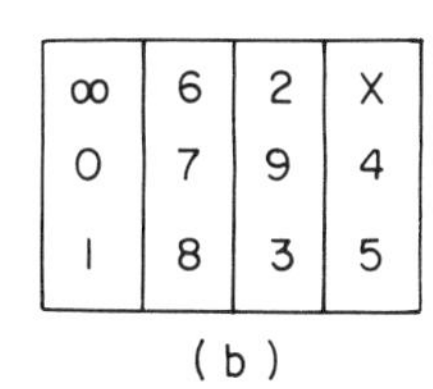

(b)

*Fig. 8*

### 3.3 Alternative version

Should the user require the Steiner system in which we take $Q$ as $\{\infty, 1, 3, 9, 5, 4\}$, that is we adjoin $\infty$ to the quadratic residues instead of 0, he should write 6, 5, 4 down the $\infty$-altitude instead of 6, 7, 8 and proceed as before.

### 3.4 Embedding of the Kitten in the MOG

$M_{12}$ is the stabilizer of a dodecad in $M_{24}$ (see [1], [2], [3]). If we take as canonical dodecad the symmetric difference of the top row of the MOG with its second, fourth and sixth columns, we find that the embedding of the Kitten is as shown in Fig. 9.

| ∞ | • | 0 | • | I | • |
|---|---|---|---|---|---|
| • | 6 | • | 2 | • | X |
| • | 7 | • | 9 | • | 4 |
| • | 8 | • | 3 | • | 5 |

*Fig. 9*

## References

1. Conway, J. H. (1971). Three lectures on exceptional groups. *In* "Finite Simple Groups" (Eds. M. B. Powell and G. Higman), 215–247. Academic Press, London, Orlando and New York.
2. Conway, J. H. (1977). The miracle octad generator. *In* "Topics in Group Theory and Computation" (Ed. M. P. J. Curran), 62–68. Academic Press, London, Orlando and New York.
3. Curtis, R. T. (1976). A new combinatorial approach to $M_{24}$, *Math. Proc. Camb. Phil. Soc.* **79**, 25–42.

# Hexacode and Tetracode—MOG and MINIMOG

J. H. CONWAY

## Introduction

The *hexacode* $C_6$ is a length 6 code over the field $\mathbb{F}_4 = \{0, 1, w, \bar{w}\}$ of order 4, which enables us to provide a very simple way of calculating inside the binary Golay code $C_{24}$ and the associated Steiner system S(5,8,24) and Mathieu group $M_{24}$, using a $6 \times 4$ array called the MOG. The *tetracode* $C_4$ is a length 4 code over the field $\mathbb{F}_3 = \{o, +, -\}$ of order 3, which provides a similar service for the ternary Golay code $C_{12}$ and the associated Steiner system S(5,6,12) and Mathieu group $M_{12}$, this time using a $4 \times 3$ array called the MINIMOG. Moreover, there are analogies and relations between the two theories which simplify calculations in which $M_{12}$ appears as a subgroup of $M_{24}$.

Let $\varphi(x) = ax^2 + bx + c$ be the typical quadratic function over $\mathbb{F}_4$. Then the typical hexacodeword $ab\ \ cd\ \ ef$ has $a, b, c$ for its first three digits (these name $\varphi$), and $\varphi(0)$, $\varphi(1)$, $\varphi(w)$, $\varphi(\bar{w})$ for its last four digits (which are the values of $\varphi$). In a similar way, if $\varphi(x) = ax + b$ is the typical linear function over $\mathbb{F}_3$, then the typical tetracodeword $a\ \ bcd$ has $a, b$ for its first two digits (which name $\varphi$), and $\varphi(o)$, $\varphi(+)$, $\varphi(-)$ for its last three (which are the values of $\varphi$).

In practice, it is simplest to *remember* all the words of the two codes. This is not too hard if we make use of various symmetries; thus, if $a, b, c$ denotes any cyclic permutation of $1, w, \bar{w}$, then any hexacodeword is obtained from one of

$$0a\ 0a\ bc, \quad bc\ bc\ bc, \quad 00\ aa\ aa, \quad aa\ bb\ cc, \quad 00\ 00\ 00$$

by permuting the three couples, or flipping the two digits in any even number of them, while if $a, b, c$ is a cyclic permutation of $o, +, -$ the typical tetracodeword is one of

$$o\ aaa \quad +abc, \quad -cba.$$

## 2 Problems 3 and 5, and Problems 2 and 4

In the use of the hexacode, it is valuable to be able to reconstruct a

COMPUTATIONAL GROUP THEORY
ISBN 0-12-066270-1

hexacodeword from partial information about it. Usually this reduces to one of two particular types of problem.

*Problem 3.* Reconstruct a hexacodeword from any 3 of its digits.

*Problem 5.* Reconstruct a hexacodeword from any 5 of its digits, *of which one may be mistaken.*

It can be shown by counting that either of these problems always has a unique solution, and so certainly the best way to solve one of them is just to write down the correct answer! If one really has remembered all the codewords, this is not at all hard.

Alternatively, we provide a beginner's way. Another definition of the hexacode is to say that it is the set of all words

$$L_1R_1 \quad L_2R_2 \quad L_3R_3$$

for which there exists a number $s \in \mathbb{F}_4$ (called the *slope* of the word), satisfying

$$\left.\begin{aligned} L_1+R_1 = L_2+R_2 = L_3+R_3 &= s \\ L_i+L_j+L_k = L_i+R_j+R_k &= ws \\ R_i+R_j+R_k = R_i+L_j+L_k &= \bar{w}s \end{aligned}\right\} \quad (\{i,j,k\} = \{1,2,3\}).$$

Now, given any three digits, the slope can be determined from one of these equations and then the remaining digits from the others. For Problem 5, we note that the mistaken digit, if any, can involve at most one of the three couples into which we habitually divide our hexacodeword. The remaining two couples contain at least three digits between them, which is enough to determine it by a case of Problem 3. The three choices for the couple containing the mistake therefore give three possibilities for the word, out of which it is easy to select the correct one.

For the tetracode, the analogous problems are:

*Problem 2.* Reconstruct a tetracodeword from any 2 of its digits.

*Problem 4.* Reconstruct a tetracodeword from all 4 of its digits, *of which one may be mistaken.*

Once again, there is always a unique solution, and since there are just 9 tetracodewords, it really is easy to just write this down. However, we give the analogous definition of the tetracode. Its typical word is

$$L \quad R_1R_2R_3$$

for which there exists a number $s \in \mathbb{F}_3$ (the *slope*), satisfying

$$s = L = R_2 - R_1 = R_3 - R_2 = R_1 - R_3.$$

## 3 The MOG and MINIMOG Arrays

The MOG is a $6 \times 4$ array whose 4 rows correspond to the elements $0, 1, w, \bar{w}$ of $\mathbb{F}_4$, while its 6 columns correspond to the 6 digits of the typical hexacodeword. The MINIMOG is a $4 \times 3$ array whose 3 rows are labelled by the digits $\text{o}, +, -$ of $\mathbb{F}_3$, while its 4 columns correspond to the 4 places of the typical tetracodeword. Each of the 24 places of the MOG is intended to hold a digit 0 or 1 from $\mathbb{F}_2$, while the 12 places of the MINIMOG are intended for digits $\text{o}, +, -$ from $\mathbb{F}_3$. We shall usually omit zero entries.

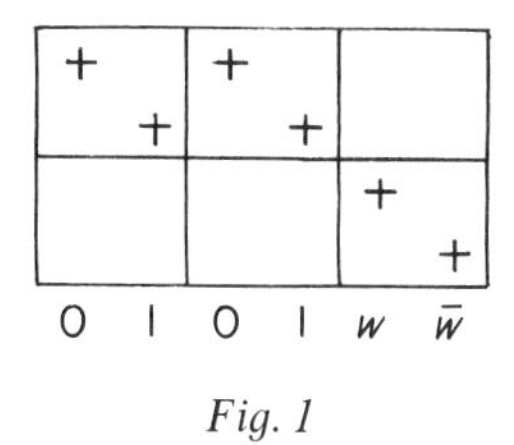


*Fig. 1*

A *hexadword* in the MOG is a weight 6 word whose 6 entries 1 are in the places indexed by the digits of a hexacodeword. Thus, Fig. 1 (where we have used + rather than 1 for vividness) is the hexadword corresponding to the hexacodeword 01 01 $w\bar{w}$.

*Tetradwords* are defined similarly for each tetracodeword. Thus, Fig. 2 is the tetradword corresponding to $+\ \text{o} + -$. (For ease of reference, the hexacodeword and tetracodeword have been written below the MOG and MINIMOG arrays.)

A *columnword* (for either array) is the word which has 1 digits (written +) in the places of a given column, and zero elsewhere. Then the *binary Golay code* $C_{24}$ may be defined in terms of the MOG array by saying that it contains all (mod 2) sums of columnwords and hexadwords in which the total number of columnwords is congruent, (mod 2), to the total number of hexadwords. Similarly the *ternary Golay code* $C_{12}$ is defined in terms of the MINIMOG array as the collection of all (mod 3) sums of columnwords and tetradwords for which the total number of columnwords is congruent, (mod 3), to the total number of tetradwords.

In terms of these, the Steiner system S(5, 8, 24) is the collection of all the

| | | | |
|---|---|---|---|
| | + | | |
| + | | + | |
| | | | + |
| + | 0 | + | − |

*Fig. 2*

8-element sets which are supports of weight 8 codewords of $C_{24}$, and $S(5,6,12)$ is the collection of all the 6-element sets which are supports of weight 6 codewords in $C_{12}$. The Mathieu group $M_{24}$ is the set of all permutations of the 24 places which preserve $S(5,8,24)$ (or equivalently $C_{24}$), while $M_{12}$ is the set of permutations that preserve $S(5,6,12)$ (or equivalently arise from monomial permutations fixing $C_{12}$).

It is very easy to check, mentally, whether an arbitrarily given length 24 binary word belongs to $C_{24}$, or similarly whether a length 12 ternary word belongs to $C_{12}$. In either case, we assign to each column both a *count* (the sum of its entries, in $\mathbb{F}_2$ or $\mathbb{F}_3$ respectively) and a *score* (in $\mathbb{F}_4$ or $\mathbb{F}_3$) obtained by summing the entries in that column times the row label.

For $C_{24}$, the condition is that the scores should be the digits of a hexacodeword and the counts should all have the same parity as the sum of the entries in the top row (the *top row count*). For $C_{12}$, the scores should, of course, be the digits of a tetracodeword and the counts should all be *opposite*, mod 3, to the top row count.

From here on, it is slightly better to split the two cases. An alternative way of expressing our conditions for $C_{24}$ words, is to say that such words are obtained from hexacodewords by *either* giving all their digits even interpretations in any way for which the top row becomes even, *or* giving them all odd interpretations in any way for which the top row comes out odd. The "interpretations" intended here are defined in Fig. 3, which also gives two interpretations of 01 01 $w\bar{w}$.

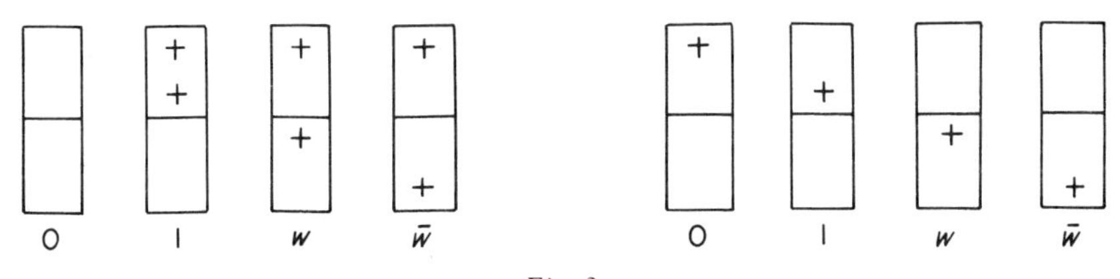


*Fig. 3*

The even interpretations of a digit are the sets on the left, or their complements. The odd interpretations are those on the right, or their complements. Thus the words on the left and right of Fig. 4 are respectively even and odd interpretations of 01 01 $w\bar{w}$.

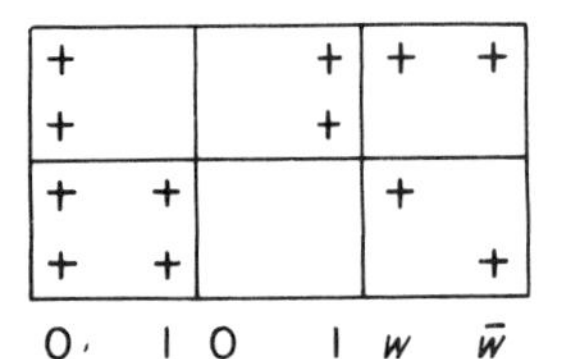


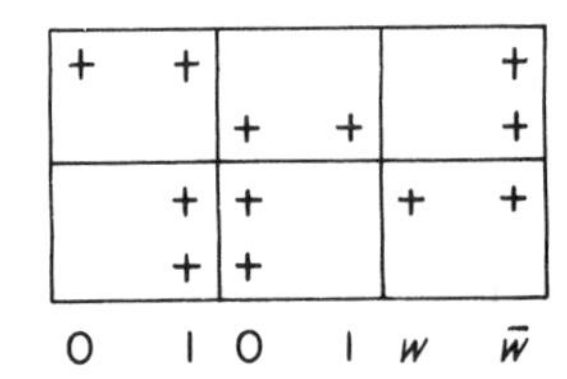


*Fig. 4*

In particular, it is easy to complete any octad from 5 of its points. If one supposes a parity for the type of interpretation required (two cases) and remembers that only 3 corrections are allowed, and that 2 are required to change a column which already has the correct parity, one finds that in every case one is left with an instance of Problem 3 or Problem 5.

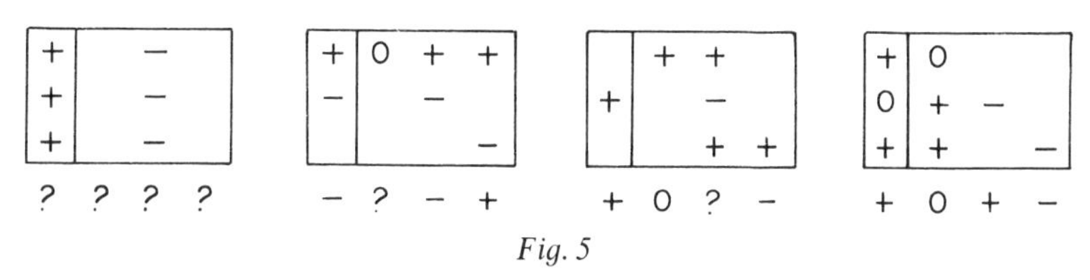


*Fig. 5*

For $C_{12}$ and S(5, 6, 12) there are additional problems concerned with signs. Using "col" for columnword and "tet" for tetradword, it is easy to see that words of the form

$$\text{col}-\text{col}, \quad \text{tet}-\text{tet}, \quad \text{col}+\text{tet}, \quad \text{col}+\text{col}-\text{tet}$$

are words of $C_{12}$ (modulo $C_{12}$, every "col" is congruent to the negative of every "tet"). Instances are given in Fig. 5. (In every case, the "tet" for $+\,\text{o}+-$ is involved and, in every case, we see that the supporting hexad is split across the columns as one of $3^2 0^2$, $2^3 0$, $31^3$, or $2^2 1^2$, while if we note the odd-men-out (when they exist), we find that they form part of the digits of a tetracodeword. (The row label corresponding to a single zero or non-zero digit in a column is the odd-man-out for that column—if all digits are zero or all non-zero there is no odd-man-out for that column.) For our examples, the odd-men-out are indicated, with ? when there is no odd-man-out for a column. Note that they can be determined merely from the support of a hexad, and are independent of its signs.

It is not hard to see that every hexad is of one of the four indicated kinds, and this makes it easy to determine a hexad from any 5 of its points and then, if desired, to adjoin a permissible sign-distribution (unique up to changing the sign of the whole word), since this is easily recovered from the expression in terms of "cols" and "tets".

## 4 Labellings for the MOG and MINIMOG

There are standard numberings for $M_{24}$ and $M_{12}$ which display the inclusion of the subgroups $L_2(23)$ and $L_2(11)$. Below, are shown such labellings for the MOG and MINIMOG arrays, with hints as to their construction.

| | | | | | |
|---|---|---|---|---|---|
| 0 | ∞ | 1 | 11 | 2 | 22 |
| 19 | 3 | 20 | 4 | 10 | 18 |
| 15 | 6 | 14 | 16 | 17 | 8 |
| 5 | 9 | 21 | 13 | 7 | 12 |

| | | | |
|---|---|---|---|
| 0 | 3 | ∞ | 2 |
| 5 | 9 | 8 | X |
| 4 | 1 | 6 | 7 |

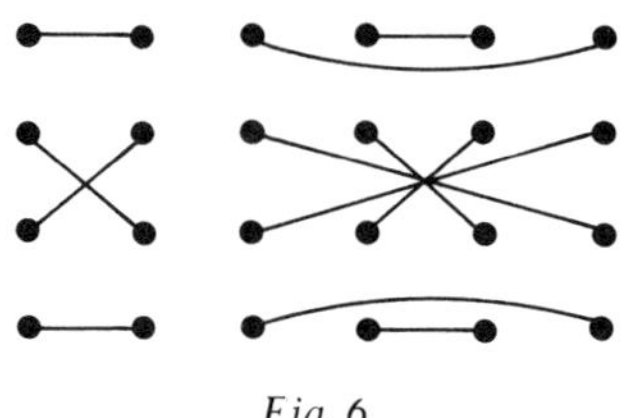

*Fig. 6*

For the MOG, enter first the quadratic residues (and 0) mod 23, according to the scheme

$$\begin{array}{ccccccc} 0 & : & 1 & : & 2 & : & \\ : & 3 & : & 4 & : & -5 & \\ : & 6 & : & -7 & : & 8 & \\ : & 9 & : & -10 & : & -11 & \end{array}$$

in which the signs are chosen so as to ensure that the given number *is* a quadratic residue. Then fill in the non-residues by the action of the permutation $\gamma: t \rightarrow -1/t$ (see Fig. 6).

For the MINIMOG, first enter the numbers from 0 to 11 in the order

$$\begin{array}{cccc} 6 & 3 & 0 & 9 \\ 5 & 2 & 7 & 10 \\ 4 & 1 & 8 & 11 \end{array}$$

Then, translate according to the scheme

$$\begin{array}{cccccccccccc} 0 & 1 & 2 & 3 & 4 & 5 & 6 & 7 & 8 & 9 & 10 & 11 \\ \infty & 1 & -2 & 3 & 4 & 5 & 0 & -3 & 6 & -9 & -12 & -15 \end{array}$$

(The signs are chosen so that 1 to 5 become quadratic residues, and 7 to 11 non-residues, modulo 11.)

## 5 The Embedding of the MINIMOG in the MOG

This is best illustrated by Fig. 7, where (a) includes (b) and (b) folds to (c).

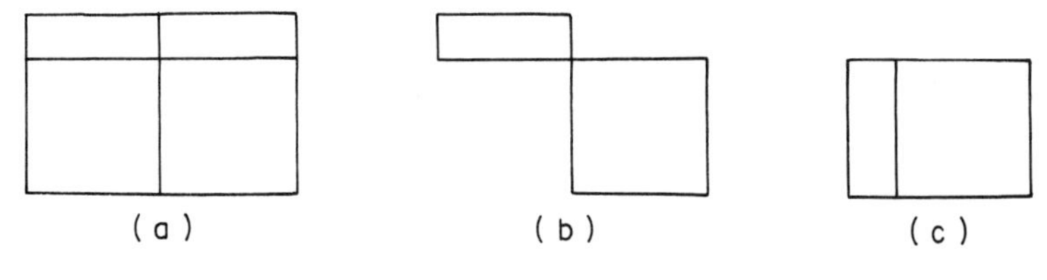


*Fig. 7*

With this inclusion, it will be found that when an octad of the MOG for S(5, 8, 24) intersects the sub-MINIMOG figure, then it does so in a hexad of the corresponding S(5, 6, 12). The complementary set of points also supports a similar structure, which is best found by reversing the order of the 6 columns and then interchanging the last two rows, since this is a permutation of $M_{24}$.

This brief note is a condensation of a large paper on these subjects, which I hope will appear among the Proceedings of the Montreal Conference, edited by J. K. S. McKay. In that paper, many examples are given to illustrate simple computations with these concepts, and it is shown how to use them to find permutations of $M_{24}$ and $M_{12}$. It is also shown how several subgroups of $M_{24}$ are easily described in MOG terms.

The MOG array was discovered by R. T. Curtis, who showed just what a powerful tool it was for computing with $M_{24}$. The reader should be aware that the version used in [1] and [2] differs from that described here, by a left–right reflection.

## References

1. Conway, J. H. (1977). The miracle octad generator. *In* "Topics in Group Theory and Computation" (Ed. M. P. J. Curran), 62–68. Academic Press, London, Orlando and New York.
2. Curtis, R. T. (1976). A new combinatorial approach to $M_{24}$, *Math. Proc. Camb. Phil. Soc.* **79**, 25–42.

# Distinguishing Eleven Crossing Knots

GEORGE HAVAS AND L. G. KOVÁCS

## 1 Introduction

Work has been done on the tabulation of knots since the last century. Perko [11] presents 552 distinct knots with eleven crossings and also provides a list of knot tabulations. In 1979 Richard Hartley drew our attention to Perko's work and to seven pairs of eleven crossing knots which Perko had not succeeded in distinguishing at that time. We indicate how we distinguished these pairs using group-theoretic calculations, not routinely applied by knot theorists. These knot pairs have now also been distinguished by Perko in the cited work and by Thistlethwaite (unpublished) using more usual calculations.

## 2 The problems

At the beginning of 1979, the following seven pairs of knots (in Perko's notation, with the notation of Conway [3], indicated in parentheses) remained to be distinguished:

| | | | |
|---|---|---|---|
| 11–84 | (3, 3, 21, 2), | 11–357 | (3, 21, 3, 2); |
| 11–173 | (8 * 30 . 20), | 11–255 | (. 21 . 21 . 20); |
| 11–220 | (21, 3, 21, 2), | 11–225 | (3, 21, 21, 2); |
| 11–427 | (3, 3, 21, 2–), | 11–428 | (3, 21, 3, 2–); |
| 11–429 | (3, 21, 21, 2–), | 11–430 | (21, 3, 21, 2–); |
| 11–433 | (3, 3, 21, 2––), | 11–434 | (3, 21, 3, 2––); |
| 11–475 | (. –(3, 2) . 20), | 11–476 | (. 20 . –(3, 2)). |

Of these, all except the pair 11–173 and 11–255 (see Fig. 1) are algebraic, and may be distinguished also by the work of Bonahon and Siebenmann [1], so we focus our attention on this pair. Fox [4] and Hartley [5] describe methods for deciding which groups from certain classes of metabelian groups are homomorphic images of a given knot group. In particular, those methods yield that the holomorph $H$ of the group of order 13, $Z_{13}\emptyset Z_{12}$ in Hartley's nota-

COMPUTATIONAL GROUP THEORY
ISBN 0-12-066270-1

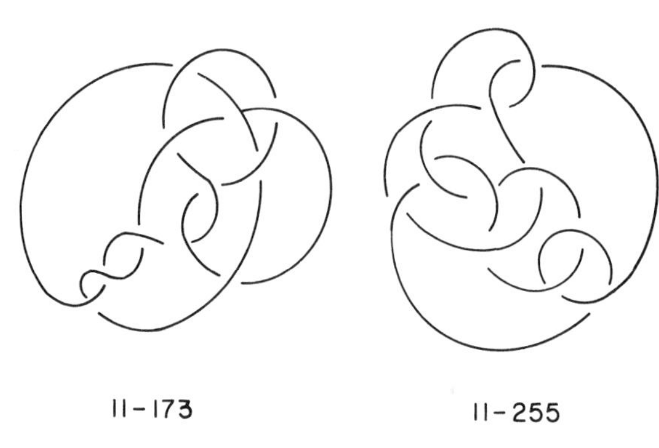


*Fig. 1*

tion, is a homomorphic image of the groups of 11–173 and 11–225. In view of the availability of various programs for computation in group theory, in particular the newly developed abelian decomposition program [**9**], Hartley suggested that we investigate subgroups $S$ of $G$ which arise as complete inverse images of index 13 subgroups of $H$, under homomorphisms of $G$ onto $H$. (For a knot-theoretic interpretation of such subgroups see, for example, [**6**].)

It is clear that, as any finitely generated group, $G$ has only finitely many such subgroups $S$, and that the family of the isomorphism types of their abelian quotients $S/S'$ is an invariant of $G$. (We speak of family, rather than of set or sequence, to indicate that the same isomorphism type may occur repeatedly and it is relevant to know the "multiplicity" showing just how often it does occur, but the particular order in which the isomorphism types happen to be listed is irrelevant.) It is also clear that these $S$ fall into conjugacy classes of 13 each, and that one may use instead just one representative of each conjugacy class.

## 3 The Approach and its Application to 11–173 and 11–255

The first task is to plan how to select a complete set of representatives of the conjugacy classes of the subgroups $S$. In fact, this is no harder—if anything, it is less laborious—than to obtain a complete, repetition-free listing of all the subgroups $S$.

Let us write $\Sigma$ for the set of these subgroups: thus $S \in \Sigma$ means that $S < G$, $|G:S| = 13$, and $G/\text{core } S \cong H$ (where core $S$ denotes the normal core of $S$ in $G$, that is, the intersection of the conjugates of $S$ in $G$). Also, let $\Phi$ stand for the set of all homomorphisms of $G$ *onto* $H$.

Without needing any special properties of $G$ or $H$, note that the composites of elements of $\Phi$ with automorphisms of $H$ all lie in $\Phi$, so the automorphism group Aut $H$ of $H$ acts on $\Phi$ by composition of maps; indeed, two elements of $\Phi$ are in the same orbit of this action if and only if their kernels coincide. From the fact that the subgroups of index 13 form a single conjugacy class in $H$, it follows that two members of $\Sigma$ are conjugate if and only if their normal cores coincide. By the definition of $\Sigma$, the normal core of a member of $\Sigma$ is the kernel of some elements of $\Phi$; conversely, if $\phi \in \Phi$ then the complete inverse image $A\phi^{-1}$ of any index 13 subgroup $A$ of $H$ is a member of $\Sigma$ whose normal core is just the kernel of $\phi$. Thus, there is an equivalence between the set of all conjugacy classes in $\Sigma$ and the set of all orbits in $\Phi$, a conjugacy class matching an orbit when the common normal core of the members of the former is the common kernel of the elements of the latter. It follows that a complete set of representatives of the relevant conjugacy classes may be envisaged as the set of the complete inverse images $A\phi^{-1}$ with $A$ fixed and $\phi$ ranging through a complete set of representatives of the orbits in $\Phi$.

In our calculations, $H$ will be taken as the subgroup generated in the symmetric group on the 13 symbols $1, 2, \ldots, 13$ by the permutations

$$a = (1, 2, 4, 8, 3, 6, 12, 11, 9, 5, 10, 7) \text{ and}$$
$$b = (1, 2, 3, 4, 5, 6, 7, 8, 9, 10, 11, 12, 13);$$

we take $A$ to be the subgroup generated by $a$ alone. Note that $b$ generates the commutator subgroup $H'$, and the cyclic group $H/H'$ is generated by $H'a^i$ if and only if $i \equiv \pm 1, \pm 5 \bmod 12$. Each of these cosets is, of course, a single conjugacy class in $H$.

Each knot group $G$ will be written down in a 3-generator "over-presentation": that is, as $F/R$ where $F$ is free on, say, $\{x, y, z\}$, and $Rx$, $Ry$, $Rz$ are pairwise conjugate in $F/R$. A homomorphism of $G$ onto $H$ must map $Rx$, $Ry$, $Rz$ to a *conjugate triple*: this will be our name of convenience for 3-term (ordered) sequences of pairwise conjugate elements of $H$. Moreover, this conjugate triple will have to be *generating* in the sense that the set of its terms must generate $H$. Clearly, a generating conjugate triple cannot be a constant sequence (for $H$ is not cyclic), and its terms must come from a conjugacy class of $H$ whose image in $H/H'$ generates $H/H'$: as that image is a singleton, the conjugacy class in question must be an $H'a^i$ with $i = \pm 1, \pm 5$. Conversely, it is easy to see that each nonconstant 3-term sequence of elements from any one of these 4 cosets is a generating conjugate triple. One can now readily count that there are precisely $4 . 13 . (1 . 12 + 12 . 13)$, that is, $56|H|$ such triples (choose first one of 4 cosets, then one of the 13 elements of that coset as first term; repeat that as second term and choose one of the 12 others as last term, or choose one of 12 others as second term and any one of 13 as last term). The image of a generating conjugate triple under a nontrivial automorphism of $H$ is an *other* generating conjugate triple: thus the set of all such triples is permuted by Aut $H$ in orbits of size $|\text{Aut } H|$. Since $H$ has trivial centre and no outer automorphism, $|\text{Aut } H| = |H|$; hence we have precisely 56 orbits. If $\phi \in \Phi$ and $\alpha \in \text{Aut } H$, the image of $Rx$, $Ry$, $Rz$ under $\phi$ is mapped by $\alpha$ to the image of $Rx, Ry, Rz$ under the composite of $\phi$ and $\alpha$; thus the map from $\Phi$ to the set of all generating conjugate triples (which maps $\phi$ to the image of $Rx, Ry, Rz$ under $\phi$, and which is clearly one-to-one) is compatible with the action of Aut $H$ on these two sets: in particular, it takes orbits to orbits. We therefore plan to select representatives of the 56 orbits of generating conjugate triples, test which of these lie in the image of $\Phi$, and use those which pass the test to define the desired complete set of representatives of the orbits in $\Phi$. The test is simple: given a triple $u, v, w$, one has to check whether each defining relator of $G$ has trivial image under the homomorphism of $F$ to $H$ which takes $x$ to $u$, $y$ to $v$, $z$ to $w$. (We have had no occasion to mention these defining relators of $G$ so far: they are, of course the generators of $R$ as normal subgroup of $F$ listed in the presentation of $G$.)

It remains to choose a set of representatives of the 56 orbits of triples. We used the union of

$$\{a^i, a^i, a^i b \mid i = \pm 1, \pm 5\} \text{ and } \{a^i, a^i b, a^i b^j \mid i = \pm 1, \pm 5; j = 0, 1, \ldots, 12\}.$$

To see that this is indeed a complete set of representatives, note that it consists of the right number of triples (56), and verify that no two of these triples can lie in the same orbit. The verification is immediate from the following facts. All automorphisms of $H$ are inner. If $a^i$ and $a^{i'}$ are conjugate in $H$, they are certainly congruent modulo $H'$ so $i \equiv i' \bmod 12$, which is only possible within the given range of this parameter if $i = i'$. Finally, if $h \in H$, $(a^i)^h = a^i$, $(a^i b)^h = a^i b$, and $(a^i b^j)^h = a^i b^{j'}$, then $j = j'$ since the first two equations imply that $h$ is central in $H$.

## 4 Calculations for 11–173 and 11–255

Let $G_{173}$ and $G_{255}$ be the knot groups of 11–173 and 11–255 respectively. We proceeded to distinguish these groups in the following way.

First we obtained presentations for $G_{173}$ and $G_{255}$. In practice, we used the knot theory program described in [**8**] which provided both the presentations for $G_{173}$ and $G_{255}$ and, for confirmation, the Alexander polynomials. Next an application of the Tietze transformation program (see [**10**]) to the Wirtinger presentation produced by the knot theory program gave simplified presentations for these groups. It was important to reduce the number of generators in the presentations in order to simplify the further calculations. These presentations (with $\bar{x}, \bar{y}, \bar{z}$ denoting $x^{-1}, y^{-1}, z^{-1}$, respectively), are

$$\begin{aligned} G_{173} = \langle x, y, z \mid\ & xy\bar{x}zx\bar{y}\bar{x}y\bar{x}yx\bar{y}zy\bar{x}\bar{y}\bar{z}xy\bar{x}\bar{z}x\bar{y}\bar{x}zyx\bar{y}\bar{z}y\bar{x}\bar{y}x\bar{y}xy\bar{x}\bar{z} = \\ & xy\bar{x}zx\bar{y}\bar{x}y\bar{x}yx\bar{y}xy\bar{x}\bar{z}x\bar{y}\bar{x}zx\bar{y}\bar{x}y\bar{x}\bar{z}xy\bar{x}zx\bar{y}\bar{x}zy\bar{x}\bar{y}\bar{z}xy\bar{x}\bar{z}x\bar{y}\bar{x}zx\bar{y}xy\bar{x}\bar{z} = 1 \rangle, \end{aligned}$$

$$G_{255} = \langle x, y, z \mid xy\bar{x}yxz\bar{x}\bar{y}x\bar{y}\bar{x}yxz\bar{x}yx\bar{z}\bar{x}\bar{y}xy\bar{x}yx\bar{z}\bar{x}\bar{y}x\bar{y}\bar{x}yxz\bar{x}\bar{y}x\bar{y}\bar{x}y\bar{x}\bar{y}xy\bar{x}yx\bar{z}\bar{x}\bar{y} = xy\bar{x}yxz\bar{x}\bar{y}x\bar{y}\bar{x}yxz\bar{x}\bar{y}x\bar{z}\bar{y}xyz\bar{y}\bar{x}yz\bar{x}yx\bar{z}\bar{x}\bar{y}xy\bar{x}yx\bar{z}\bar{x}\bar{y}x\bar{y}\bar{x}yxz\bar{x}\bar{y}x\bar{z}\bar{y}\bar{x}yz\bar{x}yx\bar{z}\bar{x}\bar{y} = 1 \rangle.$$

Now a straightforward Cayley program (see [**2**]) implementing the algorithm of §3 provided the required complete sets of representatives of the conjugacy classes of desired subgroups of $G_{173}$ and $G_{255}$, via permutation representations. In this case each group had two subgroups in the representative set.

Using the Reidemeister-Schreier program (see [**7**]) we obtained presentations for these subgroups. These are presentations for index 13 subgroups of groups with 2 rather long relators, so the subgroup presentations are not particularly palatable. In each case, we initially obtained 27 generator 26 relator presentations, with lengthy relators.

Since the next step was to find the maximal abelian quotients of these subgroups, there was no need to simplify these presentations. They were used as input to the abelian decomposition program described in [**9**], where some more details of these calculations are presented. The abelian decomposition program revealed that the abelian groups concerned all have torsion free rank 2; the two corresponding to $G_{173}$ have torsion invariants 3 and 14, while for those belonging to $G_{255}$ the torsion invariants are 2 and 3, 3. This distinguishes the groups $G_{173}$ and $G_{255}$ and hence the knots 11–173 and 11–255.

## 5 The Other Six Pairs

Each of the other six pairs of knots was distinguished in a similar way. The smallest interesting metabelian quotient of the knot groups for all of these knots bar 11–475 and 11–476 is $Z_7 \rtimes Z_6$, while for this pair it is $Z_3{}^2 \rtimes Z_4$. Accordingly we investigated the corresponding characteristic classes of subgroups of index 7 (or 9) in these knot groups. The knots in each of these pairs are distinct because their groups have different families of isomorphism types of abelian quotients for the subgroups in these characteristic classes. We calculated these families using the principles exemplified above for $G_{173}$ and $G_{255}$.

For the first five pairs of knots, we obtained knot group presentations with 4 generators and 3 relators and the required index 7 subgroups came from testing 114 possibilities for each of the ten groups. In all cases, each group had sixteen subgroups in the representative set. All of the abelian groups concerned have torsion free rank 2, and we list their torsion invariants in Table 1. For the pair 11–475, 11–476, the knot groups were presented with 3 generators and 2 relators, and we tested 10 possibilities each for the requisite index 9 subgroups. In this case, each group had one subgroup in the

*Table 1*

| 11–84 | 11–357 | 11–220 | 11–225 | 11–427 | 11–428 | 11–429 | 11–430 | 11–433 | 11–434 |
|---|---|---|---|---|---|---|---|---|---|
| 7 | 14 | 14 | 7 | 7 | 7 | 14 | 7 | 14 | 7 |
| 7 | 14 | 14 | 7 | 7 | 7 | 14 | 7 | 14 | 7 |
| 7 | 14 | 14 | 7 | 21 | 7 | 14 | 7 | 28 | 7 |
| 7 | 14 | 14 | 7 | 21 | 7 | 14 | 7 | 28 | 7 |
| 1120 | 1120 | 154 | 798 | 84 | 42 | 252 | 595 | 28 | 21 |
| 1120 | 1120 | 154 | 798 | 84 | 42 | 252 | 595 | 28 | 21 |
| 2128 | 1120 | 329 | 952 | 504 | 84 | 840 | 595 | 28 | 21 |
| 2128 | 1120 | 329 | 952 | 504 | 84 | 840 | 595 | 28 | 21 |
| 2170 | 1855 | 329 | 2, 14 | 2, 14 | 84 | 2, 14 | 812 | 42 | 28 |
| 2170 | 1855 | 329 | 2, 14 | 2, 14 | 84 | 2, 14 | 812 | 42 | 28 |
| 2, 42 | 1855 | 2, 84 | 2, 168 | 2, 42 | 2, 28 | 2, 308 | 840 | 2, 14 | 28 |
| 2, 42 | 1855 | 2, 84 | 2, 168 | 2, 42 | 2, 28 | 2, 308 | 840 | 2, 14 | 28 |
| 2, 896 | 3542 | 7, 280 | 7, 280 | 7, 14 | 7, 14 | 2, 322 | 840 | 2, 14 | 336 |
| 2, 896 | 3542 | 7, 280 | 7, 280 | 7, 14 | 7, 14 | 2, 322 | 840 | 2, 14 | 336 |
| 2, 952 | 2, 56 | 7, 280 | 14, 28 | 14, 14 | 7, 14 | 7, 252 | 14, 14 | 2, 28 | 2, 14 |
| 2, 952 | 2, 56 | 7, 280 | 14, 28 | 14, 14 | 7, 14 | 7, 252 | 14, 14 | 2, 28 | 2, 14 |

representative set. The abelian groups concerned both have torsion free rank 3. The abelian group corresponding to $G_{475}$ has torsion invariants 2, 216 while that corresponding to $G_{476}$ has torsion invariants 2, 2, 72.

## 6 Acknowledgement

We are grateful to W. A. Alford, who guided the large number of calculations involved through the various computer programs required.

## References

1. Bonahon, F. and Siebenmann, L. New geometric splitting of classical knots (Algebraic knots). London Mathematical Society Lecture Notes **75**. Cambridge University Press, (to appear).
2. Cannon, J. J. (1982). (Preprint). A Language for Group Theory. University of Sydney.

3. Conway, J. H. (1970). An enumeration of knots and links and some of their algebraic properties. *In* "Computational problems in abstract algebra", 329–358, Pergamon Press, Oxford.
4. Fox, R. H. (1970). Metacyclic invariants of knots and links, *Canad. J. Math.* **22**, 193–201.
5. Hartley, R. (1979). Metabelian representations of knot groups, *Pacific J. of Mathematics* **82**, 93–104.
6. Hartley, R. and Murasugi, K. (1978). Homology invariants, *Canad. J. Math.* **30**, 655–670.
7. Havas, G. (1974a). "A Reidemeister-Schreier program", Proc. Second Internat. Conf. Theory of Groups (Canberra 1973), Lecture Notes in Mathematics, vol. **372**, 347–356. Springer-Verlag, Berlin.
8. Havas, G. (1974b). Computational Approaches to Combinatorial Group Theory. Ph.D. Thesis, University of Sydney.
9. Havas, G. and Sterling, L. S. (1979). Integer matrices and abelian groups. Symbolic and Algebraic Computation (Ed. E. W. Ng.), Lecture Notes in Computer Science, vol. **72**, 431–451. Springer-Verlag, Berlin.
10. Havas, G., Kenne, P. E., Richardson, J. S. and Robertson, E. F. (1983). A Tietze Transformation Program, (these Proceedings).
11. Perko, K. Invariants of 11-crossing knots, *Asterisque* (to appear).

# Commutator Identities in Alternating Groups

DANIELA B. NIKOLOVA

Let $w_{m,n}(x, y)$, $m < n$, be the commutator function

$$[x, {}_m y]\,[x, {}_n y]^{-1} \tag{1}$$

when $[x, {}_m y]$ is the left-normed commutator $[x, y, y, \ldots, y]$ with $m$ $y$'s. These words can be ordered according to the lexicographic order on the pairs $(m, n)$. In view of the following theorem (see [**1**]) it is of interest to know, for a group $G$, what is the minimal $w_{m,n}(x, y)$ which is a law of $G$.

*Theorem.* Let $G$ be a group which satisfies a law of type (1) and let $w_{m,n}(x, y)$ be the minimal law. Then $G$ satisfies the law $w_{r,s}(x, y)$ if and only if $m \leqslant r$ and $n-m$ divides $s-r$.

By using this theorem together with exhaustive computer search, we have found the minimal laws $w_{m,n}(x, y)$ which hold in small alternating groups. For $A_5$ the minimal $(m, n)$ is (3, 63); for $A_6$, $A_7$ it is (4, 124), (15, 35 295), respectively.

Laws of this type have been studied in other classes of groups in [**1**], [**2**].

## References

1. Nikolova, D. B. (1983). On a class of identities in groups, *Serdika* **9**, 2 (in Russian).
2. Nikolova, D. B. (1983). Identities in metabelian varieties of groups $\mathfrak{A}_k\mathfrak{A}_l$, *Serdika* **9**, 3 (in Russian).

COMPUTATIONAL GROUP THEORY
ISBN 0-12-066270-1